# MEXICAN-ORIGIN FOODS, FOODWAYS, AND SOCIAL MOVEMENTS

**FOOD AND FOODWAYS**

SERIES EDITORS:
JENNIFER JENSEN WALLACH
AND MICHAEL WISE

OTHER TITLES IN THIS SERIES

# Mexican-Origin Foods, Foodways, and Social Movements

## DECOLONIAL PERSPECTIVES

EDITED BY
DEVON G. PEÑA, LUZ CALVO,
PANCHO McFARLAND, AND
GABRIEL R. VALLE

The University of Arkansas
Fayetteville
2017

ISBN: 978-1-68226-036-4
e-ISBN: 978-1-61075-618-1

21   20   19   18   17      5   4   3   2   1

⊖ The paper used in this publication meets the minimum requirements of the
American National Standard for Permanence of Paper for Printed Library Materials
Z39.48-1984.

LIBRARY OF CONGRESS CONTROL NUMBER: 2017942332

The Acequia Institute generously provided funding for commissions on artwork used on the front and back covers, permissions for other illustrations,
preparation of the "Indigenous Corn Belt" map featured in the introductory
chapter, and the index. For more information on the work of the institute, visit
www.acequiainstitute.org.

Cover image: *Pozole Crossing* by Anthony Ortega
Cover design: Liz Lester

This book is dedicated to our elder mentors, next-generation environmental and food justice activists, and those decolonial and Indigenous scholars who understand how the role of accomplice in service to our people is the only ethical path before us. Our colleagues in these movements have taught us how tlamatiliztli, or sabiduría (knowing), is performative and participatory—a knowing how, not a knowing that.

We mindfully and gratefully dedicate this book to the memory of our coauthor and untiring ally and advocate, Joseph C. Gallegos (1956–2016).

*La semilla es la memoria de la planta de como vivir bien en este lugar.*

*The seed is the memory of the plant of how to live well in this place.*

MARGARITA K. PEÑA, abuelita gardener and seed saver.
Laredo, Texas, ca. 1964

# CONTENTS

Part II. Witnessing: Heritage Cuisines and Decolonial Foodways

# SERIES EDITORS' PREFACE

The University of Arkansas Press series on Food and Foodways explores historical and contemporary issues in global food studies. We are committed to representing a diverse set of voices that tell lesser-known food stories and to provoking new avenues of interdisciplinary research. Our strengths are works in the humanities and social sciences that use food as a critical lens to examine broader cultural, environmental, and ethical issues.

Feeding ourselves has long entangled us human beings within complicated moral puzzles of social injustice and environmental destruction. When we eat, we consume not only food on the plate but also the lives and labors of innumerable plants, animals, and people. This process distributes its costs unevenly across race, class, gender, and other social categories. The production and distribution of food often obscure these material and cultural connections, impeding honest assessments of our impacts on the world around us. By taking these relationships seriously, Food and Foodways provides a new series of critical studies that analyze the cultural and environmental relationships that have sustained human societies.

This essay collection, *Mexican-Origin Foods, Foodways, and Social Movements: Decolonial Perspectives*, edited by Devon G. Peña, Luz Calvo, Pancho McFarland, and Gabriel R. Valle, provides a groundbreaking theoretical and empirical study of the massive contributions that Mexican-origin peoples have made toward resisting and transforming colonial food systems in North America and beyond. The sixteen essays featured here offer radical new perspectives on food and the subjective experiences of diverse Mexican-origin communities and individuals, as well as critical and precise interventions in current discussions surrounding the elevation of food as a central element of the critical theory and practice of decolonization. Not only does this volume attend to the struggles and livelihoods of human beings, but by addressing food and foodways from a critical agroecological perspective, it also considers the effects of colonialism and global food systems on the welfare of the nonhuman beings that help sustain us.

Together, the diverse voices and nuanced approaches represented here help fulfill the Food and Foodways mandate to publish research on rich and underexplored topics. They set a research agenda not only for Mexican-origin food and foodways but for the broader study of food and decolonization that should animate scholarship on global food studies for years to come.

—JENNIFER JENSEN WALLACH AND MICHAEL WISE

# ACKNOWLEDGMENTS

The coeditors convey heartfelt appreciation to our "abuelas y madres" and other relations; to Mesoamerican diaspora farmworkers and farmers; to home kitchen and guerrilla gardeners, food-chain workers, chefs and cooks, and food justice advocates who have mentored and shared *saberes* with us. You are the deep sources of the knowledge, beliefs, and practices we have shared and recounted in this book as an act of collective epistemic disobedience.

We are grateful to the many students and colleagues, fellow farmers, and seed savers who have worked with us over the years. There are too many to list here, and you know who you are. We undertook this collective project to work with and serve the environmental and food justice movements and sustain our collaboration with a wide range of grassroots organizations, including The Acequia Institute, Community to Community Development, Familias Unidas por la Justicia, Food First, Red en Defensa del Maíz, South Central Farmers Cooperative and Health and Education Fund, and many more.

The editor of the Food and Foodways series, Jennifer J. Wallach of the University of North Texas, was crucial to the development of this book and invited Devon G. Peña to submit a contribution to the series. With her assistance and counsel, Peña recruited the other coeditors, who identified and worked with the contributing authors. In all matters of preparation of the manuscript for publication, we are equally thankful for the assistance and guidance of D. S. Cunningham, editor, and Deena Owens, editorial assistant, at the University of Arkansas Press, and Mary Keirstead, copyeditor.

# Mexican Deep Food: Bodies, the Land, Food, and Social Movements

## DEVON G. PEÑA, LUZ CALVO, PANCHO McFARLAND, AND GABRIEL R. VALLE

The Mexican-origin people comprise a mosaic of millenary civilizations with diverse ethnolinguistic roots. There are officially sixty-two distinct Indigenous language groups in México today.[1] There is more diversity than this implies since many ethnolinguistic communities, like the Zapoteca in Oaxaca, have multiple regional dialects, and a good portion of these involves speakers *north* of the México-US border. We note the well-established presence of half a million Zapotecas living, working, and raising families in California and on the rest of the West Coast all the way up to Alaska.[2] This presence is exemplary of an Indigenous diaspora from Mesoamerica to multiple communities across the north. The diasporic movement of these communities signifies substantive transformation of a good deal of what scholars usually consider the Mesoamerican range in North American Indigenous foods, foodways, cuisines, and their related social movements working to protect this deep heritage.[3] Commenting on food ethnographies among Indigenous people from Oaxaca who are working and living in Alaska with their "suitcases full of mole," the ethnographer Sara V. Komarnisky insightfully notes

> how transnational lives and identities are lived out as much through what people eat—or say they eat—as through other global cultural flows . . . meals that connect places and the people

in them or foods that depend on interconnectedness and mobility across space.[4]

This volume looks at Mexican-origin foods, foodways, and social movements across a broad swath of circumstances and places—from multigenerational acequia (communal irrigation ditch) farmers in Colorado who trace their ancestry to Indigenous families in place well before the Oñate Entrada of 1598 to tomorrow's itinerant transborder travelers who will be negotiating entry into the United States across the Sonoran Desert in southern Arizona. We chart our experiences as deeply place-based and profoundly *dis*placed Indigenous peoples on both sides of the México-US border. In all these cases, movement and transmotion are evident.[5]

To these deep Indigenous roots, we add influences and syncretic fusions with other streams coming from equally diverse sets of cultures and peoples that arrived from the East. This includes so-called Spaniards—the gruff elites, aspiring merchants, privateers, and military adventurers from far-flung places like coastal Asturias or Almería, Cádiz, Córdoba, and Valencia in Andalusia; Biscaya in Castile; Galicia north of Portugal; and the high plains of León, and so on. These influxes brought people and their cultural traditions from Sephardic, Persian, Berber, Arabic, Lebanese, and Yemeni roots and branches. West Africans from Guinea-Bissau were among the arrivants.[6] These "subculture" groups—as distinct from Peninsular Spaniards and their Mexican-born but still Iberian-bred offspring—accompanied and intermarried with Indigenous peoples as they partook in movement north into Native American country in what eventually became reterritorialized as the American Southwest and to points well beyond. This contested and violent confluence of war, disease, conquest, and genocide was an unruly, wholly transformative affair changing all sides and their cultural traditions *and* ecologies of place.[7]

Indigenous foods, foodways, and cuisines survived relatively intact despite this vast, reiterative, and cataclysmic wave of dramatic change in the composition of the biota including people and the organisms inhabiting the environment. As objects of conquest, many Indigenous peoples are said to have lost their food, foodways, and cuisines—heritage erased through veritable population extinctions or forgotten in the aftermath of the collapse and expulsion of entire regional Indigenous communities by the violent forces of settler colonial empires.

A decolonial approach to critical food studies envisions the recovery and resurgence of Indigenous knowledge, belief, and practice as these are related to food, foodways, and cuisines and the methods these inspire in our agroecosystems. Decoloniality explores hidden alterNative histories of relationships between plants, animals, soil, water, and humans. Decolonial foods resurface in daily-lived acts of producing and sharing across diverse configurations of peoples— every one of us brings unique spices and ingredients to the shifting mosaic and living heritage of Mexican-origin foods and foodways gracing our fields, gardens, hearths, and multiethnic kitchen tables from Chiapas to Alaska.

Colonial observers and contemporary scholars have lamented the disappearance of Indigenous traditional and syncretic foods, cuisines, and foodways.[8] But México's Indigenous cultures are resilient. This resilience is reflected in the survival and resurgence of numerous Mesoamerican crops, herbs, and animals associated with distinct regional, national, transnational, and so-called fusion cuisines.[9] The North American Free Trade Agreement (NAFTA) and neoliberal reforms north and south of the border threaten these foodways in a second wave of bio-colonialism, only now these arise under legal frameworks imposed by investor-state treaties. These multilateral agreements privilege the accumulation and signification regimes of corporations to pursue whatever it takes to maximize profits without regard for people's own food self-sufficiency and the degradation of public and environmental health on both sides of the border. The new wave of displacements involves the spread of commercial agricultural biotechnologies including transgenic crops—which are increasingly being shown to have serious ecological, social, public health, and cultural impacts.[10]

As a result, social movements have risen to defend Indigenous farmers' agroecosystems and their associated landrace heirloom varieties, wild relatives, and intermediaries. These living agroecosystems are heritage working landscapes grounded in the complex relationships between places and their human and more-than-human inhabitants in a partnership that co-produces the bio-cultural diversity of the

Mesoamerican Vavilov Center, which for us extends from the Petén in Guatemala to today's American Southwest and points north by west and east across the northern reaches of the "Indigenous Corn Belt." This includes Northeast and Pacific Northwest bioregions, as depicted in the map of the Indigenous Corn Belt. We define this transborder biogeographical province as the appropriate frame of study for this collection of original essays.

Two other developments have altered the biogeography of Mexican-origin food cultures and politics in the extended center of origin for maize. In deference to fellow travelers: the millions of displaced farmers from México, Guatemala, Honduras, and other points south are not "immigrants." Within a subaltern history of flows in old and new Mesoamerican diasporas, many of the travelers arriving today are Indigenous farmers and environmental refugees, economic exiles, or targets of paramilitary death squads and drug cartels. An estimated 1 million Indigenous farmers were displaced by the violence associated with the first decade after NAFTA (1994–2004), and at least an additional 1.5 million left during the second decade of neoliberal-imposed austerity and widening structural violence through 2014. This context is capped by the recent disappearance of the Ayotzinapa 43 in Iguala, Guerrero, México.[11] This massive displacement has led to an unprecedented movement of millions of Indigenous people from México and points south into the United States and Canada, bringing many possible worlds along.

The people of this Mesoamerican diaspora have always brought food, foodways, and cuisines along for the journey. Plants and tubers, precious cuttings and scions, rootstocks and seeds—all this precious biotic baggage accompanies the arrivants. Defying displacement, we sustain Indigenous knowledge of farming and ethnobotany and refuse to surrender our seeds at the border. We bring our own social organization, governance, and political subjectivities. We are the living extension of the Mexican Vavilov Center farther north into the United States and even Canada. Today many milpas (maize polyculture fields) and *huertos familiares* (home kitchen gardens) in places across rural México remain for now abandoned. The native landrace varieties of corn and other crops cultivated there for millennia are now being grown in urban and rural farms and gardens across the

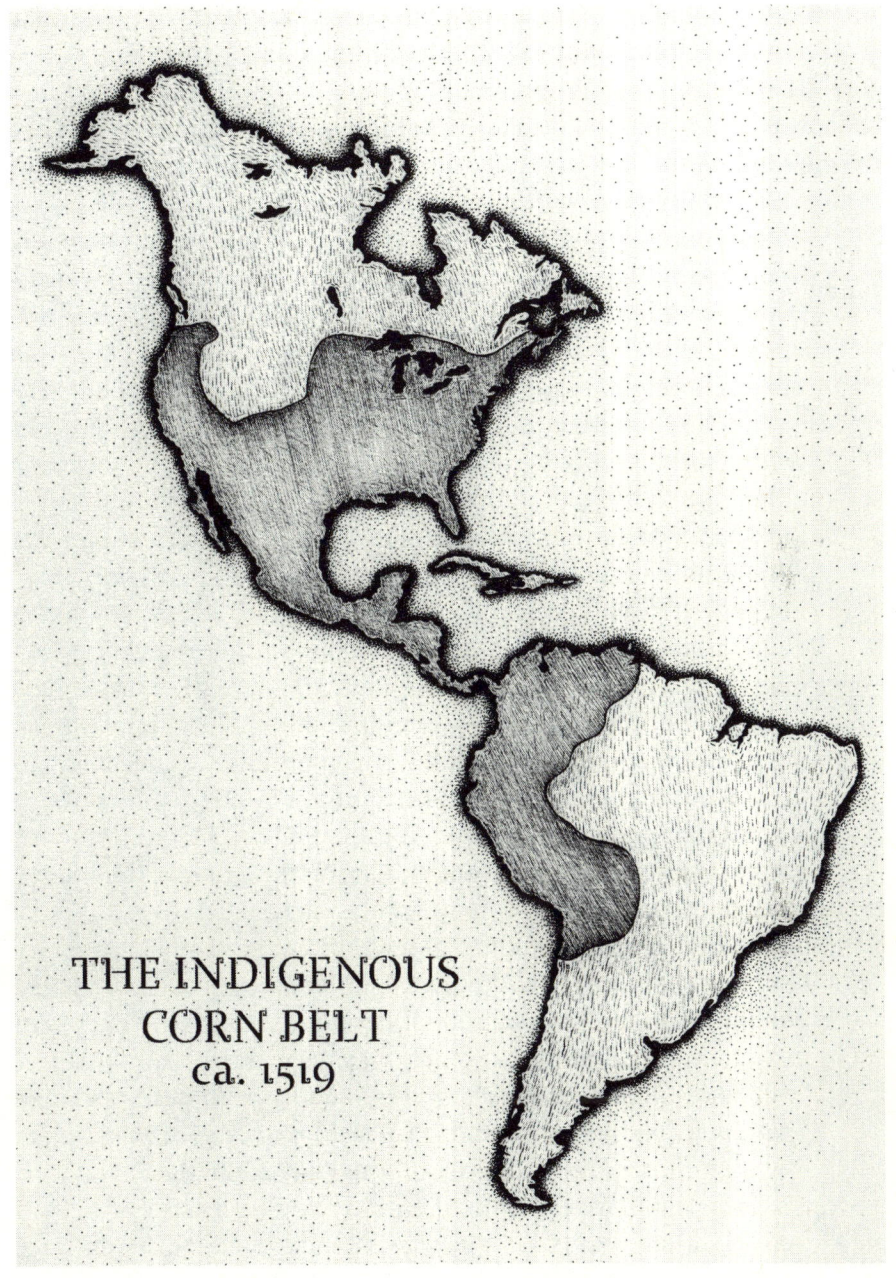

THE INDIGENOUS
CORN BELT
ca. 1519

The Indigenous Corn Belt. Map by Davidv35. Courtesy of The Acequia Institute.

United States and into Canada. Such is the nature of this vast instance of *alter*-globalization as it unfolds through the food and foodways of Indigenous peoples resisting privatization and the commodification of land, water, and life itself. The resurgence of Indigenous heritage cuisines required to nurture the crops that comprise the elements of these native foods—what coeditor Peña prefers to call "Three Sisters Plus"—are a sign that the Mesoamerican diaspora is a major force currently reshaping the crops, foods, and ingredients available to farmers, cooks, chefs, and consumers across the three nation-states comprising North America. We are at the start of a continent-wide transformation from the bottom up.

This elemental shift involves what we are calling the recovery and resurgence of "deep foods" or "first foods"—food and foodways grounded in the ancient millenary knowledge of Indigenous peoples and based on our relations with plants, animals, soils, and water and our use of these for cooking, eating, and sharing food. This knowledge ties our decolonial *comida* (food and foodways as social relations) to a normative infrastructure constitutive of ways of being in the world predating white settler colonial societies by thousands of years. These are enunciations of an epistemic refusal and delinking from colonial capitalist *dispositifs*—apparatuses of control or governmentality in the "conduct of conduct," as Foucault defines it. Deep foods are the embodiment of practices of healing and transformation. These represent "escapes" from the subjugated spaces of the global commodity chains that characterize the neoliberal capitalist agri-food system, which is dominated by six major multinational corporations.[12]

These escapes are not universal, and major challenges are posed by the decimation and erasure of heritage cuisines in many places south and north of the border. As a consequence, we are in danger of losing the advantages of the so-called Latina/o health paradox and are increasingly stricken with diabetes and cancer on both sides of the border.[13] We are also increasingly subject to the dietary cultural deprivation and senses of displacement associated with "food deserts"—or perhaps it is better to call these "food junkyards" since desert ecosystem peoples might rightly object? These challenges have led to the emergence of what we are calling the "Decolonial Food Movement"—a broadening and deepening domain of grassroots struggles reconnecting Mexican-origin peoples with their own diverse regional heritage and diasporic cuisines while encouraging food self-sufficiency through Indigenous agroecology and ethnobotany and direct agency outside the global commodity chains of the industrial capitalist food system. This strategy focuses on the long-term environmental justice principle of ecological democracy by restoring community health, personal well-being and healing, and a renewed sense of place through the resurgence of culturally resonant deep foods.

The Decolonial Food Movement seeks to end violence against food-chain workers and farmers with livelihoods vested in agri-food systems. Decolonial food activists reject any destructive agriculture that brutalizes the diversity of animals, plants, habitats, and ecosystems. Instead, our decoloniality embraces Indigenous agroecological and permaculture practices that contribute to native bio-cultural diversity, ecosystem resilience, and equity and coevalness across species, human groups, and generations.

Our collection of essays explores the origins and bioregional diversity of the food, cuisines, and foodways of the Mexican-origin people in the United States. These include chapters analyzing the environmental and cultural histories of México and those with a focus on current social movements and Indigenous mobilizations against transgenic maize and other genetically modified organisms (GMOs). The sixteen original contributions are organized into three major thematic sections focused on "Theorizing" (part I), "Witnessing" (part II), and "Organizing" (part III).

The chapters comprising part I ("Theorizing") provide various theoretical critiques addressing the intersectionalities of class, gender, race, national origin, citizenship, and sexuality and the historical origins, roots and branches, and divergent tendencies of what has become known as "Food Sovereignty." By critically interrogating this almost sacrosanct concept, Devon G. Peña, Rufina Juárez, Gabriel R. Valle, and Silvia Patricia Solís open a space for discussion of the advent of decolonial food activism and thus offer a variety of perspectives centered on decolonizing so-called alternative food movements themselves.

In chapter 1 Peña boldly exposes a set of hitherto unacknowledged contradictions embedded in La Via Campesina (LVC) principles of food sovereignty. He argues that food sovereignty remains "within the orbit of a neoliberal capitalist epistemology" because it reduces more-than-human beings to the "bare life" and fails to challenge the reduction of organisms and ecosystems to nonstatus as subjects stripped of standing before the sovereign power of the "anthropological machine" that divides humanity from other animals.[14] Peña suggests that LVC is bound to anthropocentric ideas of sovereignty. These privilege human-centered values and obscure the importance of ecosystem processes in sustaining the right livelihoods of human and more-than-human beings. He argues that food sovereignty movements need to move beyond anthropocentric sovereignty models and embrace the deep philosophy of Indigenous *autonomía*—a political project enunciating and realizing the values of coevalness grounded within a more nuanced coupling of ecological systems with Indigenous models of human rights, property, and the individual. LVC should rethink its understanding of the differences between Indigenous ways of life and the needs and practices of other smallholder and alternative but trade-oriented farmers. These differences derive from a misfit between LVC theory and local Indigenous practices and can be attributed to conflicting concepts of "property" and the "individual." Indigenous polities create value based on ideals of *property as relation*, while Europeans and neo-Europeans tend to conceptualize *property as possession*. Coloniality has propagated and imposed this property value set across the planet including in Latin America, Africa, and Asia. This meets fierce Indigenous opposition since this is primitive accumulation in continued acts of enclosure and commoditization of land. Peña explains how Indigenous languages do not have a word for *individual* and our cosmovisions reject any notion of "human rights" if this fails to account for institutions of collective and communal subjectivities obligated to take care of place(s). LVC is criticized for failing to consider innovations developed by *urban* food justice movements and especially the intersectional social justice campaigns of food-chain workers across the various constantly reorganized sectors of our "multitude." The LVC's focus on an essentialist peasant subject location is questioned by Peña, who sees this as a fundamental strategic misreading of the ongoing political recomposition of food workers (and farmers) that is occurring globally and locally, a theme further addressed by Rosalinda Guillen and C2C in chapter 12.

In chapter 2 Juárez privileges decolonial feminist subjectivities within the context of the South Central Farmers' epic and continuing struggles in Los Angeles and Southern California. This firsthand account draws from the author's experiences as a farmer at South Central Farm (SCF). She was also elected as a leader during the events she describes. Juárez elucidates the intersectionality of diaspora Indigenous women farmers and their agency in the politics of finding space to grow food and to mobilize direct action. She argues, "As *mujeres en movimiento* we faced both subtle and aggressive attacks on our persons, our bodies, and our spirits." These attacks happened when external forces outside the struggle worked against the SCF, but similar attitudes were also prevalent among the men within this mobilized community. Redeploying and modifying arguments made by other Chicana feminists, the author finds that machismo is not a thing of the past and a decolonial approach to food sovereignty must recognize and address this paramount concern. She goes on to provide instances of heteronormative and hypermasculine aggression against women's leadership and challenges the reader to consider the idea that what she calls "conflict foods," like the case of "conflict diamonds," also involve constant violence against and exploitation of workers, many of them young teens and children. Yet we seldom pause to think about the political, cultural, and economic implications of the painful and harmful conflict surrounding the capitalist production, distribution, and consumption of food.

In chapter 3 Gabriel R. Valle draws on his training as an ethnoecologist to understand and remake the theory of value through an account of working-class subjectivities in the context of home kitchen gardeners. He starts by acknowledging how home kitchen gardens have always been important to Mexican-origin peoples but until recently received marginal attention in food and foodways studies. Using original ethnographic fieldwork, Valle describes how urban gardeners in San Jose, California, have constructed home garden networks to share resources and gain mutual support. The author concludes that "the tightly woven garden network allows home gardeners to protect against vulnerability, construct mobility, and improve individual *and* collective well-being." These home gardens provide the ability for gardeners to negotiate the interstitial space between citizenship and economic life while remaining self-determining and accountable to others. Avoiding a romanticizing gaze, Valle explains how growing and sharing food strengthen social networks but also reveal their fragility. Many of the San Jose gardeners still use food pantries and other emergency food sources to avoid episodic hunger. His analysis shows that "precarity increases vulnerability" but also "offers the opportunity to find new forms of autonomy." Continued research on the adaptive strategies of home kitchen gardeners is needed because "it is . . . one site among many where the creation of new subjectivities offers the possibility of envisioning more just and sustainable futures."

The deep well and transmission of Indigenous knowledge are Solís's focus in chapter 4. Solís is also trained in ethnoecology, and she insists that deep ancestral knowledge is what helps us reconnect the everyday nature of our lives to self and communal healing practices. Solís defines *coloniality* as the modernist logic of power externally administering the lifeworlds of the subjects of conquest and empire. She is interested in revealing how this shapes and distorts our bodies, cultures, languages, and traditions. This also requires understanding how this has disrupted our relationships. Solís creatively draws from the autobiographical and pedagogical writings of Karleen Pendleton Jiménez. Her narrative ends with the place she (and Solís) both long yearned to speak to; they eloquently shared the experience of having lost and found the land of their childhoods. This opens a pathway to thinking about strategies of epistemic delinking and self-enunciation cultivated and articulated through the renewal of these deep relationships that weave the body, self, family, community, and Earth together.

Solís argues that "re-membering" (recovering cultural memory that pieces our lives back together) unveils the "colonial wound," and despite the colonizers' phenomenology of pain, the act of decolonial enunciation involves cognitive and ethical transformation. Through re-membering we carry place with us. Re-membering the place-based practices of our ancestors' traditional environmental knowledge (TEK) includes growing food and following ancestral foodways. Re-membering TEK is a tool to restore relationships and our originary senses of ecosystems and communities. There is more: Through an Anzaldúan lens, Solís wishes us to show us how *the body is place*—a site where instructions and careful observations emerge from the practice of Indigenous knowledge systems in the motion and activity of our bodies. Our bodies are corporal because sensuous, sentient, connected. This means we become ethical beings when we re-member who and where we are as one with our relations. This signifies an important contribution to Chicana ecofeminist ethics that bridges the study of TEK with the Anzaldúan decolonial theory of the body and the epistemic sublation of the colonial wound.

Part II ("Witnessing") presents six chapters focused on this practice of re-membering and of reflecting on and reformulating the conditions under which our labor makes deep food possible or renders it absent or "hybridized" in specific kinds of locations like the border as a zone of exception. These *testímonios* aid in the recovery and strengthening of heritage cuisines and foodways as revealed by the authors examining the relationship between food, cultural memory, ethnic identity, subjectivity, nutrition and health, and transborder placemaking in different geospatial contexts.

Part II opens with a unique instance of "ethno-archaeology" by cultural anthropologist Consuelo Crow in chapter 5. Crow, who has studied ethnography and archaeology, delves into the self-provisioning of food and water by transborder (entry without inspection) travelers coming into the United States from México. She opens with the classic anthropological emphasis on how food and water are charged with positive and negative symbolic and affect significance. Crow challenges us to rethink how anthropologists view evidence of traditional environmental knowledge in the travelers' survival and stealth practices. She demonstrates how Sonoran Desert crossers actually often lack knowledge of exact local conditions that may pose threats to their survival and success as unauthorized entrants. Water is clearly the most essential item carried across these desert borderlands, and bottled water is actively marketed at border bodegas to individuals looking to cross over. Arriving at a view of the border-crossing world through the eyes of the travelers, Crow argues that the "migrant meal" is in fact chosen and designed as survival gear. It must be lightweight and easy to carry, eat, and discard. Transborder foods and other travel items are coded to perform certain practical functions but are simultaneously recoded as racialized objects of both stigma and an appeal to humanitarian morals.

To illustrate this last point, Crow focuses on the contradictions faced by humanitarian relief volunteers who toil with how best to deliver water and food so the desert travelers can resupply along hazardous journeys. In one especially poignant passage, Crow notes how a humanitarian group struggled with a decision to include cans of refried beans in resupply packets. These gestures are "noble and well-meaning" but can ultimately lead to unintended essentialist and racialist constructs by those routinely attempting to combat desert deaths by self-identifying and acting as pro-immigrant humanitarian aid workers. Humanitarian relief presents the danger of spoilage and serious food poisoning and acts of sabotage (poisoning of food and water) by anti-immigrant militias and even ranchers. Crow continues by noting the frequent deaths, illnesses, injuries, rapes, and homicides transborder travelers face. These atrocities may tempt the liberal mindset "to lament the inhumanity of it all," but Crow wants us to move beyond the usual decrying of "the lack of an effective human rights regime or the failure of development programs to benefit the sending countries." She refocuses our attention on criticizing the "practice of 'medicalizing' migrants' experiences" to make them "worthy of a path to inclusion in civil society." While seemingly laudable, this discursive turn "may actually contribute to biopolitical constructions of citizenship that tie rights of membership to specific biomarkers of political suffering."

In chapter 6 Lee Ann Epstein, an interdisciplinary scholar of feminist theory and studies of placemaking, explores the concept of being *norteada/o* by drawing on her background in third-space feminism. Epstein argues that *norteada/o* is the feeling of being lost in the geospatial/geopolitical sense and identifies the feeling of "turned aroundness" as a key feature of this affect dimension of spatial disorientation. She reclaims and recasts these concepts as a "process of negotiating one's identity and making sense of the world." Through a decolonial third-space Chicana feminist standpoint, the author presents food deserts associated with many barrios as sites of identity confusion and turned aroundness. Putting landscape- and self-integrity back together again requires the radical reorganization of the production, distribution, and consumption of food in urban centers. Thompson Community on the west side of San Antonio is the locus of this inquiry because the area has been economically ravaged by the closure of a military base that resulted in the decline of many small businesses, including locally owned restaurants. Epstein explains how the resulting "preponderance of fast-food restaurants does not indicate a dearth of food, but rather a surplus of processed foods and the inaccessibility of fresh organic produce." In this context, a *norteada/o* subjectivity leads to "yearning and embodiment of hunger and identity" but is also associated with "geopolitical dis/location where the hungering to belong is pervasive." This hunger has led the author to create a liberated space in a bountiful home kitchen garden and orchard.

Chapter 7 is a contribution by Luz Calvo and Catriona R. Esquibel, the authors of the acclaimed book *Decolonize Your Diet*. Together they explore the transformational and healing powers of a "decolonized diet." Through a queer feminist perspective, the authors argue that food connects us to our ancestors, the land, and our bodies. They find that "Americanization" projects altered Chicana/o families' relationships to food. While reclaiming native foods as an act of survival, the authors stress that a return to plant-based diets is not the answer by itself. The return to native foods must also address gender oppression within its core by working to dismantle patriarchal and heteronormative constructs and relationships. The authors argue not for a "return to the kitchen" but for the "liberation of the kitchen." Food is an important site for the construction of identities, and decolonial cooking is an act of "creative resistance" against hegemonic societal norms. Calvo and Esquibel offer an eloquent call for a type of deep-food activism involving "an explosion of community gardens that function as the people's pharmacy" and "as centers where ancestral knowledge is gathered and shared." They celebrate the resurgence of "*curanderas* who can help people through the traumas associated with colonization and disease to perform *limpias* to cure us of our *sustos*." Deep food presents significant elemental spiritual dimensions of sensual pleasure and joy: "We claim food as an important site of pleasure, and decolonial cooking as a creative act of resistance."

Adobe oven-roasted white flint corn is known regionally as "chicos del horno" and is an important example of deep food. In chapter 8, the acequia farmer and environmental justice advocate Joseph C.

Gallegos details the art of sociality in the preparation of chicos in the Río Arriba bioregion of Colorado. In a plainspoken autobiographical voice, he lovingly describes how making chicos del horno is specific to place and involves practices rooted in the traditions of people who have long lived in his community of San Luis, Colorado. He explains how for generations "the chico way of life brought family and friends together." This is more than local and slow food. It is deep food. While over time there were fewer chicos-making gatherings, today families are actively working to restore this long-standing cultural practice and heritage cuisine. The chicos way of life embraces an ontological stand-point enunciating a way of living in place. The return of the chicos-making tradition is one way to "awaken the energy of our pasts" and witness how "the ambiance turns into celebration by revived souls."

In chapter 9 ethnoecologist and "artivist" María Guillen Valdovinos takes us through her memories as a daughter of Afromestiza/os by connecting home places in Washington State with the origin villages and *ejidos* in La Tierra Caliente of Guerrero, México. In her part oral history, part autoethnography, the author foregrounds food and farm-ing in the lives of her extended family from La Tierra Caliente to Eastern Washington. Guillen's narrative enunciates an Indigenous ancestry in P'urhépecha and Afro-Indigenous descent communities. She warns of "a danger in glorifying only one Indigenous past (Mexica or Maya) without acknowledging Indigenous peoples' resistance to assimilate into mainstream 'mestiza/o' society." She urges us to define our unique ethnic hybridizations and explore how diaspora families use "autotopographies" to reconnect with traditions and re-create their sense of place. Cultivating autotopography is a tool for displaced people and can nurture well-being and survival wherever a materially grounded agency enables one to identify and reconnect with ancestral knowledge.

In chapter 10, Julia Curry Rodríguez, a sociologist who studies migration, the family, and class conflict in cultural encounters, pro-vides the only contribution focused on pedagogy in the college class-room. She recounts how she came to use student inquiries on family food and foodways as a window to the study of the sociology of the family. Curry's students revealed a fascinating quality of their food-ways: A meal is as much about *where* it is prepared and eaten as it as is about the ingredients and methods. Where we eat and with whom are vital to an understanding of our foodways. Curry draws from her family's foods and foodways to explore how memories are potent paths to healing. She rejects the identity politics of "authenticity" and celebrates the creative fire of labor's hybrid circumstances. This is especially relevant given the large numbers of us among displaced people, and we note here how many of our families and relatives have been forced to flee violence and unending acts of settler colonial dispossession and capitalist enclosures. Despite this displacement, a love of food and cooking grows out of the phenomenologically sustained cultural and labor practices handed down by forebears in convivial embrace of the Other. As Gustavo Esteva and Mahdu Prakash explain in their book *Grassroots Post-Modernism*, *la comida* is more than consuming food and "nutrition." It is a social practice connecting us to the full depth of food as a culinary tradition of creativity linked to cohabitation of the kitchen as an autonomous loving space.

The six chapters comprising part III ("Organizing") explore food justice and *autonomía* struggles through select case studies in rural, urban, local, regional, national, and transnational settings in California, Colorado, New Mexico, Illinois (Chicago), Washington, and México. In chapter 11 Tezozomoc and colleagues open by reviewing the historic struggle over the South Central Farm (SCF) in Los Angeles. On the surface SCF was a place to grow heirloom seeds and continue farming traditions of ancestors. On a deeper level, the SCF was a rebellious act seeking to restore the "common." Essentially their message is: You displaced us from our communal *ejidos*? Now we are going to make new *ejidos*, right here in the heart of this city. This struggle pitted privatization interests against the *reterritorializing* practices of SCF. The eviction of the farmers and the bulldozing of the SCF *ejido* were conducted on July 5, 2006, but this was not the end. The group continues to do its work at a new co-op centered in Buttonwillow, California. The concept of "Food for the 'Hood" remains vital in Los Angeles and Kern Counties wherever SCF supports the "localized and autonomous food economy."

The authors deconstruct the neoliberal strategy of "de-subjectivation," which involves an attack on "free-range subjectivities" by means of a violent narrowing of the ontological horizon imposed under a neoliberal regime of hyperindividuality. Tezozomoc and colleagues deconstruct "food deserts" as strategically "un-signified"—by virtue of the revelation of their real constitution as a "dietary form of class warfare" waged by capitalists and producing our condition of precarity. The chapter examines "the regimes of accumulation in regimes of signification" and concludes by emphasizing how SCF embodies a movement comprising "a permanent state of economic refusal escaping spatial enclosures" and signifies a "return, through daily-lived practice, to our institutions of collective action." These autonomous spaces are fragmentary and diffused, and the authors observe how they continuously "erupt and flow across urban and rural landscapes in reiterative escapes of place-making that lie beyond the functional dominium of the apparatuses of governmentality."

Chapter 12 shares the important work of activists at Community to Community Development (a.k.a. C2C) in Bellingham, Washington. Rosalinda Guillen and C2C view food sovereignty as a liberated practice because the ability to decide gets at the very heart of our relationships. The authors enunciate ecofeminist principles to show how they link food sovereignty with immigration reform, workers' rights, and women's autonomy. Each of these elements cannot be separated from the others, which is why worker-run cooperatives have been one focus of C2C's efforts to liberate women from dehumanizing capitalist exploitation. They examine the "solidarity economy" as a form of direct democracy and explain how in participatory practice people learn best by doing. They urge us to consider that "growing food is a dignified, necessary and valued skill." The ecofeminist solidarity economy ensures the survival of all Earth communities.

In chapter 13 Tomás Madrigal, a student of labor movements, takes a hemispheric and Marxist approach to the study of struggles for food sovereignty. Laws promoting racialization have historically been used by agricultural industries and settler nation-states to control farmworkers. Madrigal presents evidence of the continued role of hypercriminalization and mass incarceration as a strategy to keep food prices artificially low and the wages of food-chain workers suppressed. Using workers' personal narratives, Madrigal explains how "migrants" and Indigenous diaspora farmworkers, including those in detention centers, are asserting agency. The narratives illustrate the myriad ways food-chain workers confront exploitation and racialization of their labor. They resist in the fields against speedup, quotas, toxic hazards, and a lack of adequate housing, sanitation, and health care. They resist inside detention centers by waging hunger strikes over inhumane mistreatment, poor food, and a lack of medical attention and access to prescriptions.

In chapter 14 Chicago-based food justice activist and sociologist Pancho McFarland focuses on "organic intellectuals" and alternative urban food justice movements. The author links hip-hop emcees to the placemaking activism of urban bearers of traditional environmental and agroecological knowledge. He investigates how racism under capitalism remains central to the "development of poverty," as was the case with the War on Poverty. Among contemporary emcees, McFarland features Chicago-based performer, "Zárate," whose songs are presented as an "insightful examination of living in a poor, Mexican/Mexican American community in Chicago." McFarland recites from "El Santuario" (The Sanctuary) and "El Sacrificio" (The Sacrifice). These songs are often heard in urban gardens and describe the multigenerational economic problems of Black and Mexican Chicago: a lack of jobs with living wages, job security, and benefits and the ubiquitous presence of gangs, drugs, crumbling infrastructure, and few educational or recreational opportunities. Despite "the conditions of political and social ecological chaos and disturbance," Zárate recognizes the barrio and the people living there with genuine fondness and welcomes us saying, "Bienvenido, compa / al único lugar que nos entiende" (Welcome, relation, to the only place that understands us). McFarland argues that emcees like Zárate and urban gardeners alike challenge the logics of capitalism. They recognize that "our economic system cannot be reformed and . . . must be replaced." Gardening and the work of emcees are emerging and coalescing spontaneously and "rhizomatically" as people create autonomous zones throughout the city.

In chapter 15 Adelita Sanvicente Tello (Fundación Semillas de Vida) and Araceli Carreón ("Sin Maíz No Hay País" campaign) present a compelling history of México as the cradle of maize civilizations. Experienced agronomists, they explore the connection between people and the sacred grain in an alterNative environmental history filled with passages from Indigenous sources. They note how "Cinteotl" (the God of the Cob) is the most important of all the deities and how "teocintle" is the wild ancestor of maize, and they explain how this "establishes another indissoluble bond: the link between the farmers' knowledge and work and the diversity of corn seeds." Their analysis and critique of the dangers posed to native seeds and civilizational cultures by the advent of transgenic maize in México are grounded in understanding the rupture of this basic dialectical relationship: "When our corn was taken out of the civilization that gave rise to it, the subject and object were split; corn lost its 'humanity' and became an object, a traveling plant that produces grain and seed."

Sanvicente Tello and Carreón chart the origins and history of the anti-GMO movement in México. They use original primary source data from Mexican governmental archives to analyze the neoliberal regulatory regime and how it conflicts with México's own biosafety and Indigenous consultation laws as well as the country's status as a signatory to the Convention on Biological Diversity (CBD) and the Cartagena Biosafety Protocols. The authors present a carefully documented argument about the importance of protecting México as a major "center of origin" for maize and multiple other food crops that constitute a major portion of the varieties grown by corporate and alternative producers across the globe. They conclude by placing "the defense of maize as the iconic heart of the common heritage of México" and emphasize how this strengthens "identity and collective organization in rural, Indigenous, and urban communities."

In chapter 16, acequia farmer, seed saver, and agroecologist Devon G. Peña analyzes soil conservation in the guise of neoliberal governmentality, which he views as an act of "epistemic violence" against Indigenous knowledges, beliefs, and practices. Peña focuses attention on farm ecosystem dimensions of food autonomy to show how Indigenous soil knowledge is situated or autochthonous— rooted from the very beginning in the ancient practices of Indigenous people who domesticated plants like the Three Sisters from extant wild relatives. This is an enlightening journey demonstrating how TEK has been subjected to erasure, displacement, replacement, nostalgic glorification, and commodification. Despite this coloniality, soil knowledge and wisdom remain active dimensions of Indigenous agroecosystems, as illustrated by the case of acequia farmers in the Upper Rio Grande watershed. Indigenous soil knowledge is a vital key to resilience and should compel us to continue working at the intersections of agroecology and ethnobotany and food/foodways studies. Describing the Culebra watershed in Colorado, Peña argues that settler colonial sodbusters "left a heavy imprint on this land." In a detailed analysis of contemporary conflicts between soil conservation districts and acequia farmers, Peña finds evidence of resistance, collaboration, and defection. This account offers insights on the future of these conflicts and possible avenues for the survival of Indigenous soil knowledge to advance "regenerative" agricultural practices in our farming communities. Peña presents an epistemological challenge to dominant neoliberal frameworks for soil conservation and grounds work at The Acequia Institute in Indigenous soil knowledge, permaculture, and biodynamics—all rooted in the ancestral civilizations of the Indigenous Corn Belt.

Each chapter here moves through time and space by embodying "Nepantla" in the multiple, shifting subjectivities we constantly inhabit and alter. Our voices blur the lines between activism, scholarship, farming, cooking, and eating. This allows for new opportunities of resistance and collaboration to emerge. The struggle for decolonial food is more than a social movement. It involves the conscious ethical practice of taking care of and healing our bodies, lands, and communities. This book offers insights into the realities of those who are most harmed by the capitalist food system and how we are resisting through personal and collective actions. As Mexican-origin people, our communities span the length of the Americas. In the interstitial spaces of re-indigenized foodscapes, we are exercising a capacity to challenge biopower. Food is part of the materiality of the human experience, and the history of how we have produced, distributed, and consumed food is layered with class, racial, gender, and sexual inequities and violence. We hope this book begins to move discussion of food and foodways studies forward in rigorous engagement with Indigenous peoples' voices because these are clarifying what food sovereignty struggles can accomplish in future possible worlds.

# PART I  ▪  Theorizing

## Decolonial Food and Movements

# From *Borderlands/La Frontera:* *The New Mestiza*

GLORIA ANZALDÚA

We are the porous rock in the stone metate
squatting on the ground.
We are the rolling pin, el maíz y agua,
La masa harina. Somos el amasijo.
Somos lo molido en el metate.
We are the comal sizzling hot,
The hot tortilla, the hungry mouth.
We are a coarse rock.
We are the grinding motion,
The mixed potion, somos el molcajete.
We are the pestle, the comino, ajo, pimiento,
We are chile Colorado,
the green shoot that cracks the rock.
We will abide.

CHAPTER 1

# Autonomía and Food Sovereignty
## Decolonization across the Food Chain

DEVON G. PEÑA

The principles of food sovereignty are associated with an influential global movement and a subject of considerable activist and scholarly discussion. These discourses have much ground left to cover, especially from the standpoint of decolonial theory and the critique of white settler colonialism.[1] Studies of the food sovereignty movement are yet to adequately address some unsettled epistemological and ontological questions posed by *decoloniality*.[2] My approach questions the concept of *sovereignty* itself by means of a critical analysis of the principles embraced by the most prominent advocate of food sovereignty, La Via Campesina (LVC). I focus on silences and implicit acceptance of suspect paradigms in the articulation of concepts of human rights, governmentality, and sustainability in the human-environment relationship: LVC's conceptualization of sovereignty remains bound to Western concepts of human rights; it traverses onto the terrain of the unique prospects and challenges involved in tribal "sovereignty versus autonomy" disputes in First Nations; and it fails to challenge the state of *economic* exception that subjects human and more-than-human beings to unabated ecological violence alongside surveillance, compliance, compulsion, and incitement to dispossession unleashed under neoliberal environmental rationalities, or *environmentality*.[3]

LVC's prominent declaration on food sovereignty presents "dominionist" and "exceptionalist" subject positions that limit and perhaps even rule out the possibility of a politics of *coevalness* among humans, other organisms, and ecosystems. Giorgio Agamben's notion of the "anthropological machine" defines this as the source of the hier-archy of human exceptionalism because it produces a divide between human and animal that subordinates the latter to the former. I depart from Agamben by rejecting the idea that the genealogy of the concept of the anthropological machine necessarily always leads back to an originary binary. I instead propose the following: Despite pretentions, the Western originary binary lacks status as a universal naturalized object. Most Indigenous cultures conceptualize plants and animals as coevals and teachers; sentient landscapes are respected as sources of knowledge and agency.[4] This leads to different results, departing company with anthropocentric concepts of sovereignty. The former reduces other species and organisms to "bare life" by miscasting these as "non-subjects" stripped of the political, social, and legal standing granted citizens.[5] I seek to extend Mick Smith's argument that "the reduction of the world to a standing reserve . . . reduce[s] humans to the status of the 'bare life' . . . [and] constitutes the 'hidden matrix' of contemporary (bio)politics." I agree with Smith on how "the natural world is precisely where the state of exception originally takes the form of the rule." This certainly pertains where the "dominant mod-ern Western philosophical and political traditions are concerned."[6] Smith does not address how this may unfold in the conflict between Indigenous and Western philosophical and social systems, but the argument can be extended to pivotal silences in food sovereignty discourses. We can then clear a way for decolonial principles guid-ing actual biopolitical practices in Indigenous place-based and other communities of resistance. For me, Zapatista decolonial concepts of *autonomía* (Indigenous autonomy) provide a more widely resonating framework for struggles integrating food sovereignty with decolonial and indigenizing methods.[7]

## The Limits and Contradictions of Food Sovereignty: Five Critical Dimensions

The principles of food sovereignty were first articulated in a decla-ration issued by LVC as a global peasant farmers' movement.[8] The statement of foundational principles was released in November 2001. Here is the preamble:

People's Food Sovereignty is a Right. In order to guarantee the independence and food sovereignty of all of the world's peoples, it is essential that food be produced though diversified, farmer-based production systems. Food sovereignty is the right of peoples to define their own agriculture and food policies, to protect and regulate domestic agricultural production and trade in order to achieve sustainable development objectives, to determine the extent to which they want to be self-reliant, and to restrict the dumping of products in their markets. Food sovereignty does not negate trade, but rather, it promotes the formulation of trade policies and practices that serve the rights of peoples to safe, healthy and ecologically sustainable production.[9]

This preamble is followed by eleven principles presented for the realization of the political goals of food sovereignty (see sidebar). These principles were later reformulated in the Declaration of Nyéléni adopted in February 2007 and so named for the site of the gathering at Nyéléni Village, Sélingué, Mali, Africa. This document is formally titled *Declaration of the Forum for Food Sovereignty, Nyéléni 2007,* and not all participants in the drafting were LVC affiliates. The Nyéléni declaration reflects profound shifts in the reframing of food sovereignty and draws a sharp contrast by emphasizing Indigenous resistance to patriarchy in food systems and insipid links to the continuing structural violence of capitalist enclosures in the current wave of global land grabs. (I will address the Nyéléni declaration in future work.)

One of the accomplishments of the food sovereignty movement has been to influence the framing of the mandate followed by the UN Special Rapporteur, who offers this definition of the right to food:

[T]he right to food is the right to have regular, permanent and unrestricted access, either directly or by means of financial purchases, to quantitatively and qualitatively adequate and sufficient food corresponding to the cultural traditions of the people to which the consumer belongs, and which ensure a physical and mental, individual and collective, fulfilling and dignified life free of fear.[10]

Recent commentary on the Special Rapporteurs visits to Canada—where Olivier De Schutter (and later Anaya, the Special Rapporteur

## La Via Campesina Principles of Food Sovereignty

The eleven original Principles of Food Sovereignty adopted in 2001 by La Via Campesina deemed critical to the realization of the movement's political goals:

1. The right of all countries to protect their domestic markets by regulating all imports; which undermine their food sovereignty;
2. Trade rules that support and guarantee food sovereignty;
3. Upholding gender equity and equality in all policies and practices concerning food production;
4. The precautionary principle;
5. The right to information about the origin and content of food items;
6. Genuine international democratic participation mechanisms;
7. Priority to domestic food production, sustainable farming practices and equitable access to all resources;
8. Support for small farmers and producers to own, and have sufficient control over means of food production;
9. An effective ban on all forms of dumping in order to protect domestic food production; this would include supply management by exporting countries to avoid surpluses and the rights of importing countries to protect internal markets against imports at low prices;
10. Prohibition of biopiracy and patents on living matter—animals, plants, the human body and other life forms—and any of its components, including the development of sterile varieties through genetic engineering;
11. Respect for all human rights conventions and related multilateral agreements under independent international jurisdiction.

Source: La Via Campesina 2001.

on Indigenous Rights) expressed concern over the "deep and severe food insecurity faced by aboriginal peoples across Canada"—led Native critics to point to prolonged discontent in Indigenous public discourse over the "imagery of a white man pronouncing upon indigenous issues."[11]

While sympathetic to most of the original 2001 principles, I have concerns over how some have resulted in the occlusion and silencing of important decolonial ethical principles and political claims. This includes misrecognizing resurgent alterNative epistemologies rooted in multigenerational Indigenous knowledge of the human-ecology nexus.[12] These missing elements involve principles that are indispensable to environmental and food justice movements and directly negate the privileging of white settler colonial *dispositifs* (governmental apparatuses, habitus) and human-centered economic systems. AlterNative epistemologies privilege Earth's ecosystems as originary sources of right livelihoods and as teachers of the rules of place-based living.

I read the principles outlined by LVC (see sidebar) as a narrative trapped within a neoliberal capitalist epistemology of the environment and labor. In this formulation human rights are still ultimately subject to the alienation of human and more-than-human entities under the spell of the commodity form. As a result, LVC's "brand" has been left open to abuse wherever affiliates reproduce the privileges of petite bourgeoisie worldviews amid the mass of legitimate struggles by precarious farmers. The 2001 LVC food sovereignty declaration avoids explicitly *anticapitalist* statements. It is critical of corporate power and industrial agribusinesses, including the biotechnology "Gene Giants," but fails to address the deeper problems inherent to the capitalist regime imposed under neoliberal investor-state trade treaties. These enclosures impinge on territories *and* local democratic spaces targeted for dismantling by neoliberal globalists. This also reflects unresolved divisions inside LVC between smallholder subsistence-oriented farmers and larger, more market- and trade-oriented commercial farmers who claim, often without proof, that they too embrace social and environmental justice ethics and standards in pursuit of "sustainable agriculture."[13] According to one gentle criticism of LVC, these problems stem from a lack of "class analysis" and tendencies toward "unnecessary localism."[14] The critical reading presented here focuses attention on five unresolved contradictions or ambiguities evident in the 2001 declaration. This involves criticism of a framework that

1. accepts and promotes Western concepts of human rights;
2. remains bound to anthropocentric forms of governmental or state sovereignty;
3. fails to recognize and engage in the active defense of the ecological and economic base services associated with

Indigenous and other traditional agroecosystems subjected to neoliberal enclosures and dispossession;

4. fails to understand how the precautionary principle is rendered largely irrelevant given the state of agricultural biopolitics and biotechnology today, a situation requiring active support for and commitment to a restoration ecology paradigm; and

5. remains uncritically committed to the "sustainability" agenda and overlooks the rise of more revolutionary confrontations with the neoliberal logic of the capitalist regime; one especially significant dimension of all these issues is bound up with LVC's failure to address the rise of urban food justice movements across the food chain.

## 1. ACCEPTANCE OF WESTERNIZED CONCEPTS OF UNIVERSAL HUMAN RIGHTS

LVC claims to promote Indigenous rights, but the original food sovereignty principles failed to clarify the differences between distinctly Western (US-led) approaches to human rights and Indigenous precepts governing all our relations. The concept of "universal human rights" has been criticized as a "Trojan horse" of neoliberal design and "recolonization" masquerading as respect for Indigenous peoples.[15] The framing of human rights under the rubric of Western nation-states, and to a more nuanced extent within the UN, has largely privileged "individual rights," and these are actually too often reduced to property rights or limited appeals for due process and equal protection. This is certainly the US position, as was evident in the work of researchers with the US Foreign Military Studies Office (FMSO) during the infamous "Bowman Expeditions" that sought to "weaponize" cognitive maps in Indigenous territories of México to compel privatization of common property (*ejidos*).[16] All this unfolds while Indigenous discourses alternately work on transforming policies through the UN Permanent Forum on Indigenous Issues and other forums.

In another sense, the problem resides in the very concept of the "individual." A vital epistemological guidepost for me, following Shawn Wilson, is the Indigenous idea of *relational knowledge*.[17] This principle is enunciated by a majority of the world's Indigenous languages that lack a word translating into "individual" let alone a legal concept that is the moral and juridical equivalent of the Western idea of a separate and autonomous (qua atomized) subject with specific rights and duties as defined by a sovereign power. Most Indigenous languages include words that translate to "person" or "human being," but these are typically coupled or nested within the epistemic tenet that to be human, to become a person, one must be in relation to others. Interconnection is being.[18] This is expressed in the Maya concept of "In Lak'ech" (You are my other me) and the Lakota concept of "Mitakuye oyasin" (All my relations), two well-known examples of AlterNative ways of knowing and being.

LVC needs to critically reflect on the epistemological challenges posed by questions over the definition of human rights, indigeneity, sustainability, equity, and the nature of the "market" and "trade." LVC should consider concepts posed by Indigenous peoples that articulate their enduring presence and voices from autonomous and collective philosophies of being. These must be understood as positions of radical subjectivity dead set against and actively delinked from the colonial politics of interpellation. This requires understanding how settler colonial intruders seek to trap Indigenous peoples within juridical and political narratives that insist we negotiate the conditions of our surrender and co-optation to the globalization demands of neoliberal capitalism. This is one among many "trickster" moves used to enact biopolitical control and impose an economic order based on hyper-individualism, disconnection, and self-aggrandizement. This configuration produces settler subjectivities willing to fulfill the role of presumably "telluric" partisans championing the cause of continued enclosure and dispossession of Indigenous territories.[19] Since it has a substantial global research apparatus, it would be useful for LVC to clarify its understanding of alterNative perspectives on so-called human rights. Decolonial delinking from Westernized precepts of human rights could inform the work of scholars of food sovereignty who rarely address the widespread but unstated epistemological influence of rational choice theories (RCT) in a barely visible and unholy union to establish a universal definition of homo oeconomicus in order to justify fundamentalist market extremism.[20]

The relationship between human rights and food sovereignty must be clarified through Indigenous voices because these enunciate their own conceptualizations of rights that lay well beyond the Western colonial emphasis on *individual* rights to possessions, due process, and equal protection. Indigenous cosmovisions and customary laws, results of *constitutive* power and the basis of actually existing autonomous polities, emphasize the principle of interrelationship including the first ethics of the obligation to take care of home places through mutual aid and social cooperation.[21] Misconstruing this to mean that Indigenous people neither deserve nor desire due process or equal protection is a grave error because First Peoples have autochthonous principles for organizing and implementing such matters especially within the spaces of institutional autonomy. This is not an essentialist or romanticist notion given the long arc of struggles against structural violence and historical trauma. It *is* a hotly contested process and *unrealized* political project in most Indigenous communities pursuing autonomy. There are no guarantees that the restoration of autonomy in pursuit of the fulfillment of obligations to all Earth communities can be fully realized,[22] and so these struggles remain salient features of movements coalescing around indigeneity and autonomy.

## 2. ACCEPTANCE OF SOVEREIGN (STATE) POWER AND ANTHROPOCENTRIC "DOMINIONISM."

Indigenous concepts of property are *relational*, and large bodies of customary law illustrate the designation of obligations to family, clan, village, or other groups holding use rights to specific territories. These embrace "Earth-care" obligations by forbidding abusive uses of the land, water, sea, and wildlife. These are originary rules for tenderly inhabiting place. Fulfillment of these ethics of Earth-care guarantee the exercise of future use rights. By invoking the so-called Seventh Generation Principle, these (home)land ethics avoid the white settler colonial subjectivity of "free riders." I have urged attention to these issues by informing environmental justice discourse on why principal Indigenous concepts of human rights are about respect for the capacity of communities to fulfill obligations as cohabitants following "Original Instructions"—understood here to involve respect for the agency of the Earth as the only true "plane of immanence" manifested in the force and substance of matter independently asserting its *presence*.[23]

LVC should take a more cautious approach to the problem of sovereignty as exercised through neoliberal governmental rationalities that are behind new enclosures seeking to extend the dominion of late capitalist globalization and even relocalization. LVC does not address how neoliberal institutions co-opt the petite bourgeoisie ideology of the nouveau peasantry. There is internal reluctance to mobilize support for effective struggles *across the entire food chain* and in active defense of Indigenous ecosystems. In addition, the struggles against racialized environmental injustices continue to challenge the alternative, sustainable agriculture and food sovereignty movements and the corporate-dominated sectors alike. Neoliberal ideologies create spaces for some family farmers to feign progressive, nonexploitative relations with farmworkers, other food-chain workers, and the environment. LVC's faith in the ability of civil society to compel sovereign powers to adopt radical political transformation in response to grassroots demands is naive at best—for capital, food will always be a political weapon in the legal civil or class war.[24]

Food justice movements are smart to redirect political work away from excess engagement with neoliberal investor-state politics and focus on civil society networks and free associations. The Zapatistas have done this to dramatic effect in southeastern México. Food sovereignty activism can also redirect resources and community assets toward the task of *Indigenous self-valorization*—the rebuilding of our traditional agroecosystems and cooperative forms of "prosumption" in both rural and urban contexts to more effectively challenge the neoliberal corporate paradigm,[25] which cannot thrive in delinked spaces. This means refocusing on actually existing spaces of autonomy and the formal and informal networks of mutual aid and cooperative labor in Indigenous ancestral and diaspora-adopted territories.[26] LVC should explicitly challenge the existing state of economic exception and the "nonsubject" status of Indigenous peoples and the Earth itself. As the Zapatistas have said, describing the ethics of resistance to neoliberalism, "Here you can buy or sell anything except Indigenous dignity."[27]

The LVC principles of food sovereignty commit a strategic error by negotiating within the fields of power/knowledge imposed by neoliberals in their formulations of state sovereignty. Discussing "peoples' right to food," LVC declares that *"Governments must uphold the rights of all peoples to food sovereignty and security, and adopt policies that promote sustainable, family-farm based production rather than industry-led, high-input and export oriented production."*[28] This appeal to benign governmental rationality is politically naive and glosses over the need for critiques of the assemblages and intersections of race, gender, class, national origin, heteronormative, species, and other structural inequities inhering to the investor-state nexus and of how these intrude on social collectivities, communities, and the associations of civil society. In the United States, numerous family farms in the alternative and sustainable agriculture movements pronounce themselves advocates of food sovereignty. Yet affiliates of LVC so far remain incapable of developing effective strategies against the notorious sites of continued patriarchal domination. Many farmers harbor reactionary and regressive attitudes toward farmworkers, women, and animals. State sovereign power is complicit in deregulating the constitution of these relations of domination and exploitation under the guise of sustainability.[29]

## 3. FAILURE TO FOCUS ON THE VALUE OF ECOLOGICAL AND ECONOMIC BASE SERVICES

LVC's principles of food sovereignty lack concern for struggles by Indigenous and other traditional smallholder farmers to revalue the ecological and economic base services provided by ethnoecological practices.[30] LVC food sovereignty advocates demand that people be able to produce their own food using agricultural methods that are appropriate to their time, place, and cultural traditions. However, what if these traditions are harmful to other members of the social and ecological community? What if other social actors, including the military, are disrupting and interfering with these practices? Situations can and do arise within dispossessed and "co-managed" territories in which presumed traditional farming systems are incompatible with the protection of native wild and agricultural biodiversity. Some LVC affiliates are engaged in monoculture production (e.g., coffee producers in México and Vietnam).[31] These farmers contribute to the decline of farmland as habitat and refuge and reproduce the patterns of exploitation many denounce as the sins of corporate monocultures.[32]

Traditional farmers often represent new generations of incoming displaced rural workers or subsistence farmers seeking refuge from encounters with enclosures in other locales. Many remain landless and are not seeking to produce local foods and are instead unwitting participants in the environmental violence of unequal geographies of development. Food sovereignty advocates need to address this by clearly identifying and defining the sociopolitical conditions and agroecological practices under which Indigenous and other place-based communities can continue to provide services to ecosystems that protect biodiversity in both rural and urban locales. This requires appraisal of the differences between established and incoming small-holder farmers. Many advocates uncritically focus only on *localizing* production without fully considering varying impacts on native bio-diversity, wildlife habitat, and landscape ecologies in the struggles for Indigenous heritage subsistence rights. LVC is concerned with protecting the diversity of landrace and heirloom crop varieties. It does so directly by calling for the protection of smallholder agroecosystems, seed saving and exchange, and opposition to transgenics and the patenting regime underpinning development of agricultural biotechnologies. Yet we still need effective policy proposals and direct-action plans to protect native agrobiodiversity and the habitat of wild relatives of heirloom cultivars in the vital centers of origin. While some scholars have documented such projects among LVC affiliates in Cuba, India, and Zimbabwe,[33] LVC itself has not appeared concerned with making this a strategic planning and direct-action priority.

This remains a major issue for Indigenous farmers, plant breeders, and seed savers and could help us reconceptualize food sovereignty to reflect our obligations toward more-than-human beings, including the threatened wild relatives of the thousands of food and medicinal plants stewarded by Indigenous farmers across the planet. In México we have the example of *Zea diploperennis* and the effective campaign to protect native varieties in a center of origin.[34] Indigenous farmers worked with civil society groups, scientists, and environmental and human rights organizations and used the courts to negate investor-state logics and reground México's binding status as a signatory to the Convention on Biological Diversity (CBD), Cartagena Biosafety Protocols, and the Indigenous consultation statutes of International Labor Organization Convention 169. This led to a ban (widely ignored with impunity by corporate growers to date) on transgenic maize and soybeans as threats to México's sixty-two maize landraces. LVC should clarify and strengthen its work on transgenics and ecosystem justice by more actively supporting grassroots struggles that strategically use select multilateral conventions and national statutes and customary law to promote biosafety and protect the genomic integrity of landrace cultivars, the habitat of wild relatives, and the autonomy of Indigenous seed savers and plant breeders in their centers of cultural and ecological origin. LVC must achieve and promote awareness of the centers of origin and diversification of Indigenous crops *within the United States* and of the struggles of Indigenous farmers seeking to protect their native maize from threats posed by gene flow and the introgression of transgenes.[35]

## 4. FAILURE TO RECOGNIZE THE LIMITS OF THE PRECAUTIONARY PRINCIPLE

The precautionary principle is one of those ethical concepts that reeks of liberal naïveté and the partial politics that arise from acting after damage has been done. For precautionary measures to be effective, one must be in a position to prevent the condition that would cause the actual harm. The Convention on Biological Diversity and the Cartagena Biosafety Protocols were meant to make precautionary regulation practical, but rules are still being promulgated and reformulated long after the environmental release of genetically engineered organisms (GEOs, a.k.a. GMOs) and the widespread commercial and experimental planting of transgenic crops. Despite success in México, precautionary measures are difficult once the gene(ie) is out of the bottle. So we might ask, What's the point? To champion an idea whose time is passed to little discernable effect? If neoliberalism throws precaution to the wind, what is the point of insisting on forms of risk management co-opted by market-steered cost/benefit analysis in the extant regime of environmentality?[36] Remember, harms are cumulative and even epigenetic. So, how are we to engage in effective precautionary regulation when many threats and their effects now involve possible hyperobjects?

A more radical declaration on the environmental, cultural, and social impacts of transgenic technologies would do well by working to (1) eliminate the legal basis of the patents on life regime (which LVC calls for), (2) emphasize a worldwide restoration strategy to eliminate transgenes from plant genomes in centers of origin and diversification, and (3) restore the integrity of landrace heirloom varieties *and* their agroecosystems. Despite the bad news occasioned by the discovery of transgenes in the landrace maizes of Oaxaca in 2001,[37] recent research suggests that transgenic lines are both "promiscuous" and "unstable," while landraces are "in-bred" and "stable" (genetically speaking). Cultural practices and sound agroecological management should allow seed savers and plant breeders to restore genomic integrity, but this will come with high costs for limited-resource farmers and perhaps some cases of irreversible harm to specific alleles in wild relatives.[38] This approach also requires land redistribution policies to address the political decomposition of Indigenous autonomy in settler colonial capitalist nation-states. The reversal of waves of neoliberal enclosures is a precondition for the resurgence of heritage landscape ecologies in Indigenous territories—we must be able to ban transgenes in the centers of origin. This requires investing in the recovery of the conditions supportive of ecosystem and cultural resilience. Getting beyond neoliberal capitalist rule is one thing; ending the logic of the commodity form and the long duration destructive effects of capitalism as prime driver of the disturbances of the "Capitalocene" is another matter altogether. In this context, the precautionary principle is rendered largely moot, and our strategic focus should encourage a shift toward direct action to dismantle the patenting regime and to promote restoration ecologies as the best-practice horizon for a more radical politics of food systems.

## 5. ACCEPTANCE OF THE FAILED SUSTAINABILITY PARADIGM

Not rendered moot is the question of how communities bounce back and overcome disturbances. The resurgence of alterNative spaces of autonomy foregrounds the protection of ecological and genetic diversity alongside fulfillment of first obligations to home ecosystems or "full habitance." This is different from "sustainability." The latter concept was made fashionable by the UN Commission on Environment and Development (UNCED) and 1992 Earth Summit. There is growing recognition of how "sustainability" has been co-opted by corporate interests and neoliberals promoting "green governmentality."[39] Also, with a shift to autonomy and resilience theory,[40] a growing number of environmental scientists, ethnoecologists, and political ecologists support Indigenous criticisms of the ethics and politics of "sustainable development."

Growing numbers of Indigenous activists, including farmers, are rejecting the concept of sustainability as co-opted by neoliberalism. We are by far more concerned with articulating concepts of resilience *that move beyond mere adaptive practices.* For food justice activists, this means espousing more than "food self-sufficiency" (*autosuficiencia alimentaria*).[41] The concept of resilience basically comes down to these principal ideas: First is the idea of the "Capitalocene," or the "Age of Capital," as one in which changes in Earth's systems still operate as a coupling of social and ecological processes just as with prior periods of "anthropogenesis." The problem now is how the linkage has become inherently antagonistic and destructive on unprecedented temporal and spatial scales. We cannot separate this coupling because conditions in one affect conditions in the other. Second, these systems are resilient only when they are able to adapt to or "bounce back" from disturbances. This ability is a key to the survival of Earth's life-support systems. The possibility is illustrated by Indigenous peoples who have weathered successive ecological revolutions unleashed by settler colonial nation-states.[42] This is why we need a food sovereignty declaration that embraces anticapitalism and resilience as key principles. Such a declaration would emphasize the "intrinsic value" of this coupling of social and ecological subsystems and assert how justice in "habitance" applies to *all* living organisms and ecosystems, not just human beings. We must overcome the tendency to presume the Earth is only our prostheses.

The United States remains relevant because it is the world's central food hub and master purveyor of "sustainable" biotechnology and industrial monocultures. But we are also at the center of the urban ecological revolution being created by food justice struggles in the cities. In the United States and some European countries (most notably Spain, Italy, and France), there is a new wave of "Great Pretenders" behind myriad neoliberal enclosures of agroecological commons. All of these play for larger market share under the brand of sustainability through organic and heritage farm-to-table production schemes. Many organic, local, and slow-food farmers are not the least bit interested in the living and working conditions of women and farmworkers and express little concern for ecosystem values beyond those commoditized as part of the newfangled agricultural tourism markets. Such illusions of sustainability are encouraged by the neoliberal devolution of regulation to the dominant actors across the varied sectors of the agri-food system.

## Food Sovereignty and the State of Economic Exception

I turn to a more detailed discussion of some problems posed by the unacknowledged anthropocentrism of the LVC food sovereignty declaration in light of a failure to challenge sovereign power more explicitly as a state of economic exception. Agamben has made the forceful argument that since 9/11, Western liberal democracies have retreated to a permanent state of exception (or emergency) in which the rule of law is suspended by the sovereign power in the interests of national security. This includes suspension of due process (habeas corpus) and equal protection.[43] But the matter is more complicated than that. Agamben states that

> modern totalitarianism can be defined as the establishment, by means of the state of exception, of *a legal civil war* that allows for the physical elimination not only of political adversaries but of entire categories of citizens who for some reason cannot be integrated into the political system.[44]

Absent integration, this "legal civil war" is leading to the wholesale reduction of entire categories of human beings (tribal peoples, undocumented workers) and more-than-human beings (endangered plants and animals, ecosystems) to "bare life"—a life without political virtue suspended in a zone of right-lessness, or a being whose life is denied biological flourishing or even survival, since nonsubjects can be killed, or, say, denied water, without such acts constituting murder, extirpation, or extinction. Agamben overlooks this as a condition that Indigenous peoples and ecosystems have experienced for more than five hundred years. Our exception did not begin with the prelude to World War II or after 9/11, and it did not result just in bare life. Instead it produced the Indigenous problem of "bare habitance." Decolonial scholar Mark Rifkin defines this "as the biopolitical project of defining the proper 'body' of the people . . . subtended by the geopolitical project of defining the territoriality of the nation."[45]

The extension of the state of economic exception to more-than-human beings and entire ecosystems is endemic. While food sovereignty advocates argue for sustainable, equitable, and place-based agri-food systems, LVC underestimates how the violence imposed on organisms and entire ecosystems as exploited objects blocks full realization of Indigenous autonomy. As long as we allow capitalism to miscast ecosystems as the stage for the unfolding of human drama, as long as we fail to enforce respect for the Earth's life-support systems with their own rules and transformative agency, and as long as we fail to recognize the intrinsic value of ecosystems independent of the economic value capital wishes to inscribe via a universal "social hieroglyphic," then radical moves beyond the institutions that colonize and commoditize all life, and all organisms, will remain elusive. We need a complete transformation of the "coupling" of social and ecological subsystems rather than reformism tweaking at the edges of selected sectors of the agri-food system. We also need to find other ways to express environmental values without reducing these to human-centered metrology. Under the state of economic exception, capital imposes nonsubject status on Earth's communities by reducing all to "bare life" or "abstract labor" while forcibly removing or alienating us from place, that is, the condition of bare habitance. This is the first line of defense maintaining the hegemony of the "Republic of Property" (see chapter 11).

Deepening critique along these lines will lead to decolonial modification of the LVC principles of food sovereignty. A chief problem is the overly eager quest for recognition under the policy- and law-making power of the state (or multilateral institutions) that reinforces the hegemony of sovereign powers to dictate to us what constitutes equitable law and policy. LVC cedes the ground to the Leviathan of capitalist environmental rationalities allowing sovereign powers the space to define what equity and fairness are and to do so in a manner that erases and marginalizes what are called the "rights of nature." Seyla Benhabib argues that sovereign states are able to more or less exercise "ultimate authority over all subjects *and objects* within a prescribed territory."[46] For me, this includes the instantiation of nature's subjugation within the construct of biopolitical sovereignty. LVC declarations on food sovereignty must be reformulated to explicitly reject the biopolitics of bare life/bare habitance and embrace a shift toward alterNative paradigms of interspecies, intergroup, and intergenerational equity, all defining obligations to the Other. This epistemic intersectional challenge can overcome the phenomenologically weak apparatuses and habitus created by settler colonial-capitalist states under neoliberal conductors who still wield control over sovereignty in the current regime of biopower.

Some critics of LVC's organizational strategy are perhaps justified for questioning a movement that spends a lot of time and energy raising funds to convene international meetings to issue proclamations while the needs of local farmers in struggle are eclipsed by the endless iteration of political demands. Jefferson Boyer is among a group of scholars who are critical of what is perhaps the most ambitious LVC policy initiative yet, a worldwide campaign for agrarian reform as a key to attaining food sovereignty. The goal of returning land to the tiller is laudable, but discursive framing by LVC has been flawed and may have actually damaged smallholder prospects and negotiations for land reform in some countries. In his detailed study of small-farmer movements in Honduras, Boyer notes:

> Via Campesina initiated a global campaign for agrarian reform. It stressed that food sovereignty includes the right, and usually the necessity, for peasants, small farmers, farm labourers, indigenous peoples, and women as well as men to shape the institutions and services of such reforms . . . [but] sovereignty is not a term

that expresses the concerns of everyday rural life in the same manner that security does. In Via Campesina's efforts to establish a universalizing alternative to food security, they clearly were seeking to balance several ideological tendencies. . . . The idea of autonomy invoked by the term sovereignty may well appeal to populists and certainly groups influenced by some anarchist traditions . . . but it also can become somewhat *confusing to the many who equate sovereignty with states and not with the rights of particular peoples* or . . . their daily lives.[47]

I share Boyer's concern and find myself growing skeptical of the manner in which we have failed—without much self-criticism or political reflection—to pursue a clearer understanding of why so many concepts in the food sovereignty discourse are objectionable to smallholder farmers and Indigenous people. Boyer surely opens a new line of criticism, but I have four key points of divergence: First is my rejection of the food security frame since, from a decolonial standpoint, we can avoid hunger and still suffer malnourishment from a lack of access to our Indigenous crops, foodways, and heritage cuisines.[48] Second, many Indigenous and other landless and smallholder farmers understandably comprehend the concept of sovereignty in light of deeper histories and lived experiences with the exercise of abusive and violent state powers. This is a realist perception since states in the Global South typically are major accomplices behind bare life/bare habitance. Third, the mostly landless farmers are adept at avoiding confusion about these matters. In Honduras, their clarity of purpose is the reason Indigenous and other smallholder farmers have remained relevant social forces for land redistribution. It is also why they participate in food justice struggles in the "spaces of neoliberal neglect," to borrow a phrase developed with my colleague Michelle Tellez. Fourth, these smallholder farmers embrace their own visions of autonomy—often understood as place-based coevalness, in a bioregional common, enacted through Indigenous traditions of community-based governance. They certainly do not confuse any of this with "sovereignty" of any kind. In the Latin American context and elsewhere, bioregional commons are the material basis sustaining long-term epistemologies based on relational knowledge. This requires understanding the difference between sovereignty and autonomy when it comes to matters of land redistribution and tenure while accounting for the differentiated political landscapes of capitalist enclosures in each colonized bioregion.

It is important to recognize how large LVC is. The organization claims more than two hundred million members worldwide. Like other large-scale organizations, LVC may experience the misuse of their "political brand." A recent example is the case of LVC Mexican affiliate UNORCA (Unión Nacional de Organizaciones Regionales Campesinas Autónomas) in Chiapas. This entity has a paramilitary wing tied to a cooperative of coffee producers known as ORCAO (Organización Regional de Cafeticultores de Ocosingo). These producers are supporting paramilitary activities by launching violent attacks against Zapatista "Caracoles" (communal villages and schools) in the uplands of Chiapas. This included one incursion resulting in the assassination of beloved Comandante Galiano.[49] When an organization becomes this large, how do leaders coordinate all the groups across the planet? How does one respond to violence unleashed by those seeking justification via association with the brand? This is altering local perceptions of LVC's relational accountability.

Many colleagues agree with the idea that LVC should offer an alternative organizational form comprised of bioregional nodes across networks linked as global feedback loops and acting as principal drivers of actual resource allocation. The extant transnational organizational form is subject to strong centrifugal forces and escapes the eye of episodic general assemblies, whose members constantly reel from one global gathering to the next, or who find themselves caught up in the endless cycles of local events that command, if only for a moment, global attention. In the meantime, groups with hidden agendas exploit the brand for aims that rely on the exercise of political violence. In México this involves an unholy alliance among narco-trafficking cartels, paramilitary groups, and municipal police acting with impunity against Indigenous communities. This was made clear in the aftermath of the September 2014 Ayotzinapa massacre and the disappearance of forty-three normal school students working on Indigenous agricultural education projects. LVC's muted response to the larger implications of Ayotzinapa has been deafening.[50] These issues are relevant to current urban agricultural mobilizations, and I address this further in the following, concluding section.

## Food Sovereignty Spans across the Food Chain

We see multiple signs of emerging alternatives to anthropocentrism and the rejection of acquiescence to a neoliberal global order whose biopolitics seek the commodification of everything related to food and foodways. The practical autonomy of place-based Mexican-origin and Mesoamerican diaspora communities relies on the culturally grounded exercise of self-governance, and this allows us to reclaim our seeds, agroecological traditions, foodways, and heritage cuisines. This is occurring in rural and urban areas through the conscious enactment of heterotopias in community gardens, home kitchen gardens, and liberated kitchens spaces, described in the chapters that follow. At the heart of many of these alternatives are organizational forms involving cooperativism inspired by Indigenous general assemblies and a consensus approach to participatory democracy. The prospects for creating true food sovereignty may come to rely on the practices, normative orientations, and relational knowledge of Indigenous farmers and other food-chain workers, including those who have been displaced from originary lands and are both transborder travelers and mobile placemakers.[51]

The strategic problem requires confronting contradictions between rural, peri-urban, and urban areas. The 2001 Principles of Food Sovereignty do not directly address this issue in any meaningful manner. This is a significant oversight since vibrant activism and innovation are occurring in urban and peri-urban agricultural movements across the planet. How do we achieve human rights and culturally appropriate, economically liberating food self-sufficiency on a "planet of slums" unless we address the metabolic disorders and political economic imbalances within and between urban centers and surrounding bioregions?[52] If what we seek is to address the ability for hungry and malnourished people in the cities to provision themselves through sustained access to fresh, organic, and culturally appropriate foods in an equitable life-affirming manner, then we should consider the insights of Philip Aerni, who observes that the restructuring of the global population via rapid urbanization poses serious strategic questions for LVC:

> The definition [of food sovereignty] implicitly assumes that local food production and consumption can ensure food security and

therefore the human right to food. It completely ignores . . . [the] process of rapid urbanization. . . . [This] means that a smaller share of the population needs to produce more food with less input [and] is focused on self-sufficiency. . . . [How is this] supposed to feed this rapidly growing urban population? Do[es] the human right to food appl[y] only to those who produce their own food within the self-sufficient community?[53]

This is a profound set of questions. LVC policies and direct actions have so far not advanced self-reflexive analyses or critical policy responses to the dynamics facing communities of resistance in the city-countryside nexus. This seems especially timely given the rapid growth since the 1990s of urban-based food justice movements across México, the United States, and other places.

Displaced farmers, often from the very same rural areas surrounding so-called global cities, and other transborder arrivants are mobilizing these movements. LVC's declaration remains silent on the struggle for the city despite the trend toward a planet of gated cities and sprawling slums pockmarked by food junkyards *and* more hidden kitchen gardens. Urban food justice movements offer a lesson here: LVC could revalorize farmers and their rural communities but also *all* food-chain workers and self-provisioning urban farmers. We all share a desire to escape bare habitance. We are all against being reduced to abstract labor for capitalist production, which is the force behind the precariousness of our condition as unwanted surplus populations forced to find a way to live off the books. The chapters that follow show us how this precarity can also be a source of the capacity to create alterNative convivial or solidarity economies.

LVC must understand why many of the landrace cultivars threatened with extinction are no longer cultivated in their original agroecological landscapes. Many rare and endangered heirloom varieties are now cultivated by displaced farmers in cities and suburbs, as illustrated by the Indigenous farmers of the post-NAFTA Mesoamerican diaspora. They are preserving these varieties in urban home kitchen and community gardens and farms across México, the United States, and even Canada. LVC needs to rethink what it would mean to become a global organization and network for *all* food-chain workers, including displaced Indigenous and other "peasant" farmers in

US cities, suburbs, and towns. The autonomy perspectives explored in this chapter are guided by awareness that our movements do not seek permission from the state or corporate acquiescence in order for us to act in solidarity. Relational accountability/solidarity is really praxis not theory; *it is a method of resistance.* We must act everywhere possible in a radical manner by refusing to submit to sovereign power as we rebuild local deep-food systems for ourselves based on relational knowledge of our place-based cultures and convivial economies. Build your own economy—one not separated from the political but converting politics into the art of cohabitation (in Arendt's sense) and dedicated to conviviality and cultural mentoring.

As a university-based research scholar I am obligated to conclude with a brief excerpt from the 2009 UN Environmental Program report *Agriculture at a Crossroads,* which identified a "growing tendency . . . in the United States, to encourage research likely to return financial benefits to the university rather than broader benefits to the public or ecological commons." The report, also known as the *International Assessment of Agricultural Knowledge, Science and Technology for Development* (IAASTD), further criticized US universities for "offering private sector partners such as the agrochemical/biotechnology industry a wider role in shaping university research and teaching priorities."[54] The future of food sovereignty movements clearly will also be shaped and constrained by epistemological politics as these unfold inside US research universities and the land-grant college complex. The future is also already being shaped by the advent of global projects by "philanthrocapitalists" like Bill Gates Jr. But the forces of neoliberal environmentality must face determined and growing opposition from our movement's "calmecacs"—the Indigenous institutions of higher learning for collective action and the survivance of agroecological knowledge in our return to full habitance.

# CHAPTER 2

# Indigenous Women in the Food Sovereignty Movement
## Lessons from the South Central Farm

### RUFINA JUÁREZ

*I wish to thank the Creator for this time in our Chicana history—I am honored as a Xicana-Indígena woman to speak to you about the importance of our rights and place within the global food sovereignty movement.*

## South Central Farmers

The farmers at the South Central Farm (SCF) began our struggle in September 2003 when a single-page notice was anonymously posted at the entrance gates of the farm.[1] It was written in English but was addressed to a mostly Spanish-speaking community with many mono-lingual Indigenous language speakers. This letter stated that the L.A. Regional Food Bank thanked us, the "gardeners," for our ongoing participation in their "gardening" program, but they were sad to inform us that the program had ended. Consequently, 350 farming families were being ordered to leave the farm. We were not only being asked to leave the land but were being told we had to discontinue growing and distributing our healthy, traditional, culturally grounded foods to extremely poor local families. The implications of this notice threatened a historic social movement that was under way; the underlying message of the notice placed our Chicana/o and Indigenous food justice and sovereignty movement directly under attack. This notice was problematic on multiple fronts. We were concerned about the threat to our struggles but also about the immediate effects on the local poor and, most of all, the complete disregard for the original conditions that drew farmers to this land.

This story is starting to sound too familiar to you all, right? Everyone here understands how the historical push-pull factor works with regard to the mistreatment of Mexicans and Latinos under our immigration and labor policies in this country. You pull in the workers when you need them, then you toss them right out when you do not. This is exactly what happened to the SCF. We were invited by the city government to create a farm, which resulted in an ancient, traditional, and sophisticated farming system, but then we were pushed off, at gunpoint, by the government and told we had no rights or legal recourse to remain on the land despite the many years of labor.

This forced us to raise several important questions regarding the mission of the L.A. Regional Food Bank: "Is it no longer the primary mission of the food bank to feed the local poor?" "Have poverty and hunger magically ended in South Central Los Angeles?" "If the farmers are effectively helping to fulfill the primary mission of the food bank, then why put an end this so-called gardening program?" Take note that we have always considered ourselves self-determining *campesinos* (farmers), but the food bank continuously insists on calling us gardeners (*jardineros*). Now, if you ask any Mexican or Latino that ever set foot on the SCF, they will tell you, based on their common sense, that what was happening there was not gardening—it was too huge of a project to be a garden. This was a farm. This distinction is important, and I will explain it further.

At the same time, this notice forced us to ask key questions regarding the Los Angeles City Council: Is it no longer a fact that the land was given to the community by the city as a form of mitigation due to the inner-city conditions of extreme poverty, racism, lack of educational and employment opportunities, unsafe parks, and widespread police abuse, which were exposed worldwide during the 1992 uprising in Los Angeles?[2]

For the next twelve years, 350 families, the majority consisting of displaced Indigenous people, worked the land and created a fourteen-acre farm in an urban setting in the middle of the most contaminated and polluted eastern section of South Central Los Angeles, at Forty-First Street and Alameda Avenue. People had been growing food for over a decade, and now it was threatened with destruction.

At the time, South Central L.A. was a dormant place, an environment where poor people did not speak out politically out of fear of police brutality or *migra* raids. However, this movement forced us to take a stand and move toward saving the farm because we felt it was the correct thing to do. This action forced us to begin to organize and mobilize around the question of land, food, and the environment.

The central concerns of our struggle depended on believing that we had a right to this land. If we had the right to land, we also had the right to continue to grow our own foods there and the right to continue to feed our families and ourselves with fresh and inexpensive organic produce, while at the same time affording a safe place for children.

Our movement consisted of 350 working-class members whose experiences and backgrounds embodied the reality that exists today in many of our poorest urban neighborhoods. Men, women, and families like the South Central Farmers were once forced to leave their place of origin and live in marginalized neighborhoods and urban cities throughout the United States, many without legal recognition or political representation. For many of the South Central Farmers, their forced migration from México and Central America was a result of megacorporations taking away their natural resources and farmlands while destroying their local self-reliant economies. This rare open space represented the only connection we had to the land, or Mother Earth, and our right to grow food that we felt was culturally appropriate. Within the fourteen acres, we created a safe place for our children and relatives to play, study, pray, and heal wounds from life in a city dominated by realtors and developers. Our elders had a place where they felt wanted and productive. This reminds me of Don Juan, who, with only one leg after amputation, had one of the most beautiful parcels on the farm. And yes, this was his only access to good nutritious food since, to this day, South Central L.A., like many neighborhoods where low-income people of color live that are not yet gentrified, does not have a single grocery store that offers pesticide-free foods.

## So, What Is Food Sovereignty?

The term *food sovereignty* was introduced in an international platform at the World Food Summit in 1996. This was only two years after the Zapatista uprising in Chiapas, México, when the world became aware of the criticism of neoliberal policies and its effects on land ownership, resource use, and exploitation of campesinos in particular. According to La Via Campesina, an international movement that coordinates farmer organizations worldwide, food sovereignty is the right of every nation and all peoples to define their agricultural and food policies. At a national level, this implies that the nation can determine how it organizes its food production and import and export policies, as well as distribution. At a more local or individual level, this implies that the community can determine the nature of food production as well as the manner of consumption and the modes of distribution.

The concept of food sovereignty aims to become an alternative to neoliberal policies since these affect food systems in a manner harmful to the majority of the people and the planet. Food sovereignty goes beyond the more common concept of food security, which is limited to ensuring that a sufficient amount of safe food is produced without taking into account the kind of food produced or how, where, and on what scale it is produced. Most food travels many miles, and it requires fossil fuel, chemicals, pesticides, and hormones to keep it "fresh" until it reaches its destination. Most regional food banks across California focus simply on food security. One of the many lessons of our struggle is realizing how these food banks are dependent on and tied to the megacorporations for food distribution. The problem that we have with our food system in the United States, according to Anuradha Mittal,

> is that food, instead of being about communities, is now about commodities. It is controlled, not by the family farm growing food for families and communities, while maintaining biodiversity, but rather it has come to mean large corporate industrial agriculture farms, where machines have replaced farmers, where corporate agribusiness has replaced family farms. What we see is the result of a disconnection between us and food, where ("donde") we have been reduced to consumers.[3]

We have all become dependent on someone else for food. As Chicana/os, who still have the traditional knowledge of farming within our extended families, we should be very troubled by this. The dominant capitalist food system and aligned neoliberal policies con-

Figure 2.1. Aerial view of South Central Farm, June 2003. The SCF was the sole green spot in a landscape of warehouses, junkyards, freeways, and railroad tracks. Courtesy of Lane Barden.

tinue to control, even monopolize, the production and distribution of food and thus undermine our struggle for food sovereignty and the right to *our* foods and foodways. The conflicted nature of food is related to the current state of immigration, as is evident from the displacement of peoples and massive movement due to the destruction of the farmland and other resources that we depend on (see chapter 13). The US government gives priority to international trade over peoples' subsistence and right to autonomous livelihoods. Most of these development and trade policies have done nothing to eliminate world hunger. On the contrary, these policies have increased our dependence on agricultural imports and intensified the industrialization of agriculture while endangering our Mother Earth and our cultural and environmental heritage and placing the health of the world's population at risk at the hands of a few capitalist corporations. These policies have driven millions of women and men, farmers and farmworkers, to abandon their traditional agricultural practices, forcing them into rural exodus and migration.

## SCF and Food Sovereignty

What happened to the SCF families working and sharing life on the farm in South Central L.A. represented what is going on in many parts of the world today: our traditional farming was brutally interrupted, and the people were uprooted from the land. You cannot see it now, but our traditional agricultural practices helped us harvest the massive amounts of food that we had on fourteen acres in an urban setting. This model challenged the institutional definition of how urban land is supposed to be used and, more important, how we are supposed to be organized to work.

The traditional knowledge we were given through the generations, and the heirloom seeds we harvested, kept, and shared created our food sovereignty. At the peak of our movement at the farm, we supplemented foodstuffs for more than three thousand families in one day. The families participating at SCF defined their own agricultural methods of growing and taking care of the topsoil, which created microecological atmospheres that reduced the temperature at least twenty degrees (Fahrenheit) on the farm compared to the surrounding hot urban concrete jungle. We were defining the structure of our food system by grounding it in socially, economically, and culturally appropriate foods that met the needs of our unique circumstances. Since there were not any natural-food stores in the area, the farmers were forced to grow serious foods and medicinal plants, which meant that all 350 families took hold of their right to safe, nutritious, and culturally appropriate food and food-producing resources that sustained us and our surrounding communities. At times, these medicinal crops were so highly sought that we had people who traveled from Nevada looking for some of the healing plants we were harvesting. This is merely one of the ways we practiced food sovereignty on the ground.

Our message has been consistent and is simple: Chicanos/Latinos should have the right to access affordable, quality, and nutritious food. When we walk into the grocery store, we believe that we have the power to purchase what we want. However, before one arrives at the store, someone else has already made the choice of food for us. In low-income barrios, that choice is usually non-nutritious food. There is no true democratic process in the selection of your food—you have given that responsibility to someone else because you no longer have that choice. Your only choice is to commute across cities to find a Whole Foods or Trader Joe's, which gives you "better" food choices, but they are expensive choices. The way food distribution works in our society is that the existing system has determined what you will buy and eat by deciding what food is shelved in distinct communities. Someone else chooses for you. Further, your money does not pay the worker who had to pick your food, since they are not paid prevailing wages. Someone is making a profit, and it sure isn't the workers.

Consider the alarm over "conflict diamonds." There is a system that has for years allowed diamonds that are mined through slave labor to enter the global jewelry market. This is clearly a significant issue, yet most people buy diamonds only once or twice in their lives. In contrast, people try to eat three times a day, and we really do not think of how our food gets to the market or who will actually benefit from its consumption. We never speak about "conflict food"!

When the SCFs talk about food sovereignty, we mean having control over the system that determines the types of foods that are available and acceptable to our community to eat. It matters to us that the food we eat does not exploit people who grow, pick, and distribute it. So, as we begin to articulate the significance of food sovereignty for our people, we begin to see why the L.A. Regional Food Bank, a food security–based program, was threatened by our existence. Food sovereignty is a threat because the work is about more than securing food in the bellies of the poor. The work disrupts the norms of securing food contracts from megacorporations in the food industry. We realized the SCF was a true threat to the establishment—we were a threat to the pocketbooks of the local McDonald's, Domino's Pizza, La Superior, and Food 4 Less. Can you imagine what our communities would look like if we had a South Central Farm in every barrio throughout the United States? So how can we achieve radical redistribution of productive resources that are fundamental to real change for an improved quality of life?

Food sovereignty is therefore, the *right* of peoples, communities, and countries to define their own agricultural, labor, fishing, and food and land policies so they are ecologically, socially, and economically appropriate to their unique circumstances. This includes the right to safe, nutritious, and culturally appropriate food and to food-producing resources as well as the ability to sustain ourselves and our societies. Food sovereignty places people's and communities' rights to food and food production over trade concerns. This entails the support and promotion of local markets and producers over production for export and food imports. But as we will see, even the struggle for food sovereignty has its contradictions and tensions.

Figure 2.2. Rufina Juárez, SCF farmer, seen here taking a much-needed rest break outside the main gate of the farm during the antieviction protests in summer 2006. Courtesy of Devon G. Peña.

## Life as an Indigenous Woman

For us, the struggle to save the farm grew from a sense of our ancestral and historical right to the land; we may live in cities, but our hearts belong to the land. SCF, a small, fourteen-acre farm, was the only place where we could teach and transfer our traditional knowledge of growing food while regaining our relationship with the land, our Mother Earth. I, as a woman, believe I carry the message for all the women in our struggle to reclaim our fire, our food justice, as Indigenous people. This is the crucial aspect of the struggle for environmental justice, as we cannot reach food sovereignty and autonomy and survival of our communities without challenging the legal system and its manipulation by real estate and other neoliberal proponents of privatization and enclosures.

My vision as an Indigenous woman involved in environmental

and food justice struggles starts with the fundamental principle that we cannot separate the environment from Mother Earth, traditional knowledge from technological advances, the right to eat from the right to grow your own traditional food, and autonomy from community self-governance of land and water. Here, I am including our four-legged brothers and sisters. Furthermore, this can only be accomplished when we have a relationship with, and responsibility for, our Mother Earth and her resources.

For us, the Indigenous women of the world, land, territories, and natural resources are the fundamental bases of our existence since we have long developed a spiritual and sacred relationship that looks for a holistic connection between "being" and "nature." We reaffirm that this sacred connection is a collective right, which is not negotiable, specifically because of our obligations to protect the Mother Earth, water, sun, and air.

## La Red Xicana Indígena and the UN Forum

In 2007 Josefina Medina, a SCF leader, and I participated in the UN Permanent Forum on Indigenous Issues with La Red Xicana Indígena, an Indigenous women's network. There we presented the following statement to the Special Rapporteur on Migration Issues:

> The corporate production of food, such as hybrid corn, soy, and wheat, has taken over the production of local high quality ancestral food (i.e., corn, squash, and beans) through the displacement of traditional agricultural communities who produce food for their own use and for trade. Corporate takeover of agricultural lands forces families to flee to local and international urban areas, where they transform from a self-reliant, highly skilled agricultural society, into poor and politically vulnerable substrata of urban society. Economically dependent on low wages for unskilled labor, men, women and children lose their relationships, roles, ancestral knowledge and practices of self-sufficiency. Their lack of economic resources makes them dependent on cheap poor-quality food produced by the corporations, which displaced them in the first place. Coupled with the lack of health education and basic health care they are highly defenseless to long term diseases like obesity, diabetes, cancer,

and asthma, which make them lifetime consumers of pharma-ceuticals. The rise of childhood illnesses produces long-term profit for corporations. Indigenous peoples in diaspora are in fact paying for their own oppression.

At the forum, we highlighted the devastating issue faced in our poor urban and rural communities that have no access to good-quality food and have no choice but to purchase high-caloric processed foods like Maruchan and Top Ramen dry noodle soups. Consequently, when the SCFs were evicted in 2006, we were denied the right to grow our traditional foods and to teach our children their relationship with and responsibility to the land. Therefore, it can be said that on a small scale the SCFs were an example of a diaspora Indigenous population that does not have any rights to practice the continuity of ancestral traditions outside of their homeland.

The SCF was physically structured in eight sections that gave individuals responsibility for communication and resolution of issues, which do arise when you bring that many people together to work on an urban farm. Before we started to organize, however, many individ-uals did not care to nor were empowered to participate in the admin-istration of the farm. As an elected representative of the farmers at our first general assembly, I was assigned this responsibility to empower and engage members in self-governance. It was not difficult for the Josefinas, Carmens, and Carolinas of the farm to step up and help. The SCF women organized, cleaned, marched, patrolled, and changed locks in the middle of the night to deter the theft of "Papalo,"[4] which sold like crazy at the local market. It was through the leadership of the women that we could organize the men and families when needed.

One of the challenges, which I am sure most activist women have experienced, was the physical threat to one's person and body. Yes, we were called names and were physically attacked by men in our com-munity who wanted to weaken our leadership. How did we overcome this challenge? By always having groups travel together and making sure we had our own individual, sometimes personal, security guards. And most importantly, by keeping our heads held high with visions of feeding our families and saving the farm. We had one goal: that the farm should benefit all of us including the men, elders, children, and the people in the community who need good deep food.

As *mujeres en movimiento*, we faced both subtle and aggressive attacks on our persons, our bodies, and our spirits and not only from external forces outside the struggle but by male members in our own community. In the SCF struggle, I and other women were attacked verbally and physically. I remember the relentless microaggressions by a young African brother from the local community who adopted what our Black colleagues at SCF called a stereotypical "gangster" or "ghetto style." He always took a hostile stance toward the SCF women and positioned himself physically close and inside our personal space. He did this to the point of appearing to threaten imminent attacks and used such posturing to manipulate the situation to weaken our expressions of political views. He was trying to intimidate us as women, which creates and feeds into the racial divide between Blacks and Mexicana/os. He consistently used words and phrases such as "You're a bitch, a whore, you have sold out the community" in allegations that sought to induce self-doubt among our group members.

Another challenge that we faced is the machismo (male chauvinism) among the farmers. At one point, women were not being allowed to grow what they wanted because some husbands controlled the family plots. In response to this problem, we created the women's cooperative section only for women who wanted to grow crops of their choice. Arguably, this action can be seen as the women conceding to male chauvinism and leaving the family plot to be run solely by the husband. However, our cooperative section offered so much more in exchange. It gave many women leadership roles, and they created a united front and eventually pressured some of the men to change their attitudes. Through the practice of women leading by doing and focusing on the main goal of growing a variety of medicinal plants and foods, we overcame some of the barriers faced in a male-dominated community and openly questioned the historical gender roles once considered untouchable. Most importantly, by having all members of the SCF become part of a participatory democracy, we were able to insist on transparency and accountability. Contracts were written not because we did not believe in the traditional oral word, *la palabra*, but because they were *a tool that documented our agreement in public* and attested to the consensus we gained in the long general assembly gatherings.

A subtler form of machismo occurred when a young Chicano

journalist wrote a one-sided piece attacking the leadership of the SCF. He copied two previous writers who took their lead from Jan Perry, the city council member who facilitated the sale of the land to a former owner at a price far below its market value. The tone of these articles was that the SCF leadership prevented democracy and stole money from the membership. They based their information on the words of angry groups that had split from the main body of SCFs either because they were scared by the threat of police attacks on the farm, were uncomfortable accepting leadership from women, or abandoned the struggle for a measly piece of land underneath high-tension electric cables in an area sixty blocks from the original site.[5] Quotes from these sources included those from upset male members who physically attacked the two elected leaders of the SCF; we had to impose restraining orders on them. The false accusations leveled by the journalist against the leadership included one targeting me for allegedly "embarrassing the humble farmers by implying that they were not men" in a speech I gave in which I placed him on the side of Jan Perry, the politician, and Horowitz, the beneficiary of her generosity with the people's money. But it did more than that. His willingness to interview and take at face value the remarks only of the people who had abandoned the struggle put him on one side of the conflict. Given the nature of his profession, he is able to mold public opinion and therefore has a greater responsibility.

If we lived in a vacuum, then we could assume that machismo is a thing of the past, but we do not fall into the trap of static thinking. Most of the more unscrupulous members of the old-guard leadership at the farm were men, and they were not willing to accept the women's self-assertion of leadership. However, in the end they had no choice but to accept this leadership whenever important decisions were made, including answering strategic questions. One time, we were offered a strip of land for relocation, and the old guard tried to divide us. The women asked, "Do you stay with the farmers and *fight for this place*, or do you accept the strip of land under the high-tension electricity transmission wires?" These efforts did not erase decades and centuries of deeply engrained machismo. Some intellectuals used machismo to hide behind cowardly support of Perry and betrayal of the residents of L.A. and destruction of the farm.

Figure 2.3. The SCF contingent at the immigrant and Indigenous diaspora peoples' protests. More than two hundred SCF members and supporters marched in Los Angeles on May 1, 2006, as part of an estimated five hundred thousand protestors mobilizing against the criminalization of immigrant and diaspora peoples. Courtesy of Devon G. Peña.

## Conclusion

We have a tremendous amount of work ahead of us. I hope that the SCF struggle can become a source of lessons and inspiration for future struggles. I hope that each of you is in the right place doing the work to liberate our communities. Food can be used as a weapon for decolonization of our youth and communities. The key questions are, Why, where, and with whom do we eat the foods we do, and who is making those choices for us? An elder gave me this personal message to give to all of you:

> Our Mother Earth is being raped, violated, and criminalized by men. They call this global warming but I say that Mother Earth is mad and all these changes that have been happening to our Mother Earth are a result of the abuse. In order for as humans to survive, we must change how we treat "nuestra madre tierra." We

all have to be in the right place doing the right work, building a movement for the survival of our Mother Earth.

I did not ask to be in the SCF struggle. Rather, it came calling when, like so many other *compañeras*, I found myself in that place back in 2003. We did the work, and we still are working, always relearning how to grow food to sustain ourselves and erasing years of conditioning that binds us to servitude in the dominant food systems and practices that are so dependent on fossil fuels. We all must close the gap that we have created in order to survive on our Mother Earth, *nuestra madre tierra*, and bring life back into balance.

CHAPTER 3

# Food Values
## Urban Kitchen Gardens and Working-Class Subjectivity

GABRIEL R. VALLE

I recently visited with an urban gardener in San Jose, California, who give me advice about growing food. He told me that growing food was not just about growing food. He informed me that it was about *comunicación*. For him, growing food meant nurturing lines of communication with the plants, the land, our bodies, and each other. In the simplest of explanations, he was talking about how food is part of a self-provisioning experience that requires strong forms of reciprocity. At a deeper level, he was explaining how the simple act of growing food requires mutual recognition of the Other. When we grow food, we are part of a larger web of social ecological interactions through which human and more-than-human actors collide and sometimes coalesce to create the possibility of new ways of seeing, being, and interacting in the world.

In this chapter, I share my experiences working with kitchen gardeners in San Jose as the context of study. There are many reasons for this; however, my primary reason is reality. The truth of the matter is that many Chicana/o and immigrant communities in the United States still lack access to the traditional community gardens articulated and celebrated by the urban agriculture movement and discourse. Regardless, through the use of agroecological knowledge, immigrant communities in the United States have been able to transform major portions of the urban landscape.

Some of the greatest contributions to the study of agroecology have come from Miguel Altieri and Stephen Gliessman.[1] Their contributions to agroecology cannot be overlooked. However, one of the greatest misconceptions is that agroecology exists only in rural settings. One shortcoming of Altieri's and Gliessman's work stems from limited efforts to conduct applied research in agroecology in US urban settings. Such a focus of analysis is becoming increasingly important in light of recent migrations from rural villages to urban cities.[2] Recently, Christine Buchmann published an essay that surveyed the diversity of crops found in kitchen gardens in Cuba.[3] She reports that while food crops are grown in these kitchen gardens, these are grown alongside medicinal and ornamental plants. While this is one study among many, Buchmann's Cuban research suggests kitchen gardens remain poorly understood as adaptive strategies, and I realized this was also the case in low-income communities in the United States.

I hope to accomplish three things. First, I argue that growing food in US kitchen gardens remains an underexplored phenomenon. Chicana/o Studies has long looked at the ways in which individuals, families, and communities have sought to better their quality of life. In fact, one of the principal goals of environmental justice was to create a theoretical framework to understand how people struggle against the oppressive nature of social and political institutions.[4] Similarly, food justice is a concept that has been developed to explain how people engage in social and political action to bring justice to the food system. As explained by Robert Gottlieb and Anupama Joshi, "food justice, like environmental justice, is a powerful idea. It resonates with many groups and can be invoked to expand the support base for bringing about community change *and* a different kind of food system."[5] One could argue that environmental justice has always included elements of food justice. Regardless, by framing the movement as food justice, rather than environmental justice, we are bringing direct attention to the way food affects our bodies, families, and communities. The point of this is not to argue the theoretical underpinning of either one but to bring attention to the many ways oppressed people continue to struggle in ways that validate their culture, bodies, and communities to improve their quality of life. It is my belief that in recognizing the importance of kitchen gardens in the United States, and particularly in immigrant communities, we can better understand how gardening is an act of self-valorization.[6]

Second, I seek to explain how kitchen gardens challenge the common logics of economic rationality. This rationality, which embraces the rational choice theory (RCT) framework, treats social exchanges in the same light as economic transactions. The principal idea is to maximize good and minimize bad. This utilitarian philosophy has its roots in theorists such as John Stuart Mill,[7] who believed that rational human actors always seek to optimize pleasure and profit. What this framework cannot explain, however, which is the basis of this essay, is the social phenomenon of trust and reciprocity and diverse subjectivity of "pleasure." While RCT seeks to come to terms with rational economic logic and utilitarian gain, it cannot explain why individuals voluntarily join groups where collective benefits are practiced and celebrated over those of self-gain.

As explained in great detail by Marx, value is a social phenomenon, which means the value of an object is produced only when a social group recognizes its value. This is where kitchen gardens make a significant contribution to the theory of value. The kitchen garden is nurtured by individuals and families, but the family is supported by a larger network of kitchen gardeners who are guided by similar goals. In most cases, kitchen gardens are small. They can range from an eighth of an acre to a few square feet. In other words, a family cannot produce all of the food it may need. To meet these needs, families rely on social networks. What one family has too much of, another may need: lemons for peaches; blueberries for chilis. Marx argues that the value of a commodity derives from the human labor that is put into producing it. Does Marx's theory of value work in the context of kitchen gardens? Or is value in the kitchen garden closer to that of Nancy Munn, who believes that value derives from the ability to act or create change? Or further yet, is value in the kitchen garden closer to that of David Graeber, who believes value appears when individual acts become socially meaningful?[8] While these ideas of value have differences, the similarities appear to be grounded in the realization that individuals and groups make use of value in ways that generate positive social transformations.

If kitchen gardens are part of the food justice paradigm, then they are, as Gottlieb and Joshi explain, about creating a "different kind of system."[9] Does this mean the food grown, consumed, and shared in kitchen garden networks is valued for the potential to change or bring justice to the food system? For many people, this different type of system has led to the concept of food sovereignty. But food sovereignty as a theoretical enterprise for *urban*-based food justice movements across the food chain falls apart very quickly, as delineated by Peña in chapter 1. Perhaps we need to view food sovereignty as an ethical lifestyle?[10] My pursuit in this chapter is to present kitchen gardens as I have experienced them—*as a way of life.* Kitchen gardens are not simply a means to produce fresh, culturally appropriate herbs, vegetables, grains, legumes, and fruits. The convivial labor and social relations of the kitchen garden networks I worked with produce values beyond any form a mainstream or neoliberal economist would willingly recognize. Instead, kitchen gardens are a pathway-and-practice complex created and used by persons seeking to maintain open lines of communication that produce subjectivities (ways of being in relation to others) that challenge the logic of the dominant economic rationality and embrace an ethical lifestyle of generosity.

In the third section, I seek to bring attention to a decolonial approach to food sovereignty. In academia today, courses on "decolonial" or "decolonized" food, health, medicine, the body, and so on are ascendant. My goal in this section is not to discredit any of these approaches. It is simply to recognize that decolonial struggles involve more than simply removing the shackles of oppressive institutions that have historically hindered colonized communities. Decolonization is an intimate journey that is highly personalized and requires the ability to heal the historical trauma caused by colonization, so that a decolonial approach to the study of food sovereignty is guided by much more than a critical perspective.[11] It is also inspired by the desire and willingness to liberate the production of knowledge,[12] while generating knowledge that is useful to the people it seeks to liberate.[13]

Linda Tuhiwai Smith explains decolonial research projects as documenting how peoples and cultures survive. She also notes this is about "the struggle to become self-determining, the need to take back control of our destinies."[14] Simply because alternative food movements declare themselves as alternatives to settler colonial and globalized capitalist food systems does not mean all food movements embrace a decolonized approach. Alternative food movements are often highly hierarchical and loaded with class, ethnic, racial, and citizen privileges.[15] A decolonized approach to urban agriculture requires a rethinking of the value of food, labor, and health in view of settler colonial forms of domination and enclosure. Food is essential to language, culture, community, and health, and the ability to grow, produce, and consume food is in itself a self-determining project. Food has the potential to heal our bodies, communities, cultures, and relationships. Home kitchen gardens are a window into understanding how families heal through the collective experience of growing, sharing, and consuming food.

This essay focuses on three central issues important to practitioners and advocates of urban agriculture and more specifically home kitchen gardens. First, by bringing attention to the importance of home kitchen gardens, this chapter encourages further research on how kitchen gardens contribute to a family's ability to navigate interstitial spaces. Second, the analysis brings attention to the ways in which kitchen gardens embody practices that gardeners use to continuously transform and remake the value of food and labor. Lastly, this chapter draws from over three years of participant observation, interviews, oral histories, and focus groups with gardeners in a mostly Mexicana/o and Chicana/o community in San Jose. I collaborated with them to learn how they use their gardens to heal physically, spiritually, psychologically, and environmentally from the damaging effects of capitalism. Food and gardening mean different things to different people, and this chapter seeks to enunciate how local, often displaced and diaspora people address global problems through humble acts of gardening and food sharing.

## The Home Kitchen Garden

Christine Buchmann writes, "worldwide, home gardens are a community's most adaptable and accessible land resource and are an important component in reducing vulnerability and ensuring food security."[16] This alone underscores the importance of kitchen gardens, yet very little collaborative and action-oriented research has been done to assess their impact on food sovereignty struggles in US urban communities. Low-income communities across the United States are often burdened with dilapidated infrastructure, inefficient and costly public transportation, lack of access to fresh and quality food, unsafe neighborhoods, unsafe domestic water, and few open spaces. The urban environment created by low wages further perpetuates economic, social, and environmental inequality. Kitchen gardens can ease these pressures because when communities are able to control local food systems, they can reduce their vulnerability and create more resilient households and communities.[17] In addition to these contributions, one study conducted in Cuba found that home gardens make a significant positive contribution to household budgets. Moskow found that families with kitchen gardens "reduced food bills [ . . . and in addition,] gardeners could earn money selling their garden products."[18] The study also found that the average participant was able to produce 60 percent of the household's produce needs from the garden alone. Furthermore, these kitchen gardens benefit the community in five distinct ways: greater food supply, contributions to the country, that is, nationalism, neighborhood beautification, improved safety, and enhanced urban ecology.[19]

Recent research demonstrates that home gardens found in urban or rural settings "are characterized by a structural complexity and multi-functionality which enables the provision of different benefits to ecosystems and people."[20] This means that kitchen gardens serve different functions depending on the location, and those functions cannot be generalized across all kitchen gardens. Kitchen gardens are unique to a family because they are culturally constructed as a place where knowledge and customs are preserved.[21] While this research provides support for claims about the importance of kitchen gardens, all of these studies take place in the Global South or so-called developing world. Some of the most robust theoretical inquiry, seeking an understanding of kitchen gardens in the US context, has been done by Teresa Mares and Devon G. Peña, who advocate the concept of "autotopographies." According to Mares and Peña, autotopographies occur when groups culturally reinscribe public and personal spaces as a form of "self-telling through place-shaping."[22] Gardeners re-create the familiar landscape of their homeland by planting and cultivating crops that are culturally significant, which helps "the grounding of self and communal identities through place making."[23] In this collaborative study, Mares and Peña were less concerned with home-based kitchen gardens and more interested in community gardens. In their ethnographic study of Seattle and Los Angles community gardens, they found that food "encompasses a set of deep social and cultural relationships that foster community, cultural, and place-based identities."[24] Just as the concept of community is defined by those who are included, so too are community gardens. A community garden is always selective and requires the group to be open and exclusionary at once. Kitchen gardens are not exclusionary despite being part of a "private" home space. Kitchen gardens allow households the opportunity to create an identity that strengthens existing and fosters new social networks that build a collective experience of mutual reliance.

Many studies have found that the "establishment of food producing gardens, often based on local seed systems and traditional crops, in areas of explosive urbanization, is becoming an important tool for making cities more sustainable."[25] Gea Galluzzi, Pablo Eyzaguirre, and Valeria Negri argue that the kitchen garden is a tool to help against hunger, negative health outcomes, and low wages and for the preservation of cultural identities. What is particularly interesting about this statement is that the authors explain that kitchen gardens are often based on local seed systems and traditional crops.[26] How does this correlate to kitchen gardens in immigrant communities? The immigrant experience is often one that requires families to navigate between the formal and informal economy, and the kitchen garden can help families negotiate this unsteady terrain. In my experience, working with gardeners in San Jose, kitchen gardens challenge inequality in a variety of ways. First, by having the garden in the backyard, families take up the struggle for transportation justice. Second, the social networks created by sharing food, resources, and labor foster a working-class subjectivity and thus encourage cooperation and trust, which makes neighborhoods safer. Third, with direct access to food, families can take control of their personal health. Fourth, these gardens transform food from a commodity to a political tool of self-determination. Gardeners use food as they see fit: they trade, sell, eat, compost, and so on; gardeners give food meaning *and* value. In Mares and Peña's account of Seattle and Los Angeles community gardens, they argue that immigrants plant and harvest traditional crops from their homeland to re-create landscapes, cuisines, and celebrations reminiscent of their homeland. Kitchen gardens are no different, and research must acknowledge their role in the persistence of local knowledge, traditional foods, and heirloom seed saving.

## Garden Values in Uncertain Times

In his central work, *Capital* (volume 1), Marx explains that "value . . . does not have its description branded on its forehead; it rather transforms every product of labour into a social hieroglyphic."[27] By "social hieroglyphic" Marx is referring to value's ambiguity. A social hieroglyphic presents itself for what it is on the surface as "natural," but there is nothing "natural" about it. Marx spent a lifetime attempting to denaturalize or demystify how, why, and for whom value is produced. As George Henderson explains, "value is not self-evident. It calls out for rediscovery and reinvention."[28] Value is always interpreted and reinterpreted, resignified, and thus the living determination of value is always a contested, shifting, and unfinished political project (see chapter 11). Henderson argues that Marx was not a master of the life of value; rather Marx scattered the *lives* of value throughout his work. Kitchen gardens are a vital part of the many lives of value because gardeners call out for its rediscovery and reinvention in ways that are meaningful and useful to the given circumstances of production and social reproduction.

Gardens have been central to people of Mexican origin for centuries. When the Conquistadors arrived in México in 1519, Bernal Díaz del Castillo attempted to explain the splendor of the gardens of Tenochtitlan but was left speechless. In amazement he wrote, "I do not know how to describe it, seeing things that had never been heard of or seen before, not even dreamed about."[29] More recently, historians have sought to understand the importance of *chinampas*, or floating gardens, in the urban design and architecture of the Valley of México.[30] Other researchers have sought to explain the importance of urban agriculture and community gardens in ancient México as an adaptive strategy to the changing environmental and political conditions of the region and depict the residents of the Valley of México as "an innovative population, facing the problems of serious environmental deterioration brought about by urbanization."[31] In the past, kitchen gardens, family orchards, and animal husbandry were common. It should come as no surprise that these gardens and their uses are still part of the everyday lives of Mexican-origin peoples in México and elsewhere.

While the presence of kitchen gardens in ancient and present-day México is important for the context of this essay, what is more paramount is how kitchen gardens have historically helped people in pursuit of a more sustainable lifestyle. As we have seen, the kitchen garden is an adaptive strategy where individuals, families, and communities utilize the spaces and resources they have to pursue the quality of life they wish to see. The world today is an uncertain one, and kitchen gardens are a buffer against this uncertainty. My experiences have led me to believe that kitchen gardens are socially constructed spaces that produce food, culture, tradition, knowledge, and language. These benefits have been cited by scholars in a wide variety of contexts.[32] For the families in my study, kitchen gardens help to produce a way of life that is socially meaningful.

The food networks existing within the community are not based on some sort of "gift economy,"[33] because the act of gifting is not the central social structure of the community, nor are they "moral economies,"[34] since the livelihood of gardeners is no longer rooted in subsistence agriculture. For many Mexican immigrants living in the United States, kitchen gardens are the foundation for social networks because they allow people to navigate between the formal and informal economies. Kitchen gardens connect individuals to their food by gaining them direct access, and to their community through the sharing of resources and knowledge. Sharing is both a liberating and subversive act because it helps to foster autonomy, on the one hand, and diminishes the role of the state, on the other. One gardener recently told me that he never saw gardening as a political act until he realized the transformative potential embedded within it. He never really wanted to be a part of the "food things," as he said, but he recently told me that gardening can help construct pathways for people to shape the neighborhood into the type of community they wish to live in and to transform themselves into the persons they wish to become.

In a recent essay, the Chicano urban planner and geographer David R. Díaz observes that there is nothing new about "New Urbanism." His point is that the Chicana/o community has historically practiced a lifestyle that is today going mainstream. Partly as a result of design preferences reflective of cultural desires and partly as a result of necessity, Chicana/o communities have most often been "walkable," and people have often gathered in common social spaces; they have grown their own food and participated in all forms of collective action. In other words, Chicana/o communities have historically practiced a form of multilayered citizenship, which Díaz calls "the life of the 'ciudadano.'" According to him, "the conceptual framework of the 'ciudadano' (city-dweller), encompassing active citizenship in all its manifestation, is a product of barrio everyday life."[35] An active citizenry is not simply about people passively paying taxes and voting; rather, active citizenship is about collectively creating a sense of place, and gardens are just one of the many ways people act in affirmative ways to improve self and community well-being.

In pursuit of improving the quality of life, participants in the San Jose study use gardens to transform the world around them through their relations. According to Munn, in Papua New Guinea value is relational. Value is not, and has never been, particular or substantive. In her research, she finds that "actors create communal value through effecting positive value transformations." In this way, people actively pursue positive value by transforming the world around them. In the negotiation between positive and negative value transformations, a "community creates itself as the *agent of its own value creation*."[36] While kitchen gardens are often separated by household, they are tightly woven together through social networks. An individual gardener will act upon her own garden to produce a value that is meaningful to the individual. However, because the food in kitchen gardens is often shared through social networks, that value of the garden is socially recognized as well. Munn believes that value derives from the ability to create social relations. When her understanding of value is applied to kitchen gardens, we can see how communities are agents of their own self-determination.

Echoing Marx and Munn, the anarchist anthropologist David Graeber argues that value is the result of a social process.[37] What makes Graeber's interpretation of value insightful for understanding kitchen gardens is his emphasis on social creativity. He believes that creativity is always a social experience and that the creative potential of everyday life possesses the possibility of opening pathways to revolution where new forms of value and social interaction may emerge. Value comes from pleasure but not pleasure in the utilitarian sense as individual satisfaction. In this anarchist conceptual view, pleasure arises in a more nuanced manner that includes elements that are biological, social, and aesthetic. Graeber does not restrict value to any one of these elements, because he believes all three are required for something to reach full value potential. Seen through the lens of this paradigm, the kitchen gardens I documented reveal all three elements. The San Jose kitchen gardeners address biological factors by promoting positive health outcomes and saving their own seeds and rootstocks. They address social aspects because kitchen gardens have the potential to lead to the creation of networks that create communal safety nets through solidarity. Lastly, kitchen gardens are rooted in aspects of aesthetics because they may function as autotopographies allowing families to culturally reinscribe space to create a sense of place.

The interpretations of value offered by Munn and Graeber illustrate the way that kitchen gardens challenge the logics of capitalism by creating things that are socially meaningful. But where does Marx fit into this contemporary political composition of our communities? Marx believed that the value of an object derives from the labor put into it. His goal was to demonstrate that value appears from the exploitation of labor. From this exploitation, class conflicts arise. These antagonisms result in opportunities for the emergence of new social relations capable of responding to the political recomposition of the working class. It is here that kitchen gardens hold the unique position to creatively and affirmatively transform the *social recomposition* of a group. In many oppressed communities, urban agriculture has provided a means to gain access to the production of food and health. We must not forget the role of labor. Marx pinpointed labor because he believed it to be the one thing we all shared—the ability to work. Marx did not *dislike* work; rather, he believed labor to be the "living, form-giving fire."[38] Through labor we give form to life because all of life is defined by activity, and it is activity that gives life meaning. The labor of home gardeners transforms space into place by creating autotopographies,[39] and brings form to the social, cultural, economic, and environmental worlds of the gardener. The ability and desire to enact one's own labor provide the "form-giving fire" to transform worlds in pursuit of well-being.

Similarly, Bruno Gullì envisions labor as embodying creativity. In his view, labor is "neither-productive-nor-unproductive, but simply creative."[40] His argument is that accepting labor as a creation of capital is to accept the metaphysical reality of capital. This understanding does not allow us to transcend the restraints that capital imposes on labor. This is what Hardt and Negri mean in their critique of the "Republic of Property" (as the constituted regime of the USA) when they emphasize the relation between capital and law, between economy and politics. Gullì believes that when labor is freed from the manipulation of capital accumulation, it exists in its free and "natural" but always "social" state. This creative possibility is what allows kitchen gardens to provide a family access to the means of their own production and reproduction. With this, people become agents transforming their worlds because, as explained by Marx, labor is "in itself a liberating activity."[41]

Marx, Munn, and Graeber all provide useful insights on how we can view the value produced in kitchen gardens as a complex relational process. Gardeners are active participants in transforming their social world because active participation creates meaning. The informal economy created by kitchen gardens is not one devoid of responsibility; rather, it is the recognition of one's responsibility to others. The ability to embrace an ethical lifestyle called upon by the food sovereignty movement requires us to continually reenvision and reinvent the meaning of value, especially as enunciated by those leading the struggles for food justice.

## Decolonial Gardens

On a warm fall afternoon, I sat in the kitchen with a gardener. She made me a traditional dish of chayotes, tomatoes, onions, and garlic, accompanied with roasted stuffed jalapeños, tortillas, and homemade lemonade. Everything we ate was produced in her garden. We talked about what she was growing in her garden, how she was trying to save water, and how difficult it was to eat healthy. Many of her friends and family members suffer from what are seen as diet-related illnesses,

Figure 3.1. *Huerto familiar* with nopal cactus, San Jose, California. Raised beds, orchard trees, and cactus are visible in this small kitchen garden, September 2014. Courtesy of Gabriel R. Valle.

predominantly type 2 diabetes and obesity. They are well aware of their conditions and understand the risks of unhealthy eating, but that doesn't make their decisions any easier. The reality is that many immigrant and diaspora families live in similar situations across the United States and México. The food that is available and affordable in many low-income and immigrant communities often further perpetuates negative health outcomes.

In a detailed text about diabetes in the Mexican community, Michael Montoya explains that "the diabetes genetic research enterprise would lead us to conclude that Mexicanos/as are *biologically* predisposed to diabetes."[42] His argument is that food-related illnesses have been framed in such a way that the Mexican community has been singled out as being at "high risk" for diabetes. However, throughout the text, Montoya pulls apart this assumption to reveal that the complexity of the disease is a result of many factors, including biological and environmental. He concludes by stating, "as a study of diabetes, the larger project links the politics of sugar . . . to the colonial and thus racial histories that produce it."[43] Similarly, Yvette Flores argues that the mental health of Chicanas/os must be situated within a historical context. The intergenerational trauma of racism, discrimination, and environmental neglect experienced by many Chicanas/os and immigrants greatly influence their overall well-being. According to Flores, by taking into account cultural and social factors, we create a better understanding of how Chicana/o "life experience may nuance their mental health."[44] Just as culture does not exist in a vacuum, neither does health. The conditions that produce (un)healthy communities are the by-product of larger systems of power.

Food deserts and the negative health conditions associated with them are the result of a historical process.[45] The shifting policies of urban development have moved investment from the city to the suburbs. This process of uneven development in many urban centers has shaped the lived experiences of the urban poor. Nathan McClintock argues that "the physical experience of hunger, malnutrition, or the body's biochemical metabolic process cannot be treated as disconnected from the larger-scaled processes determining the availability of food."[46] The health of immigrant and low-income communities is the result of the structural inequalities that have historically influenced urban development and planning. When jobs left many postindustrial urban centers, wages dropped, and rents rose, leaving low-income communities with the need to find or make alternatives. According to Marx, "Men make their own history, but they do not make it as they please; they do not make it under self-selected circumstances, but under circumstances existing already, given and transmitted from the past."[47] Marx was suggesting that the dominant powers establish the circumstances under which the working classes must make history. Thus, the working class is the result of the choices made by their ancestors under racist, capitalist colonialism. The immigrant families growing home kitchen gardens involved in my study face many challenges that often go unrecognized by the general public. Yet, their story is not one of victimization but of self- and communal determination where the humble act of growing food creates alternatives to enhance their collective well-being.

When I speak with gardeners in San Jose, they are aware of the context shaping their daily lives. While many of the families lack easy access to healthy, affordable food choices, they do have their own home kitchen gardens. "Walking into my backyard," one gardener said, "is like walking into a supermarket." In her backyard she is growing medicinal and ceremonial plants to insure the health of her family and community. When I last visited, she was preparing for the winter by gleaning plums and nectarines throughout the neighborhood and canning them. Fall is a busy time for her. She makes tonics, canned fruits, and dried herbs and preps the soil for winter. Her work ethic is tied to cyclical rhythms, and she is always getting ready for the next season. The 2008 economic crash left her and her husband without jobs or health care and dependent on food banks. Without affordable health care or a job to buy sufficient food, they both became ill. The following spring season she began growing food. She worked with neighbors and organizations to ensure they would not suffer this vulnerability again. Today, she is healthy, and she gladly shares garden surpluses with those in need. She also shares knowledge about gardening and health with anyone with an open ear and tirelessly advocates for the transformative power of kitchen gardens.

There are two contrasting frames from which to view labor within the context of urban agriculture. Gullì argues that the return of labor is necessary for the full development of the human experience, but neoliberal logics block the ability of labor to be widely realized as the pursuit of collective class self-determination.[48] The concept of "self-care" has been used within the urban agriculture discourse to present labor as a site of self-governance. Mary Beth Pudup explains that early community gardens were "intended to substitute for the inability of unemployed workers to purchase their means of subsistence by allowing them to grow their own food."[49] This ideology supports the basis of the neoliberal marketplace that positions responsibility to the individual. As a result, "gardening puts individuals in charge of their own adjustment(s) to economic restructuring and social dislocation."[50] Similarly,

Alison Alkon and Teresa Mares critique the food justice movement for following neoliberal logics.[51] They acknowledge that while alternative food movements often object to neoliberalism, they often unintentionally reproduce it by shifting responsibility from the state or civil society to market mechanisms and entrepreneurialism. Alkon and Mares demonstrate how a food justice perspective can identify the ways in which economic and racial inequality is embedded in our food system. The authors argue that while the food justice movement is critical of institutional racism and inequality in our food system, it is less critical of how capitalism perpetuates and normalizes these inequalities. Food justice can be both liberating, because it can help to explain how inequality is created, and constraining, because it reduces inequality to isolated instances rather than larger racial projects.[52]

Food sovereignty moves beyond both food security and food justice because it focuses on the community right to produce for itself rather than continuing dependency on food aid. What sets food sovereignty apart, according to Alkon and Mares, is that embedded within its framework is the need to transcend the local to consider how industrialization, globalization, and centralization impact communities around the world. In doing this, the food sovereignty model rejects the "high modernist" corporate regime of agriculture,[53] and places the people most negatively affected by our food system at the core of the movement.[54] Alkon and Mares argue that many involved in the food justice movement believe "government, not capitalism, is largely responsible for the racist policies and programs that produce hunger."[55] Without recognition of the ways in which capitalism perpetuates inequality, labor itself cannot be freed from neoliberalism, because it cannot escape its self-exploitative historical baggage.

In contrast to self-care, "the care of the self" aims to do good for others. Foucault explains that "care for the self is ethical in itself, but it implies complex relations with others, in the measure where this ethos of freedom is also a way of caring for others."[56] This is because while Foucault believed liberty to be political, he also acknowledged that it must be practiced ethically. The care of the self is the recognition of the knowledge of the self, which implies rules of conduct in relation to others. One person's liberty cannot come at the cost of another's. Home kitchen garden networks function in a similar manner. The success of one garden is reliant on the success of others because the resources and knowledge used are tightly woven together to create a broad mosaic of people engaged in a "network struggle."[57] These networks are made from the individual choices of home gardeners, but the choices are not devoid of recognition of the network as a whole. In other words, the garden network influences the choices of gardeners by connecting them to a web of relationships. The tightly woven garden network allows home gardeners to protect against vulnerability, construct mobility, and improve individual *and* collective well-being. This network struggle enables gardeners to engage in collective action against the oppressive nature of industrial capitalism by facilitating challenges to the meaning of legal and social citizenship.[58] It also promotes negotiating entry across formal and informal sectors of the economy. The strategies of gardeners are less about the complete rejection of capitalism and more about the construction of something new and decolonized.

When we finished lunch, I helped the gardener clean up, and then we walked in her garden. She explained that having a home garden strengthened her network of friends and at the same time provided her more freedom. For her, gardening was not simply a form of self-help, but rather it meant that she had control over her own well-being. For her, gardening transcends the alternative food movement because not only is her food more than a mere commodity, but the labor that she invests in her garden is self-constituting.[59] Growing food allowed her to begin to heal the problems (i.e., health and well-being) associated with the structural inequalities embedded in our food system by providing the means of her own cultural, environmental, and biological (re)production. She understands how her actions influence the choices and opportunities of other gardeners and, as such, how these are embedded within shared values guided by what Native American scholar Shawn Wilson calls "relational accountability." This worldview emerges from Indigenous ontological and epistemological precepts that recognize that daily lived experiences sustain long-duration ancestral ties only by "maintaining accountability to these relationships."[60] Our worlds are constructed by the relationships that link us together, and accountability to those relationships allows for people and communities to remain or become self-determining.

While clearly much work still needs to be done to dismantle a food system flawed to the core, it is important to recognize the small victories of self-affirmation. Eric Holt-Giménez and Annie Shattuck explain that "ending the [food] crisis requires more than simply producing more food or making healthier choices. Ending the food crisis is a political project requiring social, economic, and political organization and transformative change."[61] Mainstream food movements can learn from the adaptive strategies vital to the survival of immigrant communities because these strategies are proactive. Home kitchen gardens are autotopographies where gardening is a form of "self-telling" and embeds gardeners in the social and ecological relations of a place they are creating.[62] While gardeners transform their gardens by adding crops from homelands, they are transformed by their gardens. Garden subjectivity celebrates place because traditional agroecological practices and social networks adapt to meet the needs of each individual, family, and community. These adaptive strategies seek autonomy to actively heal the historical trauma of colonization through cultural affirmation. When we parted, the gardener gave me a handful of seeds and told me how the neighbors all want to control their lives and food. She understood the effects of larger political and social institutions but assured me they would continue pursuing autonomy. Her last words were, *arriba y adelante* (upward and onward).

## Creative Spaces

Julie Guthman argues that the alternative food movement presents itself as inclusionary and transformative; however, discourses and practices often fail to recognize how privilege continues to exert its hold on the movement. Her point is that "the discourses of alternative food hail a white subject and thereby code the practices and spaces of alternative food as white."[63] In pursuit of remaking a more sustainable food system, these logics act as exclusionary projects that hinder the transformative visions of food justice activists who emphasize equity and sustainability. Farmers markets, CSAs, and community gardens are only a handful of the ways marginalized people are moving toward food sovereignty. Yet, in many cases these options reinforce neoliberal logics and undermine transformative outcomes. In my experience, home kitchen gardens are far different from mainstream activism. The success of home kitchen gardens relies on a broader grassroots network of agroecological exchanges involving knowledge, seed, labor, and food. This means that kitchen gardeners are bound together by sharing garden and nutritional knowledge in the form of recipes and gardening techniques. Many gardeners will also share seed or seedlings they have collected, and others may share their labor for harvests, planting, or canning. Everyone I have worked with shares food from their garden because it serves as a medium to express oneself, while the exchanges deepen both the complexity and importance of the social networks. Gardening is a simple act, and perhaps because of its simplicity, its transformative power has been overlooked. Yet, as San Jose home gardeners demonstrate, gardening is a buffer against vulnerability and a catalyst for self- and communal determination.

In this chapter, I have attempted to accomplish three things. First, I have brought to light the importance of home kitchen gardens. In my experiences, these gardens are often hidden out of sight yet highly integrated into the exchange networks of a community. These garden networks help to cultivate a working-class subjectivity that is positioned as both self-reliant and responsible toward others. When gardeners culturally reinscribe their gardens to meet their needs, they possess the potential to make this activity transformative and inventive. Home kitchen gardens are often a family's first line of defense against the unpredictability of climate change, economic instability, and illness. Yet, in spite of all of these benefits, home kitchen gardens remain an underresearched phenomenon in the United States.

Second, home kitchen gardens challenge the logics of the dominant economic rationality because they produce a value that is constantly in the process of being reinvented. Marx recognized that labor is a liberating activity, and when gardeners pursue the self-valorizing aspects of gardening, they too can achieve liberation. For gardeners in my study, the act of gardening is neither productive or unproductive but rather creative. As Graeber argues with respect to a general capacity for agency, the creative potential of gardening allows growers to reinvent forms of social interaction and organization. Those I have worked with commonly use the phrase *poco a poco* (little by little) to refer to the small victories of gardening. They are well aware that gardening has its limitations, yet those limitations appear to be flexible. When I first began interviewing gardeners, one grower stated, "I just want to grow food. I'm not all that interested in the so-called food system." The interview with him revealed a personal journey in which he discovered his own value and social meaning by growing food. At the end of our conversation he said, "You know, with one given crop, I probably come into contact with fifty people I wouldn't have otherwise talked with. There is something to that." He is right, there is something to that.

Third, I have argued that the value produced in home kitchen gardens allows gardeners to decolonize urban agriculture by healing the wounds caused by the historical trauma of uneven development and environmental racism. This is important because trauma moves the act of resistance outside of the political space and into a space of healing. This is not to so say that healing is not political. When we discuss decolonization, we are referring to a political project that seeks to remove forms of oppression. Yet, healing does not simply refer to the spiritual, emotional, and psychological wounds caused by trauma; it refers to our ability to take care of our lands, communities, bodies, and places in ways that are regenerative and restorative. Decolonization must allow for healing to occur and therefore involves cultivating spaces for personal exploration and self-determination. Home kitchen gardens open positive frames, which are important for creative and positive engagement.

There is unease about the way society is headed. Global politics often embrace the myopic logic of economic gain and wealth accumulation. Yet, there are spaces where this logic no longer works. One gardener told me, "I feel like I'm headed upstream, against the current. But I don't want do go with the current because it seems to me like it's headed to some sort of cesspool." Julian Agyeman and Duncan McLaren argue that "smart cities" are actually quite dumb.[64] The logic used in planning cities often separates people through the ideology of neoliberalism and market fundamentalism. A "sharing city" is truly a smart city because "sharing establishes a precondition and motivation for collective political debate."[65] The creativity and political will that Agyeman and McLaren claim are lacking in governance institutions are alive and well in the informal networks of home kitchen gardens. The people in my study know what life in precarious conditions is like. Many have survived precarity for generations. They continue to survive not at the whims of some broader system, but rather because they are innovators and proactive participants in their pursuit of a desired quality of life.

I do not claim that the lives of these home gardeners are easy. I do not wish to romanticize the garden networks. Life for many of the gardeners remains precarious and unpredictable. Despite growing food, many are still forced to use food pantries. Some have been evicted from their homes because of rising speculative real estate values in the Santa Clara Valley. While growing and sharing food greatly strengthens social networks, my collaborative research with the gardeners also demonstrates that these social networks are extremely fragile. Gardening strengthens a community's resilience, but it is not the only answer. While precarity increases vulnerability, it also offers the opportunity for people to develop new forms of autonomy. We need to continue collaborative, community-based, action-oriented research on the adaptive strategies of home kitchen gardeners because it is one site among many where the creation of new subjectivities offers the possibility of envisioning and enacting more just and sustainable futures.

# CHAPTER 4

# Del alivio y coraje la tuna nacera
## A Re-membering of Land and Place

### SILVIA PATRICIA SOLÍS

*I am in a journey to re/member my body to those of my*
abuelas *and* nuestra tierra.
*I use the tools passed down to me to sew myself together*
*again.*
*I grasp the roots of* el maguey para aguantar el ardor y dolor
*I prick my flesh* con la espina del nopal cardona
*to pull through* el hilo de la palma saba.
*I place* albahaca *on my bleeding flesh*
*Listening to the silence the body makes to heal the wounds.*
Del alivio y coraje la tuna nacera.[1]

## Indigenous Lands

I echo Karleen Pendleton Jiménez's advice to "start with the land."[2] Reflecting upon Indigenous epistemologies and their contributions to issues of land ethics,[3] I am struck by the fact that I am landless. Important issues such as sovereignty, treaty rights, and the telling of creation stories are not part of my own understanding of land or place. From my experience, acquiring a sense of place—or a way of being that connects me to "land, community and traditions of reverence for nature"[4]—comes from visiting the lands my grandmother grew up on or by remembering the beauty in my mother's graceful presence

whenever she listens to Facundo Cabral.[5] The land Indigenous thinkers speak of does not feel like the same land I was born on and should
still inhabit. The land I grew up on is a territory full of gravel, cement
blocks, toxic and tired soil with beautiful gardens growing from it.[6]
The land I know has a long metal fence—called a border wall—staked
on a line drawn on a geopolitical map by elites, patrolled by men and
women carrying guns, disregarding children and the places where
they play. The lands I am now learning from are Indigenous homelands to many, bordered—"*entre norte y sur*"—between two places:
"Heroica" Matamoros, Tamaulipas, México, and Sarita, Texas, United
States. These are political markers of how we are "territorialized" by
settler nation-states.

I honor the work of Indigenous thinkers. My intent here involves
more than a desire to seek any form of recognition or flight toward
a romanticized Indigenous genealogy from other peoples' renderings
of land and place. Nor is it a "comfortable reflexivity"[7] of symbolic
lands and places to consciously or unconsciously seek "settler moves
to innocence," which are "diversions, distractions, which relieve the
settler of feelings of guilt or responsibility, and conceal the need to
give up land or power or privilege."[8] This is an intimate approach
to deepen my own connections to land and to better understand
notions of place by taking into account the materiality of everyday
bordered lives and bodies and the forms of violence these experience.
My humble pursuits, the support of and alliances with decolonial and
anticolonial projects, begin with the unsettling of my own intentions
to eliminate all possibilities that may emplace colonial logics brought
about by my own embodied story.

I did not grow up understanding my family's stories as Indigenous
creation stories placing me outside of the time/space of colonization,
empire, and capitalism.[9] The only creation stories I can speak to are
the one my mother tells of my birth and the one my grandmother
tells of my mother's birth. My grandmother was not allowed to be
present at my birth to assist her own daughter, and my father was not
by our side. These stories of creation do not link me to an ancestral
empire and cosmology, but they are real and come directly from the
materiality of our telling flesh; that they are born out of my mother
and my grandmother's flesh also makes them stories of colonization

when placed within a story of diaspora that grapples with borders, violence, longing, and hope.

To rely on such overarching and imposing discourses as diaspora or border theory to explain my personal and family's experiences defines our existence in overly determined absolutes. When I thread through the experiences within the stories of my family, yes, one theme would be of diaspora experienced as entangled embodied experiences of loss, longing, death, returns, borders, crossings, inspections, social and economic exploitation, and terror of the other. But these experiences include land- and place-based healing knowledge surviving generations, and travels that are escapes within the scope of diaspora. One of the potential pitfalls of diaspora discourses is the tendency toward a native/migrant binary—who is Indigenous and who is a settler? I am interested in engaging with this binary only to elaborately articulate its rejection. Trying to find a place as either native or migrant, or even in the interstitial spaces of the line running through them, only serves to distract and distort any attempt to deepen my connection to land, and it only serves to further entrench colonial violence within our relations.

What I am bringing to the forefront are entangled embodied experiences as the very constructions of land and place: The making and unmaking of land and place require us to unlock the way we carry land and place in and through our bodies and in our relations to the Other. This opens new ways to think about the embedded possibilities within these as paths toward healing from historical traumas and to understand that *dolor* (pain) has myriad ways to express itself. Cherríe Moraga and Gloria Anzaldúa describe these entangled embodiments as our "theory in the flesh" and a place "where the physical realities of our lives—our skin color, the land or concrete we grew up on, our sexual longings—all fuse to create a politic born out of necessity."[10] It is precisely these entanglements that render land a messy place from where I stand. This messiness is the beautiful "persistence of a memory of embodied wildness living in the interstices of degraded land" that seeps in and through our porous flesh to secrete its messy corporeal fluids and knowledge.[11] Messy is to wake up at five or six in the morning to catch a glimpse of my grandmother praying with her *rosario* (rosary)—a beautiful vision of persistence that grounds me by rejecting any logic—and then later that day at the hospital caressing and paying my respects to a beloved relative who had just died due to medical negligence. The land I have experienced has peculiar ways to make me feel departed, and then at once a walk through my grand-mother's garden, *el jardín de Rosita*, brings me home.

## Western Eurocentric Place

In the process of starting with the land, I have also come to reflect on Western Eurocentric environmental education and immediately question the intentions defined within the spaces scholars of place construct. To remember the violence over and on the lands they speak of, "place" must be rendered racially cleansed, treated with a sadistic gaze, and commodified. To rescue place from colonial and capitalist zombies, some scholars suggest an apocalyptic future. Perspectives of land and place that do not account for the entangled colonial violence between land and people do not speak to my experience. Western Eurocentric perspectives do not relate to the memories I have of my loving uncle and his tragic death. They do not relate when I try to imagine that bit of land his dying body must have rested on after the car blew a tire and went off the shoulder. For years, I fixated on the hope that if any grass grew from that bit of land, it touched his flesh one last time. Maybe, just maybe that feeling spared him some pain. He was only thirty-eight years old. We carried his ashes in a small wooden box across a border. His ashes needed to be in his mother's arms in Brownsville, Texas. He died on the *carretera* (freeway) taking him into Reynosa, Tamaulipas, México. I sensed him standing by the kitchen sink as I handed him to Rosita, his mother, my grandmother. He made it home. It struck me that I am also placeless.

## The Departed in Need of Place

A few years later, I was introduced to a reflexive autobiography by Karleen Pendleton Jiménez, "'Start with the Land': Groundwork for Chicana Pedagogy" (2006). I cry each time I read this extended essay. Her way of bringing attention to the layered and entangled colonial histories we embody, and her telling of displacement stir in me deep sentiments. Finding herself driving around Los Angeles, music blasting, crying, desperate and angry, she realizes that once her mother had died, there was no longer a place left for her in her childhood land. She shares with us her pain:

> Then it hits me last summer on one of those freeways about my mother's grave. How it is actually the only land in my name. I find the 210 freeway, find Myrtle Street. I exit and buy flowers and a Guadalupe candle. In the foothills of giant mountains, I lay on the ground above her. Maybe to begin to build a relationship with the land it has to come to something as simple as my body and a few feet of earth. Hot winds hum as they blow through the trees. The wet from the soil seeps into my jeans. The smell of cut grass comforts me, reminding me of playing soccer as a kid. I dirty my hands, wiping clean my mother's tombstone, to see her name. I cry for her and write for hours. Nobody bothers me here.[12]

Perhaps I relate to the displacement that sets in as Pendleton Jiménez grapples with the lack of connection to land and how something as simple as feeling a few feet of earth can begin to rebuild that relationship. Or perhaps the panic that sets in when struck by the loss of her mother along with her childhood home, and then feeling repelled from her childhood lands, resonates with my own lived experience of hope in the midst of such deep loss. Pendleton Jiménez speaks to the constant prodding to displace, through genocide and epistemic violence, generations of Indigenous peoples from their ancestral lands and to dismember any kinship they have to those lands.[13] These enclosures brought about by centuries of violent land theft involved the racial classification of people as defined by colonial and capitalist invaders. Centuries of violence and the struggles and rebellions to recover stolen lands and knowledge(s) created a deep "colonial wound" on our relations with land and with each other.[14]

## Let us talk about the violence of place and how we bear our colonial wounds

The colonial wound is so deep that as children we must learn about violence. I was taught about violence as part of the *saberes* passed down to me.[15] Luis Urrieta draws from Indigenous knowledge systems to describe *saberes* as involving complex understandings of the world and how to survive in it. They are tied to *familia* and *comunalidad* (knowledges of communality) but also encompass larger social, natural, and spiritual beliefs about well-being. *Saberes* are how children learn "indigenous conceptual understandings of belonging, responsibility, and integration into familia and pueblo (community) life."[16]

For my family and those closest to us, it is important for our children to be aware of the many forms of violence surrounding us to enact strategies to keep us alive and thriving. *Como me dijo un pariente* a few weeks before her own premature death, "Por si un dia no estoy aqui, por lo menos saben como hacerse un plato de comida." (If on some day I am not here, at least they [her daughters] know how to prepare a meal.)

I have learned different ways to identify violence. One way is to simply name it: *La oscuridad* (the darkness) is one name we have for it. *La oscuridad* is spiritual violence that becomes lodged in the fabric of the places we inhabit; it materializes and lingers unless it is treated. As a child I learned to see it and sense it in people's bodies, in their homes, and in their relations. I was taught to sense its presence when our *muertitos* (little dead) roam our lands due to the violence they experienced in life or at time of death, or in the stench coming from a sick body needing to be treated from the environmental conditions they are exposed to daily. It materializes in the toxic soil and water, and in the torture and death of *animalitos*. It materializes in the torture, murder, and disposability of so many people whose bodies are found along roads, ditches, or levees or *en el monte* (in the woods). There are many understandings and manifestations of *oscuridad*, but what remains constant is its relation to power, the ecological degradation of place, and how intimate it can become. Violence is always up close and personal. The danger and the terror are in the power to produce and reproduce the violence that lingers, as it becomes part of the ordinary. Franz Fanon stated it well: "each individual represents a violent link in the great chain, in the almighty body of violence rearing up in reaction to the primary violence of the colonizer."[17] We know that violence has lingered in a place far too long when death becomes a spectacle—lynchings, mobs, skinning people alive, public and now cyber decapitations. This is the reality we must all contend with day in, day out as Indigenous, Mexicana/os, Chicana/os. Violence is embedded in the very fiber of our intersubjectivities and materiality. The reality is that we are all embedded in the violence, even as we go about our day. We are being made fools of if we think otherwise. We need to tend to our wounds.

## A Place to Tend to Our Wounds

There is also an intimacy to our *saberes* that makes them a place to treat our wounds. Despite the colonial/oppressive conditions many Indigenous, Mexicana/os, and Chicana/os endure, we creatively and strategically hold on to and transform our relations to land by militantly defending land and land-based wisdom. Thinking about *saberes* and *la oscuridad*, it came to me even within the entanglement in which we live that I also experience ways to treat the wounds we carry. I got a stronger sense of this as I continued to read and cry with Pendleton Jiménez. She finds a place amid all the pain—the plot of land her mother's body is buried in. Her own body lay above her mother's grave to feel the connection to her mother/land through the wind, the wetness of the soil, the smell of cut grass, and the dirt on her hands. In this moment, she exposed a reconciliation of sorts. She exposed our intimacies to heal with the land.

The connection she made with land, her body, and her mother illuminated a familiar and very intimate place for me—all the healings Rosita carried out over the years on overworked and beaten bodies. I remember my grandmother walking outside to her garden with such ease and confidence to gather water in a small tin bucket as she picked flowers and basil—some for her altar and the rest to treat a friend or relative. As a child I remained close. I watched, learned, and waited for some indication to fetch her something, anything. I knew to remain close enough to protect the intimacies of the process, and to keep enough distance to allow for those she cared for to share and release some of their pain. Rosita is my maternal grandmother, and I am her oldest granddaughter. I lived for the time when we traveled to her *pueblo natal* Salinas, San Luis Potosí, México. Rosita's relatives knew when she was in town. They would visit and ask for a *curación* (cure or remedy) or *bendición* (blessing). I remember the pain. I sensed when they were about to leave as soon as Rosita walked into her room and reached for her purse to pull out some money. She folded it carefully to fit the palm of her hand. Walked back out and gently placed it in their hands. They quietly tucked the money somewhere on their clothes, out of sight. I remember the tears. She is now in her eighty-third year of life, and I am in my thirty-ninth. We still remain close. Now, I watch how she carries her aging body outside to her garden to gather flowers and *albahaca*. She lays out her fresh cuttings on the kitchen counter-top and fills a vase with water from the faucet. She arranges her flowers and basil and sets some aside for her daughter's house. She goes into the quietness of her room, places the vase on her altar, sits down, and with a rosary in her hands, closes her eyes. I also find my mother in this very intimate place. I was present when my mother, Armandina, after a long day at work paid visits to relatives or friends. I watched as she reached out to hold many hands to give them a sense of presence. No words, only the sounds silence makes. After their tears, there were deep sighs, perhaps to release all the tension and sometimes the pain. There is beauty and possibility in the moments when the body exhales to release. Later, in my grandmother's company, my mother reflects on her visit, on her friend or relative, and their situation.

"Re-membering" made me understand that in these intimate, mundane moments, my mother and grandmother *me enseñaron*, taught me to carry on and defend our relations with the land and with each other. I am also thinking of Linda Tuhiwai Smith and her notion that remembering is a decolonizing methodology, which "relates not so much to an idealized remembering of a golden past but more specifically to the remembering of a painful past and, importantly, people's responses to that pain."[18] Remembering, as a decolonial practice and methodology, requires us to learn, talk, reflect, and act upon the violence that does not differentiate between lands and bodies, making them inseparable as well as mutually defendable. Remembering this way works to expose our colonial wounds and begs of us to strategically use the tools our grandmothers and our mothers fight to carry on and to suture and heal. Remembering is a deeply painful process "because it involves remembering not just what colonization was about but what being dehumanized meant for our own cultural practices. Both healing and transformation become crucial strategies in an

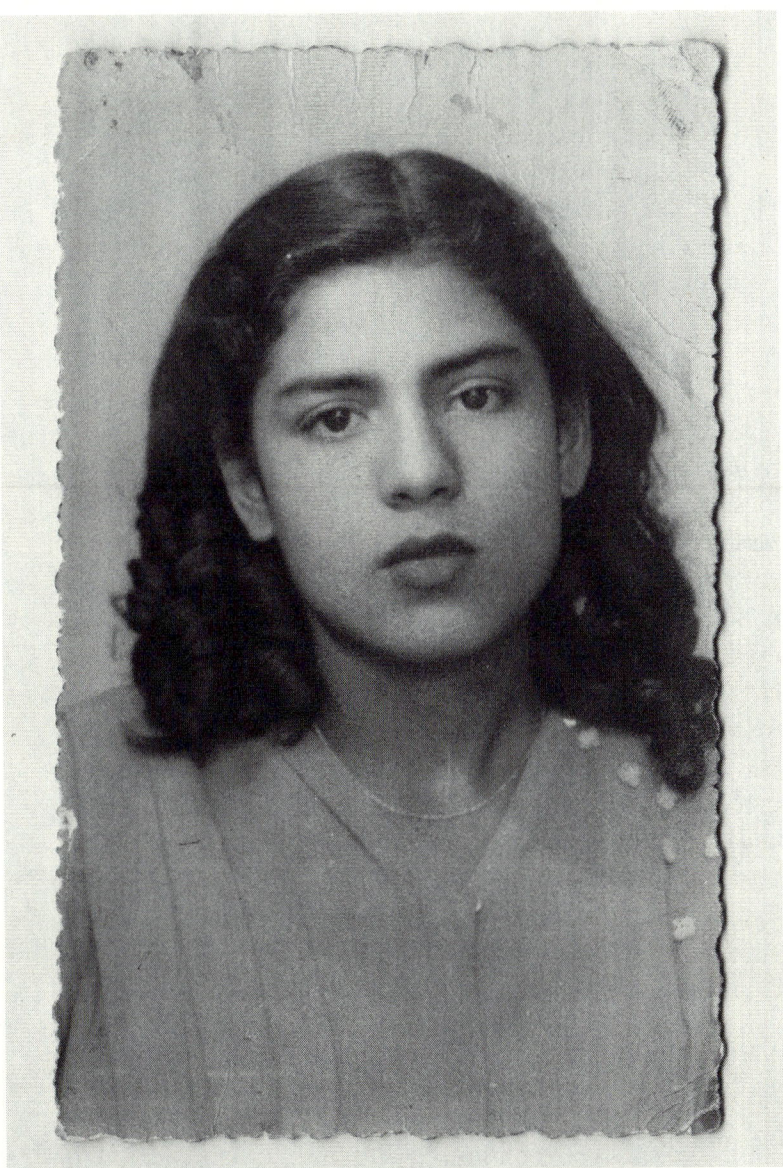

Figure 4.1. The author's maternal grandmother, Rosita Pantoja Hernandéz. This postcard photograph was taken when she was sixteen years old in Salinas de Hidalgo, San Luis Potosí, México (1948). Courtesy of Silvia Patricia Solís.

approach which asks a community to remember what they may have decided unconsciously or consciously to forget."[19]

Re-membering (putting back together) helps us heal our wounds in a way that is familiar. My grandmother's respect for her *plantitas* (beloved little plants) and her knowledge of their vitality to help heal the body are a strong and vital connection to land. My mother and my grandmother taught me that I have a connection with land that is deep rooted and cultivates ways to relate with each other that does heal our wounds. It became clear to me how deeply rooted our intimacies are within our *saberes* and that my mother and grandmother hold on and even transform this knowledge based on our needs so it continues to thrive. This is a profoundly subversive act—to educate our children of the ways to remain connected to land through the mundane and to educate ourselves in old ways to stay alive and thrive. I was never placeless.

Argentine feminist philosopher María Lugones pushes us to think of intimacies beyond exclusivity or mainly about sexual relations. Intimacies are part of the everyday interwoven relations among people not acting as "representatives or officials" within the most insidious and "central dichotomy" of coloniality/modernity—human/nonhuman.[20] Our intimacies to heal are moments and practices where we recognize the fractures and limits of coloniality in our places, our relations, our bodies, our lands, and our everyday life and create, carve out if need be, dwellings for healing that emerge from our relations with land.[21] In that moment, touching the land that embraces her mother's body, Pendleton Jiménez illuminated for us those intimacies that are beyond the fractures/limits of coloniality, a place where she/we can heal the colonial wound. The intimacies to heal were illuminated thinking and remembering my mother's presence and my grandmother's garden.

## "From a body as place . . ."

I have been writing this essay since 2011, when I was first introduced to eco-pedagogy. In my initial thinking, I considered the body as place. This was based on my grandmother's movement from Salinas to Brownsville, Texas, where she settled. I thought of the moment she had to leave her mother's side at sixteen and how that day remains clear in her memory. She remembers the pain. I wanted to express how people, particularly racialized sojourners who are forced to leave their home/lands due to social, political, and economic repression, have the ability to carry and transform land-based wisdom. I find meaning in my grandmother because she helped raise me. I went everywhere she went. Her entangled embodied experience gave me a different understanding of land and place that did not relate to the particular conceptualizations I was exposed to at that time.

My grandmother left her *tierra natal* in Salinas de Hidalgo in 1949 at sixteen to find higher-waged work. This led her to *la frontera*— Heroica Matamoros, Tamaulipas, México and Brownsville, Texas. In her movement she also lived in Valle Hermoso, Tamaulipas, México and Monterrey, Nuevo Leon, México, where her parents and siblings migrated to, also seeking higher-waged work. She worked picking cotton for a short period and then sought out domestic work. She met my grandfather, Juan, which led her to finally settle in Matamoros. I was born to her oldest daughter, Armandina, in Matamoros in 1975. Four years after I was born, we moved and settled in Brownsville. *Mi abuelo*, Juan, was US-born. This granted us the privilege to facilitate a constant transborder movement even after we migrated to Brownsville. My grandmother, my mother, and her siblings grew up traveling from Salinas, to Matamoros, then to Valle Hermoso, and to Monterrey to visit relatives. I spent winters and summers in Salinas, running the streets, experiencing the meals, celebrations, and the everyday. I hated to leave. I also remember visiting my great-grandmother— Margarita—in Monterrey. I sat and watched her massage her granddaughters' legs or arms or anywhere they needed attention. I experienced this movement along with them.

For Rosita, Salinas is her womb. She talks about the mountainous desert surrounding her and *mi bisabuelo* Francisco's *ejidos*, El Con and Ferritos, where he labored to harvest maize and tend his animals. She remembers the desert plants, particularly *sangre de grado*.[22] She describes the plant as almost cactus-like, with a long, green stem and a white root. *Mi bisabuelo* gathered it for her from the *monte*. She cleaned, cut, and pulped it to make herself a shampoo. She remembers her long, soft hair. Decades later, living in Brownsville, when my grandmother wanted to make her family a special recipe, we caught the bus early in the morning, crossed the border, and caught several *peseros* (taxis) heading to the *mercado* to buy all the food she needed. She wanted all of us to continue tasting *comida* Matamorense with a Salinense regional fusion. She remembered Salinas and Matamoros, and through her embodiment she taught us about how to carry place. So much so that her place(s) never made me feel departed.

I consider my grandmother's movement and all the *saberes* she carries with her an epistemological embodiment of land—land as she experienced it in all the places she traveled through, settled, and dwells.[23] She carries all these different places with her, and these places materialize around her. She reproduces these in her everyday life through her cooking, her sewing, her knitting, her gardening, and in the way she raised her family, and in the intimacies to heal. These materialize in the way she has fought against suffering and dying from diabetes among friends and family. Her body is a place for me to understand land through her knowledge. She taught me to dwell in multiple places.[24]

## "... to the lands we inhabit"

I was no longer placeless, but what about landless? How is being landless, or for that matter, being "land," constituted within an entangled embodiment within our flesh? Alice Walker tells us "that often the truest answer to a question that really matters can be found very close."[25] For Walker, close is her mother's flesh—her body, her labor in the cotton fields, her children, her quick and violent temper, her children's schooling, her sewing all their clothes, the towels and sheets they used, and the quilts to keep them warm in their beds, and her summers spent canning. It is when Walker is in the garden with her mother that she sees beauty in the flesh and remembers: "I notice that it is only when my mother is working in her flowers that she is radiant, almost to the point of being invisible—except as Creator: hand and eye. She is involved in work her soul must have. Ordering the universe in the image of her personal conception of Beauty."[26] Walker illuminates the epistemological embodiment of land in my mother and my grandmother's spiritual, material, and ecological production of everyday life, but it is when I am close—dwelling in our intimacies—that I am able to contemplate the beauty that radiates from their flesh, the beauty in their embodiments, which become the lands I inhabit. There is beauty in their respect and commitment to want to heal our relations, our bodies and sustain our connection to land.

I consider myself a child of a colonial form of diaspora who has experienced the tragic and false sense of dismemberment from the lands and the places our mothers and grandmothers carry with them, all those lands and places they remember. Even if the dismemberment is false, it is felt, and in those moments of pain, the bodies of our mothers and grandmothers become for us the constant and closest place to Indigenous lands we know. Their/our bodies—their/our flesh are the lands we inhabit. If their bodies are not there to tell our stories, to pass down our *saberes*, and educate us in our old ways, then the dismemberment becomes real.

This is why I cry with Pendleton Jiménez. She speaks about the body as the "primary tool for sensing the land, for establishing intimacy."[27] In my case, sensing land means sensing the bodies of my mother and my grandmother to establish an intimacy with the land. It is as simple as holding their hands, as simple as feeling the *albahaca* touch my flesh. In that intimacy, I begin to feel the weight of history and knowledge subdue, and I can take a deep breath. "It is a pleasure to be held by the earth."[28] There is pleasure in feeling the sand on my feet. There is pleasure when my mother holds me and when I walk through my grandmother's garden. And so I began to think of the importance of our flesh—the presences, excesses, and absences of our materiality. I began to think of how we sense land and not only how we make sense of our lives through it: Pendleton Jiménez asks, "when is the physical connection [to land] replaced by the weight of history and knowledge?"[29] What will happen when our mothers and grandmothers are no longer present to share their *saberes*, to make our intimacies part of the mundane, and to educate us about keeping our connection to the land? What then?

In my landlessness, I turn to the body of my mother. She showed me the importance of our presence and our relations with each other, of the respect we must give each other and the righteous rage to defend ourselves, our bodies, and our lands from further exploitation. Most importantly, she taught me to see our flesh as the lands we inhabit.

There is no separation; we are connected to land in the most mundane ways "through the water we drink, the air we breathe, the food we eat, and our own bodily processes."[30] For those of us who have experienced the tragic and false sense of dismemberment, land is a mother's embrace, our grandmothers' home kitchen garden, the spaces where we labor, sleep, play, pray, where our children go to school, and where we tend the gardens, care for our *animalitos* (animal companions), and engage in our dangerous migrations. The land is the clothing we put on our bodies, our relations with each other by way of healing practices, including struggles with violence, insulin injections, high blood pressure medication, cancer treatments, burying our loved ones, birthing our children, and remembering the silenced abortions we endure. Cherríe Moraga reminds us that

> land remains the common ground for all radical action. But land is more than the rocks and trees, the animal and plant life that make up the territory of Aztlán or Navajo Nation or Maya Mesoamerica. For immigrant and native alike, land is also the factories where we work, the water our children drink, and the housing project where we live. For women, lesbians, and gay men, land is that physical mass called our bodies. Throughout las Americas, all these "lands" remain under occupation by an Anglo-centric patriarchal, imperialist United States.[31]

To know and feel both beauty and pain with the lands we inhabit go further than just understanding land in any more narrow scientific ecological sense. The task is to dig deeply to eradicate the human/nonhuman binary that continues to plague our relations with land and with each other. We need to move toward unsettling place and continuing to defend land. This requires us to deeply understand how bodies and lands are inseparable. Tears roll down my face each time I think of the beauty my mother and my grandmother embody and how it keeps us/me alive and thriving despite the violence we are immersed in. Continued violence puts it all at risk.

## Start with the Land

From time to time I get a sense of Rosita's silence as she concentrates on chores or as she sews or knits. Her body moves, but I know she is remembering. She sometimes breaks the silence by telling me with a sigh that she misses her Salinas and her mother. I asked her recently what would be the most selfish act she would take if she won the Texas lottery. She stayed quiet, looked up at me, and said, "*Regresar a Salinas*" (return to Salinas).

I echo the call to "start with the land" with the understanding that land is also the physical mass we inhabit all the time—our flesh. Needless to say this is not a metaphorical use of the body as land or place. This is growing up learning that I have an intimate connection to the lands we inhabit in the most radical and mundane ways, through the *saberes* my mother and grandmother carry with them. This is not ontology, so it does not place the body first and above land, such as "I am, therefore place is."[32] This is learning that you respect the lands we walk and run on and with, as a vital member of our corporeality, our family and community. To a child of a colonialist-induced diaspora, to "start with the land" means to start with the closest and most constant feeling of land we know, our flesh and the flesh of our mothers and grandmothers. Therefore, it is incumbent on us to ethically and respectfully protect land and the tools our mothers and grandmothers carry with them to suture and heal. In my case it involves an intimacy that "starts with the land." We have to contend with the complicated and layered colonial histories we embody, as Indigenous Chicana/os and Mexicana/os. I offer that we acknowledge, support, and work alongside scholars, activists, pedagogues, thinkers, and workers who "start with the land" in their work and in their cosmologies and that we take our bodies into account in our renderings. I was never dismembered; I just allowed the Eurocentric mirror to distort my flesh.

# PART II · Witnessing

## Heritage Cuisines and Decolonial Foodways

# El Quelite

## TERESA VIGIL

"¡Allí! Quiero hallar un fresco quelite," said "Mana," or Manuelita, in early spring. Quelite (Lambs quarters, Chenopodium Album L.) was the prize of the search, the wild spinach of the village. It was considered almost sacred as the long cold winter lacked fresh farm-made produce.

It meant new life, nourishment.

Every cook had a special way of preparing it. As spring went into early summer, the children and adults were gathering the tasty leafs to dry for the winter. Certain greens were gathered, dried, and put into recycled flour sacks. In the winter when greens were needed, the dried leafs were placed in tepid water, and new life came to this wild spinach.

It was then cooked and enjoyed.

People, who were productive, knew they must gather quelites to appreciate the gifts of spring during the long cold winter.

Manuelita, now walked along the dry garden, and an earthy fragrance of onions reached her nose. She was elderly now and did not move too fast or bend easily. She took a long stick and poked the ground until an onion came up. It had remained in the warm ground waiting for spring.

She pushed away the dried leafs and weed. Suddenly, she saw a bright green quelite and then a few more. Manuelita bent carefully and gathered a small bag of the wild spinach.

She clutched them tightly to her chest as if the plant was a precious child. In the kitchen, she cleaned the quelites to cook and said:

"How delicious this will be!" Out came a piece of salted pork that was chopped with the onions. Next, these were sautéed until translucent and softly crisp, then the quelites were added.

The wonderful aroma made her feel young again.

She danced a gentle whirl around the kitchen.

"¡Que delicioso!" Her humble little bowl of quelites disappeared, folded into her warm tortillas. Manuelita was satisfied.

¡Allí! Que buen dia. Gracias Señor.

# CHAPTER 5

# Tracing Food Packs and Tuna Cans on *La Línea*

## Food, Water, and Foodways during Transborder Travel

### CONSUELO CROW

The US government continues to spend billions of dollars on "invasion prevention" technologies, forces, and tactics to control and stop unauthorized entry into the United States by brown bodies. The deployment of this border control apparatus is evident in isolated rural areas of the great Sonoran Desert where the US Department of Homeland Security (DHS) has strategically erected tactical infrastructures designed to funnel brown immigrant bodies into spaces where nature can enforce human laws that restrict movement across geopolitical boundaries. Feminist political ecologist and geographer Juanita Sundberg questions why DHS agencies like the US Customs and Border Protection (CBP) continue to request monies for enhanced border enforcement infrastructure to police nonhuman entities that are constructed by the border control apparatus to "belong to the realm of nature as opposed to society." She asks whether the Border Patrol is battling undocumented immigration or the desert, because the environment of the Sonoran Desert "inflects, disrupts and obstructs" the policing of the border.[1] I have overheard an echo of this in a phrase used by border control agents in my presence: "The desert is deterrence by death."[2]

It is supremely ironic that transborder travelers are so confined in the open space of the desert, like *animalitos encerrados* (caged animals), wherein movement and physical and social activity are governed by their embodiment within a state of exception and status of "illegality." In addition to revealing a biopolitics of citizenship and governmentality, zones of confinement represent opportunities for the accumulation of capital through exploitation and subjugation of migrant bodies, as in the case of the supply bodegas in migrant-staging border towns serving the smuggling networks. However, this still seems like a conventional take on the travelers' experience with the heat death chamber of the desert in that it emphasizes control and enforcement over the living (albeit battered) subjectivity of the travelers whose bodies are reduced to objects of control or humanitarian rescue.

Bruno Latour points out that all social and political expression is mediated through "instruments" and "mechanisms":

> Hence, all actors can be said to leave traces, whether these take the form of texts, oral narratives, footprints, or feces . . . landscapes . . . [that] tell stories through specific configurations of vegetation, soil types, and myriad other traces. . . . The notion of traces transcends humanist understandings of talk and text as the only mechanisms through which politics can be registered . . . traces created in the process of group formation.[3]

What are the nontextual traces left behind when the apparatus objectifies both nature and immigrants in the Sonoran Desert? Sundberg suggests this condition is set when the Border Patrol "treats rivers, mountains and deserts as objects of geopolitical calculation and instruments of enforcement."[4]

Does anthropomorphizing landscapes as inanimate objects for border control and enforcement accompany the stripping away of humans of the qualities that make us human? Is this an instance of the materialization of what Leo Chavez calls the "Latino Threat Narrative"? As Chavez explains,

> The Latino Threat Narrative posits that Latinos are not like previous migrant groups . . . assumptions and taken-for-granted "truths" [that] Latinos are unwilling or incapable of integrating. They are an invading force . . . bent on reconquering land . . . destroying the American way of life.[5]

The word *trash* is interchangeably applied by anti-immigrant activists, some environmentalists, and federal and state border control and policing agencies to the material remains left at the migrant desert "relief stations" and to the humans that have left these traces, resulting in the objectification of Mexicans and other brown bodies as an invasive form of human trash.[6]

The clandestine and corporeal lives of migrants who cross the Sonoran Desert each day leave behind traces that reveal much about the act of "unauthorized migration" and inform us of the travelers' enactment of his/her material culture(s). Transborder travelers enter into a state of exception where constituent human rights are suspended and border enforcement agents perform in a manner that resembles wartime in a foreign country. The administration of the border zone under the current regime creates a state of liminality for travelers where they enter into a condition of invisible nothingness in a place that is policed but not watched over. They are neither here nor there, less than human or object, not just the enemy or a potential ally, but are regardless presumed to lack "communitas" or structure; they are transient, unstable subjectivities suspended in a zone of anonymity that resides beyond the realm of conventional nomenclature.[7]

In an earlier work, Latour explores the mediated nature of technique and begins by recalling the myth of Daedalus to remind us that when it comes to technical knowledge, all things "deviate from the straight line." He goes on: "The clever technical know-how of Daedalus is an instance of *metis*, of strategy, of the sort of intelligence for which Odysseus (of whom the *Iliad* says that he is *polymetis*, a bag of tricks) is most famed."[8]

This essay explores the flip side of suffering and marginality by examining the praxis of brown "migrant" bodies adapting their "bag of tricks" as they work to survive the life-threatening journey across the Sonoran Desert. I do this by focusing on the artifacts they leave behind as traces not just of their journey but shifting subjectivities and material conditions that constrain such acts of desperate defiance. I should note that dialectical materialist tensions exist between symbolic action and power relationships. These may offer insights into the process of the performance of foodways during the act of unauthorized entry. Abner Cohen defines symbols and symbolic action as

> objects, acts, concepts, or linguistic formations that stand ambiguously for a multiplicity of disparate meanings, evoke sentiments

and emotions, and impel men to action. They usually occur in stylized patterns of activities like . . . eating and drinking together. All social behavior is couched in symbolic form.[9]

One of the most important functions for such symbols is the "objectification" of relationships between individuals and groups, and we see this deployed in the exact context of border control and immigration enforcement. Such symbolic politics are also involved in what Fredrik Barth sees as the subjection of ethnic(ized) groups to a system of social organization in which the most politically significant characteristic is self-ascription or ascription by others: "A categorical ascription is an ethnic ascription when it classifies a person in terms of his basic, most general identity, presumptively determined by his origin and background."[10] Barth sees the cultural content of ethnic dichotomies as comprised of two kinds: (1) "overt signals or signs—diacritical features that people look for and exhibit to show identity, often such features as dress, language, house-form, or general style of life"; and (2) "basic value orientations: the standards of morality and excellence by which performance is judged."[11]

How food is consumed, where it is consumed, and who is consuming it can all be charged with emotion and significance, both positive and negative, depending on who is doing the charging. Foods are transformed from quotidian household staples to items that emanate philosophies of race, class, gender, religion, and politics. The arrangement of what is eaten, and when, how, and what it means are closely tied to ethnic identity. There also exists a theme from ethnic to regional and political identity. Cultures invest heavily in themselves whenever they engage in the act of eating. The decoded "migrant meal" summarizes a rigid and tragic subsistence inasmuch as it reminds us of what is to be eaten, with what, and in what order as a matter of survival. What fictions are created for clandestinely crossing geopolitical landscapes, and how do we imbue them with value? How do foods develop the "power of identification"?

Existence or absence of social group mediation, when selecting from a menu of fixed options for food items before beginning the journey across *La Línea*,[12] creates a treacherous balance. How and what is inside the (learned and preferred) expected bag of tricks of the unauthorized transborder traveler's diet that is consumed on the trail?

In this essay I show how eating patterns are socially mediated, transmitted, and reinforced based on the presence or absence of knowledge of the journey and the environment.

Food is symbolically encoded to perform certain functions to and for the consumer and is consumed in very specific times and places, which are not necessarily predictable while crossing. Calculations are made to determine when food is to be consumed while walking or running through the desert, and these must also weigh the need to avoid detection while seeking protection from extreme elements by locating well-hidden and shaded areas that hide migrant bodies within the landscape. Does this necessarily create an alternate reality in that time and space? I am reminded of a commentary by Sidney Mintz:

> It is the special character of the substances consumed that, like sugar, they provide calories without nutrition; or like coffee and tea, neither nutrition nor calories, but stimulus to greater effort; or, like alcohol and tobacco, respite from reality.[13]

Travelers typically begin this portion of their journey at the US-México border with one to three gallons of water for the entire passage and sometimes less. The recommended amount of water that should be carried by one person for a summer day hike in the desert is 2.5 gallons, which weighs approximately twenty pounds.[14] Travelers purchase or procure water at the start of their crossing in migrant-staging towns. Water for sale in prepackaged plastic bottles is readily available in small bodegas. Occasionally, bottles that once contained bleach or motor oil have also been found in migrant sites in the desert. What has been the most fascinating aspect of the Mexican water bottle industry is the transborder traveler economy that is manufacturing and marketing specifically designed bottled water for the act of migration. Two examples of this phenomenon can be found in the form of black plastic water bottles and bottled water containing landscape feature depictions on the product label.

For humanitarians, principles of prestige and exoticism created by their place in the world underscore the symbolic character of these consumables. Utilitarian factors override symbolism in the process of food choice for transborder travelers. Humanitarians determine "migrant meal" options by weighing impressions from science and religion, whereas travelers select foods based on availability and portability. It seems hardly surprising that humanitarians offer food options to travelers that are Americanized ethnic foods.

Megan Carney's research on food and apprehended persons in detention centers examines how "state practices around food contribute to the militarization of the migration experience."[15] Carney does not address, however, what happens during the experience of travel to and into the United States that occurs prior to detention, and focuses only on what happens during a specific type of undefined detention. She also fails to define the location and type of detention that she is highlighting. These issues matter to the traveler and her community.

For example, depriving detainees of food and water—in effect, physically and emotionally abusing them by using food as a weapon of discipline and punishment—is a consistent and well-documented practice in both governmentally and privately operated facilities of the Immigration and Customs Enforcement (ICE) and US Border Patrol (USBP) agencies. There are different types of detention with segregation and administrative classification based on sex, age, and country of origin, for example. The type of food provided in detention, and the manner in which it is made accessible or not, is vastly different from one facility to another. As if to illustrate this exact point, a man who was deported to Agua Prieta, Sonora, handed me a foil packet of Hormel Chili. He asked me, "Why would they feed us dog food?" In another incident, a woman deported into Nogales told me that her frozen burrito was thrown at her onto the floor, and she was instructed to pick it up and eat it without using her hands. It is difficult for me as an anthropologist to find a way to retreat toward a positive life-affirming spin on incidents like this. Certainly this matter of food as a weapon is not a semiotic nicety for the individuals who have had this experience with systemic daily and unending acts of microaggression and immiseration. Both individuals informed me that they intended to cross again within the next couple of days, which seems suggestive of a form of cultural resilience few of us acknowledge or recognize.

Relying on Foucault's theory of governmentality, Carney describes detention as a technology of the self that produces a particular kind of subjectivity. She argues that "depriving detainees of food is a primary mode of constructing detainee subjectivity."[16] In other words, depriving captive travelers of food and water during detention presumably produces a compliant and acquiescent detainee yearning to return home and to the comfort of familiar foods. Of course, for many such subjects there is no such comfort in food back home; avoiding so-called food insecurity is one reason that they came to the United States in the first place. Carney offers a proposal for change in current detention practices that relies on a discourse foregrounding detainees' trauma, yet, she says, "we may find migrants enacting resistance to the larger structures in which a system of detention is embedded through reinterpreting everyday expressions of affect."[17] But this appeal to affect seems like a dead end given the suspension of these bodies as homo sacers, reduced to the bare life. Is it possible for detainees to set themselves free from the oppressive conditions of detention through acts of emotive resistance in an era defined by streamlined fast-track deportation policies that reveal a racialized administration of the so-called rule of law?

Before the current manifestation of the border wall and the increased number of "boots on the ground"—language used by Obama in his 2014 State of the Union address, invoking the same narrative that further militarized the geopolitical border shared with México—migrants were not trekking through the parched heat and death chamber of the desert. Travelers entering the United States in an unauthorized manner utilized different methods to cross the US-México border, but most of these were deployed primarily across urban areas. Male travelers were more likely to climb through holes in the fencing and walk into metropolitan areas or to be picked up by an associate waiting close by. Women were less likely to take such a risky approach to border crossing and might instead purchase counterfeit documents to attempt crossing at an official port of entry to then be picked up by friends or relatives.[18] The fortification of the border with new infrastructure concentrated in populated urban centers, along with the expansion of the CBP after 9/11, changed unauthorized entry patterns and points of entry into the United States.[19] These patterns shifted deeper into unmaintained and less populated desert areas, where resources, such as food and water, are far less available.

My central aim in this essay is to begin to fill a gap in the literature by focusing on food and foodways during unauthorized migration through a desert environment by transborder travelers in the Sonoran Desert, an arid ecosystem that covers most of southern Arizona. I will focus on how the militarization of the US-México border discussed above has led to changes in what the travelers bring before they are apprehended and detained for an undefined period of time in an ICE detention facility in the United States.

## Food and Foodways on the Journey across *La Línea*

There are fundamental items that travelers routinely bring with them on the journey to *el otro lado* (the other side) if they are properly prepared for the journey.[20] Sustenance and survival are essential considerations during the crossing, and entire economies have been created around the business of unauthorized migration with the sale and marketing of border crossing–specific foods that fit into the small backpacks used to transport food, water, clothing, and other personal items.

The demand for transborder travelers' goods has spawned an entire economic sector in México as well as in the United States. Goods are purposefully manufactured and marketed for unauthorized migration through the Sonoran Desert, creating an increase in small business enterprises that sell these goods directly to travelers in border-town bodegas. I am currently conducting research on whether the manufacturers of goods not produced specifically for migration are aware that their products are now routinely being marketed and sold to would-be travelers preparing to cross into the United States.

In 2009, I was introduced to the newly designed black-plastic water bottles manufactured by multiple Mexican bottling companies and marketed to transborder travelers for their desert passage beginning at the US-México border. These bottles are extremely common at migrant resting sites and migrant stations.[21] Traditionally, travelers would coat white or clear plastic water bottles with black or green spray paint. This tactic is used to prevent detection in the desert by keeping the water from reflecting light, even at night, when most clandestine groups are walking.

Over the past two years, I have noticed that the quality of the plastic has changed from a stiff plastic to a more pliable plastic. One veteran humanitarian volunteer explains that the booming industry found a way to make more money by cutting material costs by using less packaging material. Black-plastic water bottles are manufactured by several companies in México; Las Montañas is bottled in Sasabé, Sonora (the label ironically depicts snow-capped mountains and a glassy lake), and Sun Water is bottled in Altar, Sonora, a staging town that exists almost entirely by supporting would-be travelers with supplies, including foods, for their journey into the desert.

Nestlé Waters—which is headquartered in Vevey, Switzerland, and linked to the Mexican corporate conglomerate Grupo Modelo in 2006—has established a strategic bottled water market presence in México. Grupo Modelo owns, produces, and distributes Corona brand beer. This corporate alliance further developed Nestlé Waters' existing bottled water business in México, which includes brands such as Santa María, San Pellegrino, Perrier, Nestlé Pureza Vital, and Pure Life. This alliance allowed Nestlé privileged access to traditional distribution channels in México. Nestlé USA came under fire recently for its water extraction and production practices; it was exposed that a water-harvesting permit from a California forest on reservation land had expired twenty-seven years previously, and a CEO publicly proclaimed that water is not a human right, all while California was experiencing its worst drought in its recorded history.

In October 2014, Nestlé México expanded its dairy factory in the state of Jalisco to add to its Cero Agua water-bottling operations.[22] This is the company's first "Zero Water" manufacturing site in the world located in a water-stressed Mexican region. Cero Agua converts cast-off water from its dairy operations into potable bottled water to be sold in México. Discarded Nestlé Pure Life water bottles are a frequent sight on trails and in migrant hiding places. While Nestlé is not intentionally manufacturing bottles and bottled water specifically for migrant economies and migrant consumption, it is one of the leading bottled waters consumed by travelers in the desert.

Water bottles and electrolyte bottles have proven to be the most consistent and most consistently shifting food-based artifact in the desert.[23] Bottles are discarded on the journey because they are no longer useful or are reused to collect water from bacteria-infested cattle tanks and then discarded. Discarded bottles have been located and identified as having cattle tank water in them with clumps of green algae. Humanitarian volunteers have informed me that, in the past, cattle tanks have been poisoned to intentionally harm unauthorized travelers walking through the desert.

Jason De León conducted an experiment on a variety of water bottle types, analyzing the temperature of the water contained within those bottles based on type, size, color, and outside temperature over the course of several days. De León found that the water temperature differed by 13.5°F between the black and the white water bottles during the test period with average daily temperatures of 100.7°F. The white bottle reached a maximum temperature of 112.8°F, while the black water bottle reached a maximum temperature of 126.3°F.[24] The problem with drinking hot, putrid water on a hot and dry day when dehydrated is that the human body's internal regulator is unable to reduce the temperature of its internal organs, which can lead to rhabdomyolysis and renal failure.[25]

During a time when concentrations of border crossers were located in the Altar Valley of Arizona, the Baboquivari Bottling Company based in Sasabé, México, began manufacturing small, twenty-ounce bottles of water specifically for the act of unauthorized migration and entry. The label contains an image of the Altar Valley's famous Baboquivari Peak,[26] which can be seen for miles from several vantage points in the Sonoran Desert valley. One traveler said, "I check the picture to keep going north." Since 2010, unauthorized migration through the Altar Valley still remains frequent but has subsided from its 2005–8 peak numbers,[27] making this bottle less common in migrant sites than previously.

Migration-based artifacts in the desert are subject to an array of decay and destruction. The decay rates of artifacts in this arid environment are empirically unknown. The decay of materials is now being analyzed and considered for correlation when found with other materials like food cans with imprinted expiration dates and batch numbers. HDPE 2 is the most recognized recyclable plastic in production and is used to make detergent bottles, bleach bottles, milk cartons, shampoo and conditioner bottles, motor oil containers, and other nonedible goods.[28] Some black-plastic water bottles made in México are embossed with this symbol. These plastics will degrade in just under a hundred years depending on the thickness of the plastic used. This, however, is a generalized US landfill-based assessment of the plastic's decay.

The expiration and decay rate of the contents in canned foods is rapid, according to the manufacturers recommended handling of their products. They recommend storing low-acid canned foods such as vegetables for two to four years in a dry place with a cool temperature, and ensuring that that they are not stored in a warm place exposed to direct sunlight. Canned foods should be kept dry and cool to prevent the can itself from rusting, which may cause leaks and eventually spoil the food inside.[29]

Humanitarians slowly came to realize that their efforts to prevent migrant death and reduce suffering were failing to do so in any truly meaningful manner. In summer 2013, during a scheduled, weekly Tucson Samaritans meeting, it was revealed that certain canned foods provided by this group had presented several problems to those walking in the desert. Earlier, in spring 2012, the group collectively made the decision to add one-gallon water jugs and 234-gram individual-serving cans of refried beans to their standard preassembled food-pack offerings that are cached at predetermined locations on known-to-be-active migration trails in the desert.[30]

Prior to this new decision, the group carried only preassembled food packs in one-gallon ziplock bags that contained salty snack crackers, granola or cereal bars, salty chips, candy, apple sauce, and a variety of other high-salt, high-fat, and high-protein snack foods obtained through public and private donations. Volunteers previously handed out food packs to travelers encountered walking in the desert or as the apprehended travelers were being loaded into confined Border Patrol paddy wagons. Now instead travelers could locate food and water at predetermined sites twenty-four hours a day since the volunteers had placed these in canisters in an effort to provide more assistance to those in need.[31] As I will discuss later, preventing tampering or destruction of goods and materials left for travelers in the open desert is problematic and often unavoidable. Sites are visited and replenished once a week or once a month depending on the current activity documented at a particular drop site; it is believed this may encourage a sense of security among travelers who are dependent on these items.

Avoiding detection by border-policing entities is a primary goal starting at the border wall and into the desert. Packing supplies for mobility and portability is crucial to avoid detection and apprehension in the desert. Humanitarian aid groups leave abundant materials on well-mapped and well-traveled trails that are intended for consumption while moving through the desert to avoid detection. When providing supplies, humanitarians consider the comfort, nutrition, and mobility of each item.

Much of these materials are (overly) determined through direct experiences with travelers and interactions with migrant stations. Spoiled canned foods can develop botulism and other food-borne toxins. Debates about unintended racism and possible food-borne illness, such as botulism, were a particular focal point during a Tucson Samaritans meeting in January 2013.

According to the Food and Drug Administration (FDA), improperly canned foods can cause exposure to *Clostridium botulinum*, resulting in the food poisoning more commonly known as botulism. The symptoms of botulism begin within twelve to seventy-two hours of consumption of contaminated foods, and include vomiting, diarrhea, blurred vision, double vision, difficulty in swallowing, and muscle weakness, which can result in respiratory failure and death. This kind of violent death adds to the myriad of other forms of violent death that hundreds of travelers have encountered in the desert while crossing into *El Norte*.[32]

The group decision to offer small cans of refried beans was noble and well meaning but ultimately reflective of an unintended reductionism—no less essentializing and racialized a construct—by those who routinely attempt to combat desert deaths by self-identifying and acting as pro-immigrant humanitarian aid workers. The decision to offer canned refried pinto beans involved deliberate consideration leading to a conscious decision to provide high-protein and hermetically sealed offerings, which is good enough. However, the decision had a subtext—a largely unarticulated assumption—that the migrating population is mostly Mexican, and this was compounded with the submerged notion that all Mexicans eat refried beans. This, of course, implies the odd essential coupling that migrant equals Mexican and all Mexicans are alike in their diet and cuisine. Unfortunately, this sort of essentialist humanitarian construct of the migrant Other also potentially increases a traveler's risk of violent death in the desert with the possibility of never being identified or found—the unnamed and unclaimed ghosts of death by food poisoning.

Tucson Samaritans began researching this issue in part by contacting the manufacturer of this food product. The manufacturer gave the volunteers their recommendations regarding packing, storage, exposure, and shelf life of their refried bean products. It was determined that canned foods exposed to the desert elements, particularly during the hot summer months, presented a danger to travelers, and the group has now ceased to offer this canned food item. There was no discussion of the diversity of foods among a diverse flow of travelers.[33] However, canned Vienna sausages are still offered in prepared food packs in the desert, as has been traditionally the case for pre-assembled food packs.

Baskets full of garlic bulbs next to shelves lined with water jugs can be found in border-town bodegas. Finding a garlic bulb in a discarded backpack that I was inspecting allowed me to make a connection between garlic and migrant artifacts as traces of ethnobotanical and ethnomedical lore. Desiccated garlic bulbs or cloves are often found in backpacks that have been discarded at the migrant stations. Humanitarians and travelers later explained to me that garlic is eaten on the journey because it helps to counter the effects of a dry mouth. Travelers also rub garlic on their skin in an effort to keep animals and insects away. Some members of Indigenous groups believe *ajo* wards off bad spirits. One deported traveler in Nogales, México, told me that the connection of garlic to spirits and Indigenous people was probably a "myth about *Chinos*," a derogatory term for Indigenous Mexicans.

The introduction of traditional ecological knowledge (TEK) in the performance of unauthorized migration presents a specific set of questions. The predominant demographic in migrating populations is Indigenous Maya traveling from rural and/or remote regions within their home countries. Many are traveling outside of their place of birth for the first time, and many are from subtropical climes. Whatever knowledge of flora, fauna, landscape, and climate they may have is often not applicable in the Sonoran Desert context. TEK is more than a collection of information about objects in a given place; it is also an active epistemology that includes a cognitive mapping ability that is embodied through the observation of relationships between different objects. Does TEK in fact generally allow the transborder traveler to become an itinerant placemaker?

The journey presents a range of challenges to the uses of point-of-origin TEK when the need to adapt is invoked in an unfamiliar and militarized environment functioning as part of a designed zone of exception. These challenges present travelers with the need to deploy their cognitive mapping methodology and to collect and transmit information about the biophysical challenges of the environment during their journey to other travelers. That information can be found in comic book–style brochures produced by the Mexican government that warn of the desert's dangers and in the communicative praxis between previous travelers and would-be travelers. A question to be answered is whether TEK that is believed to be effective back home can, instead, be a detriment to the transborder traveler in the desert.

For example, consider the use of fresh garlic as an insect repellent in relation to one of the most common insect threats in southern Arizona, Africanized honeybees.[34] Some travelers have died while crossing due to massive injection of venom from these especially aggressive bees. Does garlic help repel the bees from attacking? The secondary chemical compound known as allicin is one of the principal active ingredients in freshly crushed garlic. According to Serge Ankri and David Mirelman, "Allicin, in its pure form . . . exhibit[s] . . . antibacterial activity against a wide range of . . . bacteria . . . includ[ing] most of the bacteria normally found in the gastrointestinal tract that can be responsible for disease."[35] The authors propose that allicin is also effective against bacterial meningitis, cholera, and bubonic plague. It also has antifungal properties useful in treating Candida albicans, a common yeast that is associated with superficial infections of the skin and mouth. Finally, the authors suggest in their meta-analysis that this compound has antiparasitic and antiviral properties.

Did the people whose traces I have encountered, like the owners of the backpacks left behind in the desert, receive this ethnomedical knowledge from previous travelers about the possible benefits of carrying and using garlic in the desert? Is garlic an implied necessity for a desert crossing, as suggested by the widespread inclusion of this item along with other "migrant supplies"? Did traveling with garlic align with travelers' potential expectations and conceptions of preparedness versus the actuality of desert dangers?

None of the travelers I have encountered have reported, denied, or confirmed the effectiveness of using garlic while crossing through the desert, especially to treat Africanized honeybee stings. One informant, living and working in Colorado, suggested that garlic repels snakes and other venomous reptiles. However, considering the intensity of some of the most common external injuries that travelers incur in border crossing,[36] the blood thinning that could be caused by garlic is life-threatening when attempting to control bleeding from an open, gaping wound. (For other problems, such as blood clots and cholesterol, blood thinning is extremely beneficial.) Using garlic as an external antimicrobial on open wounds in a hot, dry, dusty environment on a person who is profusely sweating or bleeding may be ineffective. I therefore suggest that garlic has the potential of creating serious negative corporeal side effects for desert travelers who may be attempting to transfer TEK out of context.

Beyond these ethnobotanical and ethnomedical traces, travelers leave other artifacts that are suggestive of the techniques adopted to circumvent the state of exception and illegality while improving the odds of survival during the desert crossing. Hand-embroidered cloths often created by wives, mothers, and daughters are given with the intention to protect travelers' foods, including tortillas, when they leave home for *La Línea*. The tortillas will most likely be replenished at a bodega in one of the many migrant staging towns before crossing into the United States. Tortilla cloths are adorned with brightly colored flowers, birds, and words, stitched with crocheted filigree and string, or handwritten with paints and ink. Among the cloths I collected were those that read *Para ti, Mama* (For you, Mother), *Siempre estoy pensando en ti* (I am always with you), and, simply, *Suerte* (Good luck). These artifacts are typically found in larger migrant sites along with discarded water bottles, empty tuna cans, clothing, and even entire backpacks and may have been quickly discarded (perhaps unintentionally) by "coyotes" as they arrived or as border agents descended upon them in an apprehension event.[37]

Humanitarian-based artists in the Tucson area have been collecting these cloths and creating quilts and other found-object art
installations with the materials. Some of these quilts and artistic renderings have been publicly displayed in the Tucson area as an embodied art practice. The objects are produced to enunciate, from an etic
standpoint, the struggle and the identity politics of representation that
the artists produce of a traveler leaving home and their subsequent
journey of hope, despair, and suffering. From the vantage point of
the travelers, I wonder, do these particular traces reveal how the act
of migration and transborder crossing is linked to vast networks of
familial and friendship ties that seek to support and console the traveler and her family?

During crossing a person may come in contact with a variety
of contaminants. Contact with animal or human feces and other
fluids are common. Food is handled, prepared, and consumed in
an environment where water to wash hands is scarce. Exposure to
contaminants while traveling increases the risk of obtaining food-
borne illnesses. These illnesses can be mild with flu-like stomach
symptoms or involve more severe physical effects, like bodily con-
vulsions that lead to dehydration and even rhabdomyolysis. It is not
uncommon to find the human remains of travelers with their pants
pulled down to their ankles as evidence of a final desperate attempt
to overcome the horrific intestinal effects of food-borne illness and
rhabdomyolysis.

Clandestine migration performance is learned through doing
and prior experience employing selected tools that directly shape this
process. These selected tools can be redeployed during the journey as
a method of classifying threats and dangers presented in the desert,
which can lead to the various innovations and adaptations found in
the discarded material remains of the journey.

It has been suggested to me that humanitarian groups might consider employing the use of low-cost water filtration systems alongside
other supplies provided to travelers in the desert. There are several
issues that arise when leaving any supply out in the open desert. I will
highlight humanitarian-provided water versus open-source water as
an example to discuss these issues presented on federal, state, private,
and reservation lands.

## Tracing Water in the Desert: The Bare Life of the Traveler in a Zone of Exception

First, it is unlawful to leave or provide supplies to what is ostensibly a known felon under the pretext of littering and aiding and abetting. In 2009, federal agents caught No More Deaths (NMD) volunteers leaving bottled water openly in the desert. They became known in humanitarian circles as the Basura 13.[38] Each member was convicted in federal court of littering and ordered to pay steep financial penalties with possible jail time. Eventually, the members of the Basura 13 group were acquitted of their charges. This case set the stage for discussions with federal agencies, including Buenos Aires National Wildlife Refuge, on obtaining permits to leave water on state and federal lands without retaliation. However, the condition of receiving and maintaining these permits was unsatisfactory to the NMD group. They argued that tethering stationary one-gallon water bottles to trees would create a climate in which travelers would be highly prone to detection and apprehension.

Second, the sabotage of supplies left by humanitarian groups has been a habitual issue. Water jugs have been routinely punctured, smashed, or removed from their locations to prevent their use by travelers. Border agents, self-appointed border militia groups, and cattle ranchers have been well-documented stealing or destroying water, food, and blankets left at predetermined points in the desert.[39] In 2011, Tucson-sector Border Patrol agents were directed by their sector chief to not remove or destroy supplies left by humanitarians. Shortly after this directive was issued, Border Patrol agents were secretly video-taped kicking and smashing water jugs and removing blankets, socks, and food items left in canisters on trails. Humanitarians have accessed social media sites, such as YouTube, to openly report such occurrences to the general public. Another issue is presented when livestock and birds eat or damage water jugs and canisters when searching for food and water during the hottest months. Cattle ranchers have reported and filed complaints with authorities that their livestock have become sick or have died due to consuming articles such as fluorescent-orange emergency blankets. Confrontations are common between ranchers and humanitarians about the legality and illegality of brown migrant bodies infiltrating US boundaries and borders, and leaving supplies on open grazing lands.

Third, as previously discussed, bottled water donated to humanitarian groups is already provided to help deter travelers from drinking water from insect, algae, and bacteria-infested cattle tanks and other open-source watering holes that are primarily used for livestock. This may also reduce the likelihood of a crisis moment when travelers start drinking their own recovered urine. Unfortunately, not all travelers are walking trails where high traffic has been previously identified, leaving their options limited to accessing cattle water or drinking their own urine.

Cattle ranchers tap into natural freshwater springs throughout the desert to supply water to their livestock. This involves using plastic hoses to run lines to the cattle tanks to keep a constant supply of water. These supply lines generally follow low-lying arroyos (steep-sided gullies) that when dry, are favored by migrating groups and their coyotes to avoid detection and apprehension. Area ranchers have issued complaints that travelers access their piped water by cutting or piercing the hoses. The lines are initially buried into the sand but are exposed by powerful flash floods during monsoon season, which overlaps with the peak of unauthorized crossings. Locating natural springs and other water sources requires a degree of geologic and hydrologic knowledge of this stark basin and range terrain that is not necessarily presented to travelers and leaves them visible and vulnerable to detection and apprehension.

Fourth, humanitarian groups like Tucson Samaritans and NMD function largely on public and private donations of very specifically requested supplies. There is a cost-benefit analysis that is considered when asking for donors to supply more than extremely low-cost, specific food items and water or new and to-be-recycled clothing and blankets. Based on the amount of accessed supplies left in the desert by humanitarian groups, acquiring the number of water-purifying apparatuses required to serve this migrating population is not cost-effective for donors or volunteers. It also carries the added risk of making an even more attractive target for sabotage or theft once left in the desert.

To stay within their budget, NMD volunteers dumpster dive at local grocery stores and restaurants to obtain discarded food to feed to incoming extended-stay, out-of-town, seasonal volunteers. They have attempted to rely on donated and volunteer-prepared foods to accommodate feeding multiple and rotating volunteers at their base camp, particularly during spring break and summer. However, in 2013, NMD returned to dumpster diving as a means to provide food to their base-camp volunteers in the desert to reduce organizational costs. Tucson Samaritans and NMD also request and accept donated vehicles to patrol the desert for lost or sick travelers and to perform water drops on trails. The cost of maintaining vehicles driven on rugged terrains also contributes to NMD's budget needs.

Fifth, Altar, Sonora, is utilized as a migrant staging town, where would-be transborder crossers prepare to make their journey through the desert by hiring coyotes to guide them on the trails, replenish their gear and personal staples such as food, water, and backpacks, and rest at one of a handful of migrant shelters before crossing over. Altar is located just south of Nogales, Arizona, and the Tohono O'odham Indian Reservation. Depending on their intended final destination, travelers can be directed into either area for crossing through the desert. The US-built border fencing is designed to funnel travelers—who aim to arrive in Phoenix or surrounding areas—onto the Tohono O'odham Indian Reservation, the third largest in the United States. This portion of the transborder landscape is the most extreme, barren, and remote of the Sonoran Desert and is surrounded by mountain peaks that soar from seven to twelve thousand feet across Arizona. Recently, the US Border Patrol established two base stations at points frequently traveled by unauthorized migrants and drug smugglers. From those stations, the agency monitors the open desert areas and roads and the San Miguel border gate, where tribal members traditionally walked across freely to access their tribal lands on both sides of the geopolitical border.

Because of the remoteness of the Tohono O'odham Reservation, the death rate of unauthorized travelers is extremely high.[40] To help combat death on the reservation, tribal member Mike Wells began distributing large quantities of one-gallon jugs of drinking water close to the border fence in highly traveled areas. This action has been in direct violation of the cease and desist order issued by the Tohono O'odham tribal government citing that this action only encourages people to walk through and then die on their lands. When a body is found on the reservation, the tribal government is financially responsible for the disposition of that body, which ultimately finds its way to a county coroner's office for identification and disposal (qua anonymous burial or cremation). Despite this order, Wells continues to leave water out on his reservation for people making the journey. Because it is unlawful to be on the reservation without the escort of a tribal member, requests from humanitarian groups for permission to leave water and supplies on the reservation have been routinely denied; so too are the requests for access and permission to patrol Tohono O'odham lands to assist travelers in need.

## The Dangers of Deception

Travelers unfamiliar with the desert terrain and landscape, whether it is their first time or fourth time, are subject to the myths and misconceptions elaborated by coyotes or depend on misinformation from other travelers. Often they are told that "Phoenix is only a few hours walk just over that hill right there." I have found many people walking in the desert alone. They ask, "How far to Phoenix?" or "How far to Tucson?" When I find them, they are devastated to learn that they are still a minimum three- to five-day walk from their destination after having already walked one to three days with little food and no fresh water.

Travelers are often woefully underprepared for the heat, insects, rocky terrain, absence of potable water, high elevations, and lack of roads. Some travelers have found themselves on the Barry M. Goldwater Range during war game exercises with live shelling. In recent years, coyotes have taken to abandoning entire groups of people only a few hours into the crossing, leaving them stranded in the desert with no idea of where they are and where they need to go. Recently, the border-crossing groups have grown from an average of five to ten to more than thirty people, wandering around lost after having been abandoned. More recently, cyber coyotes are being encountered in the desert, using technology to continually pass groups off from one coyote to the next or to perform reconnaissance and avoid being detected by Border Patrol forces.

Due to distance, physical strain, and the extreme elemental conditions of the desert, travelers must attempt to carry enough food and water to sustain them during their multiday crossing. Unfortunately, no one is or can ever be properly prepared. Some would-be travelers are given deceptive or false information by their coyotes about what to expect during crossing.[41] The amount one can comfortably and reasonably carry while making the journey through the desert in extreme temperatures and on uneven terrain, based on information transmitted by experienced and inexperienced others, translates into one to three gallons of drinking water, a bottle or a few packets of powdered electrolyte-replacement drink, some salty snack chips, cans of tuna and jars of *mayonesa* (mayonnaise),[42] and some *pan dulce* (pastries) or tortillas. People quickly become exhausted and depleted and begin to discard unnecessary and extraneous "luxury" items from their packs onto the desert floor.

What is clear is that managers and staff at bodegas in Mexican border towns have capitalized on transmitted information from previous travelers, even if they are not themselves experienced travelers. It cannot be underscored enough that a person migrating to the United States from home does not begin their journey at the US-México border; rather it is one more treacherous leg of their journey north. They have been traveling for days, or even weeks or months, and are tired, hungry, destitute, scared, and confused. The limited, predetermined, and encoded choices available in border town bodegas give would-be travelers a restricted freedom *from* choice, which can help alleviate some anxiety over making sound personal choices; this often occurs to their detriment.

## Conclusion: Transmotion and the Limits to Biopower

From a purely biopolitical vantage point, desert crossing involves above all a precariously dislocated human body, a living self that nonetheless maintains ties to family, community, and nation(s). However, the state of exception and the corresponding recoding of such bodies under the spell of the regime of illegality—attenuated by the militarization of the border—clearly and unambiguously impose on the transborder traveler the status of a body reduced to bare life.[43] In practical terms, this condition imposes the need for travelers to constantly reevaluate how to best conserve water and food supplies to survive their journey undetected while prolonging life (biological functions) under the most dire of circumstances. Those who find themselves out of food and water will turn, if presented with opportunity, to replenishing water supplies with the green open-source cattle tank water or resort to drinking their own urine when no other source has been located.

This creates a life-threatening predicament for travelers caught up in a precarious and deadly situation: should they begin conserving what little water they have and risk organ failure, or replenish what is gone with tainted water or urine, which may also cause organ failure? The millions of empty water and electrolyte bottles that are dispersed across the desert are evidence of both successful and failed attempts to conserve lifesaving fluids during crossing. Knowledge of the shared migration experience as a technical performance can mean the difference between success and failure in crossing the border undetected and saving one's life in an unforgiving desert ecosystem. It is difficult to convert compelled geo-mobility into effective "transmotion."

If walking on the correct trails, travelers may find humanitarian-created "dignity bags" made of material sewn together and fitted with cord straps to carry like a backpack. These contain preassembled food packs and other determined-as-necessary supplies for clandestine desert crossings. The contents of these so-called dignity bags change with the available donated goods. I have located the remains of these dignity bags on my desert patrols and have spoken to people on both sides of the border who have also come across these bags on the way north. They appear to be an effective tool for providing assistance to some travelers.

Given the number of deaths, illnesses, injuries, rapes, and homicides these desert travelers face, it is tempting to dwell on and lament the inhumanity of it all, to decry the lack of an effective human rights regime and the failure of development programs to benefit the sending countries. The practice of "medicalizing" migrants' experiences to make them worthy of a path to inclusion in civil society, while laudable, may actually contribute to biopolitical constructions of citizenship that tie rights of membership to specific biomarkers of political suffering, like the widespread notion that "illegals" are undeserving of basic protections and provisions to sustain life. Orientalist imaginaries of Latina/o immigrants have manifested brown bodies as a "depraved class" and a "cancer" threatening the presumably more civil United States and its freedom-loving way of life. The border control apparatus detailed in this essay seeks to equate the effectiveness of control with the removal of pathogens.

The sovereignty embodied in individual acts of transmotion involves the exercise of the freedom to move across physical and conceptual boundaries, or between what Vizenor calls "communal tribal cultures and those material and urban pretensions that counter conservative traditions."[44] However, in the absence of that freedom of physical movement, many other freedoms become impossible, as the traces of transborder travel illustrated here indicate. The act of procuring food supplies before beginning the journey across the "wire" is an act of resistance against geopolitical borders and boundaries. It implies a defiance of the DHS and its border-policing entities and challenges the construction of the illegality of unauthorized entry and migration into the United States. All the traces examined and discussed in this essay can thus be read as artifacts of subaltern resistance under the most extreme conditions of biopolitical marginalization.

Biopolitics merges with the political economy surrounding unauthorized transborder travel to create constraints that increase the risk of death and injury. Devon G. Peña, coeditor of this volume, recently asked if I had ever found discarded evidence of the adoption of more lightweight foil packets of tuna versus the heavier canned tuna that is a common staple of transborder travel packs. I have not. The absence of foil packets is related to their cost and availability and shows that the techniques of survival are further constrained by political economic conditions. This means that the risky act of exercising the freedom to travel successfully by optimizing travel pack items is actually constrained by the lack of choice to determine which objects and items are physically more practical. Asserting food dignity is difficult under such circumstances.

The precarious migrant body most certainly possesses elements of trauma and suffering, and narratives seeking to address this structural violence and historical trauma invoke a liberal call for immigration reform or better medical attention to those in need. This risks isolating brown bodies in a state where the Other must remain precarious and powerless in order to remain exotic, that is, of interest to anthropologists, many of whom remain bound to a colonial gaze when dealing with the techniques of survival used by transborder travelers in a zone of exception. Recognizing and valuing the agency as a given quality of the Other—even if this involves the act of finding a way to stay safely hydrated in the desert or rubbing one's body with crushed garlic—imply that we are no longer bound to the undercurrent of essentialist liberal and humanitarian frames.

The clandestine and corporeal life of migrants crossing the Sonoran Desert leaves traces that reveal much about the act of "unauthorized migration" and the transborder travelers' material, symbolic, and even ethnobotanical and ethnomedical knowledge. They also tell us a lot about humanitarian work to relieve the travelers from these threats to life and limb. However, these traces also reveal much about the political economy that frames the limited choices for survival techniques that can be deployed to avoid detection while prolonging biological functions. This knowledge should encourage food justice and Indigenous rights activists to confront this dismal life-threatening condition by working to end the underlying state of economic exception and illegality that undermines transborder struggles for a more civil and democratic future.

# CHAPTER 6

# Norteada/o en el barrio

## Decolonizing Foodscapes in South Central Texas and Reclaiming Belonging

### LEE ANN EPSTEIN

While I was driving in the car with my stepmother, she commented on our route and mentioned, "*No uso los highways, porque me quedo bien norteada.*" I asked her, "Did you just make that word up? What does *norteada* mean?" She assured me *norteada* is a real word, and we continued with other conversations. Not long after, I asked native Spanish speakers in my English conversation class what *norteada/o* meant. The student from Colombia said the word did not exist in Spanish, and another student from Puerto Rico said it did not really have meaning the way I explained it. I was pretty sure I was confused and misunderstood what it actually meant.

According to the Real Academia Española (RAE), the standardized Spanish dictionary definition states, "norteado, da.: 1. adj. vulg. *Méx.* Desorientado, perdido," or disoriented and lost (RAE). This entry notes how norteada/o functions as a descriptive in adjective form, and also labels the word as "vulgar" in Mexican vernacular. The word *vulgar* indicates that norteada/o as used is pedestrian and informal slang. Other sources offer insight into the word's alternate denotations. The verb *nortear* has nautical and navigational connotations: "to steer to the north" or "to veer northwards."[1] The RAE simply defines *nortear* as directing oneself toward the north, especially by sea.[2] These seafaring Castilian terms undoubtedly reference European sailors navigating the waters as part of empire-building and colonizing projects.

Finding north on a compass, by the stars, or from the wind meant the Spanish colonizers could travel "east" in pursuit of commerce. These navigational technologies were not reliable, since the colonizers got lost. Their "conquests" brought about violence, rape, disease, and land theft to those already inhabiting the lands.

As a reflexive verb, *nortearse* is predominantly of Mexican regional usage and translates into "to get lost." Another definition of *nortearse* states, "(De *norte*.) Desorientarse, perder la noción del rumbo," or to be disoriented; to lose track of direction.[3] Once the verb is reflexive, it reifies the self in the word's meaning and brings the subject into being. Later, while reading Gloria Anzaldúa as a graduate student, I came across her use of norteada/o. She reclaims the concept of being "norteada/o" or "estar norteada/o" as a theoretical framework and interrogates issues of Chicana/o or mestiza/o identity. She claims, "Estoy norteada por todas las voces que me hablan simultáneamente."[4]

When I realized my stepmother had not made up the word, I experienced a difficult moment of cognitive dissonance as I had to confront how I privileged a standardized Spanish language and was socialized to downplay meaningful borderland epistemologies and TexMex dialects from home.[5] Later, I sat down with my stepmother in her living room to recall our conversation and to apologize for questioning the collective and generational knowledges she carries. In my experience, part of the process and the ruptures in coming to consciousness can be informed by moments of confusion experienced through estar norteada/o. Part of the process in coming to consciousness about our borderlands subjectivity can be informed by moments of confusion experienced through this condition of estar norteada/o.

To engage in the geospatial/geopolitical sense, the "turned aroundness" as a Chicana Tejana can in part be explained through a historical shifting of surroundings, a reference to Texas's changing political designations and boundaries ranging from being indigenous, to being Spanish, French, Mexican, to being an independent nation, to being part of the US Confederacy, and finally to being a US state.[6] An additional way to examine estar norteada/o is to engage with how Emma Pérez weaves diasporic movement with issues of identity. She posits:

A kind of colonial diaspora emerged, created by populations dispersed through a land named, renamed, bordered, measured, mapped, and fenced to restrict more movement, whether dictated by Spanish colonialists, Mexicans, or Euroamericans—all have mapped and demarcated with artificial lines land where travel persists through time. Identity itself transforms as diasporas weave through historical moments. The unmarked identities of the diasporic become ordered and categorized according to the named and renamed geographic spaces on which they travel.[7]

In addition to situating the self on a historical and geographic map, or through autotopography, "the grounding of self and communal identities through place making . . . [where] the cultivation and celebration of meaningful food [is] central,"[8] norteada/o can refer to the confusion found in coming to consciousness in the process of working through one's cultural identity (inclusive of the meaning of food) and finding belonging in the context of place. And while autotopography refers to agential "self-telling through place shaping," estar norteada/o is indicative of the making-sense process of one's surroundings and identity when boundaries are unclear.[9] It is in these interstitial and hyphenated spaces where meaning is made: between Mexican and American; speaking English, Spanish, TexMex and searching with the tongue for erased Indigenous languages; between the family lore laden with celebrated colonial histories and the cognitive dissonance of an emerging consciousness that challenges what these stories value, and what these stories erase.

## Estar Norteada/o: Mapping a Theoretical Framework

This work examines applications of estar norteada/o through reclaiming an understanding of my Chicana Tejana identity and of an urban, working-class neighborhood in San Antonio, Texas, in order to contextualize locations where food choice is confusing. Reading the landscape is what I call a decolonizing project, in that I argue against deficit terms like *food desert* and challenge Chicana/o environmental in/accessibility to food. By using a decolonial, Chicana, third-space feminist theoretical approach, I will unpack the manner in which a yearning for understanding one's culture emerges in the experience of being norteada/o, that is, through reclaiming identity in the face

of geopolitical dis/location and as a means to connect to food in/
accessibility.

## DECOLONIAL THEORY

Several scholars have engaged in decolonial theory in their respective
regions to contest structural and epistemic colonial matrices of power:
Walter Mignolo in Argentina, Anibel Quijano in Peru, and Emma
Pérez in Texas.[10] Decolonization, as defined by Eve Tuck and Wayne
Yang, "brings about the repatriation of Indigenous land and life . . .
Because settler colonialism is built upon an entangled triad structure of settler-native-slave, the decolonial desires . . . can similarly be
entangled in resettlement, reoccupation, and reinhabitation that actually further settler colonialism."[11] Linda Tuhiwai Smith challenges the
dehumanizing negations of inadequacies projected onto Indigenous
people as a result of colonization. She asserts that

> imperialism and colonialism brought complete disorder to col-
> onized peoples, disconnecting them from their histories, their
> landscapes, their languages, their social relations and their own
> ways of thinking, feeling and interacting with the world. It was a
> process of systematic fragmentation.[12]

I will suggest that this also leaves people disoriented and stranded
from their histories and cultural memories.

## CHICANA FEMINIST THEORY

Chicana feminist scholars also challenge imperialism and elitist academic thought as they engage in theory that applies to working-class
women of color who have been left out of previous feminist thought;
they address issues of race/ethnicity, class, gender, sexuality, language,
and nationality and challenge dominant patriarchal epistemologies
of power.[13] Chicana feminist theory informs my discussion of estar
norteada/o in my effort to decolonize systems that exclude, repress,
and censor all other knowledges, and consider how US third-space
feminist theory applies within South Central Texas to those whose
discourses and knowledges have been marginalized.[14]

Decolonial and US third-space Chicana feminist theory present a framework to lay out the concept of yearning for erased epistemologies through being norteada. Emma Pérez defines the decolonial imaginary as "the time lag between the colonial and postcolonial,"[15] indicating that our state of understanding has not yet reached the point of being beyond colonial restraints. Pulling further, in Pérez's discussions of cultural erasure through perpetuated colonial histories, Chicanas seek refuge in identifying ourselves culturally: through trying to reclaim the Spanish language (albeit touching colonial wounds), hungering for the mother's tongue, hungering for the foods from Earth that feed not just the body but also the mind and soul. Pérez negotiates part of the remembering process and the reconstructions of history as she states, "Memory as history, as social construction, as politics, culture, race—all are inscribed upon the body. Inscriptions upon the body are memory and history. The body is historically and socially constructed. It is written upon by the environment."[16] In the case of estar norteada/o, the environment, particularly the in/access to food, is contributing to historical bodily memories, and being norteada/o can also be theorized as contributing to the hunger for the histories and knowledges literally and figuratively starved out of us, both in terms of identity and in terms of a visceral hunger. Thus, through a Chicana feminist and decolonial theoretical approach, I reclaim the concept of estar norteada/o as being lost in a geospatial/geopolitical sense (referring to historical and geographic location), as experiencing a turned aroundness, and as the process of negotiating one's identity and making sense of the world.

## Norteada/o "bodymindspirit" Epistemologies

Indigenous epistemologies are ways of knowing that are both antecedent to and have survived the logics of coloniality. Understanding this is important because the production of meaning is a process we all seem to use to make sense of the world. One problem we can address is the creation of binaries. Challenging the erasure of the spirit within the Cartesian mind/body split, Gloria Anzaldúa deliberately centers her theory on Indigenous epistemologies. These ways of knowing (and being) connect the cultural and the physical to the spiritual. She states, "With the loss of the familiar and the unknown ahead, you struggle to regain your balance, reintegrate yourself (put Coyolxauhqui together), and repair the damage . . . call your spirit home."[17] The remembering of Coyolxauhqui is part of piecing back together the stories lost to us in our colonized imaginations. It is in this confusing process of repair where we experience a politicized turned aroundness.[18] In the same vein, the concept of estar norteada/o pulls from how Anzaldúa indicates the sense of finding belonging and home *within* one's spirit as well as recognizing how the spiritual, the visceral, and the locational are all woven together as overlapping parts of the whole being. Recovering the spiritual as part of "bodymindspirit" allows for wholeness and openness to belonging and connecting to others.

Not only is the sense of belonging a spiritual state of being, but our selves can also embody the theory of "differential consciousness" that extends beyond what is cognitively privileged in Western ways of knowing. Chela Sandoval states that "the radical form of cognitive mapping that differential consciousness allows develops such knowledge into a method by which the limits of the social order can be spoken, named, and made translucent: the body passes through and is transformed."[19] To define third space as part of Chicana feminist theory, Sandoval engages with finding a place to belong, albeit marginal and interstitial, as "surviving in that in-between (silent) space."[20] It is in these in-between, lost, and erased spaces where we can recognize what we yearn for and why we have been systematically denied an understanding of ourselves.

## From Pathologizing Brown Bodies to Reading Food as Cariño

Examining food accessibility for many Chicana/os shows that historical starvation stems from systemic processes of "industrialization and forced urbanization of the Mexican-origin population."[21] These processes historically have contributed to colonized diets consisting of processed (fake) foods, such as white bread, bologna, and shrink-wrapped cheese, or existing in the forms of genetically engineered corn, soybeans, and wheat.[22]

Public health research scholarship widely pathologizes Mexican American and Chicana/o obesity as health problems and blames a lack of education and a failure at assimilation rather than acknowledging accessibility or taking cultural histories and the effects of structural violence into account. While nutritionists, dieticians, and others in the field of public health are quick to point out the physical effects that processed foods have on the body (diabetes, heart disease, obesity), the use of a cultural deficit paradigm only further places blame onto brown bodies through "lifestyle trends" and "familial influence" for what is assumed as Chicana/os inability to feed the self and the family properly, resulting in generational patterns of obesity and other medical conditions.[23] To contest the deficit paradigm theory that places blanket blame across a whole culture, estar norteada/o acknowledges the agency of those who are turned around by economic and food accessibility limitations. Further, obesity occurs not as an act of deliberate gluttony but rather as a result of social and environmental surroundings inscribing illnesses on and preventing food access to brown bodies.

Writings such as the recent book by Luz Calvo and Catriona Esquibel, *Decolonize Your Diet*,[24] call for plant-based diets and reject processed foods as acts of healing and self-love. Even "cooking a pot of beans is a revolutionary act" as it is simple, requires plant-based ingredients full of healthful benefits, and also functions as a way to express love, given that cooking and eating together constitute *cariño* (loving care).[25] Within my multigenerational cultural memory, the way elders communicated and projected love and compassion was through preparing and sharing a meal; this love and compassion was reciprocated by eating together. A participant in this space could not deny the offered foods, as it would signify the rejection of the preparers' love. My grandparents expected us to eat a second helping to ensure that no one left the table hungry for food or *cariño*. The hunger to feel whole and to belong in the world can be furthered by how Anzaldúa uses the term estar norteada.[26]

In "La conciencia de la mestiza: Toward a New Consciousness" Anzaldúa writes, "Estoy norteada por todas las voces que me hablan simultáneamente,"[27] which is a poetic reflection of the internal *nepantlism* she experiences, or the navigation between cultural and spiritual borderlands. According to AnaLouise Keating, "Nepantla indicates space/times of great confusion, anxiety, and loss of control."[28] Unlike nepantla, with its spiritual and mythological connections as the in-between space, estar norteada calls additionally for a concretized geo-political recognition of a northward orientation. Estar norteada represents a migratory pull that not only displaces people in the Mexican diaspora but also continues to displace Mexican American diets and food options. In urban San Antonio, where many of the food options are limited to corporatized chain restaurants, selecting foods based on price and convenience makes practical sense for working poor and precariat groups. Fast-food restaurants are accessible in terms of location, while grocery stores and food options found within are less so. This seems to reproduce emotional states of disorientation, dis-location, and feeling lost and confused when trying to make decisions about food.

The concept of norteada reflects that same confusion and chal-lenge of navigating through borderland spaces that are unfamiliar but necessary for survival. The theory behind norteada indicates the inter-stitial spaces of diasporic labor, of food decisions, and of displacement in terms of cultural identity. The intent of the larger social structure is to seek a cultural assimilatory process, which many Mexicans and Mexican Americans succumb to through migration or displacement. The process of cultural assimilation erases the prominence of tradi-tional foods. These diasporic struggles linger in the cultural memory of Chicana/os, who continue to suffer from displacement resulting in the feeling of norteada/o, an overlapping element of yearning for that which has been erased and the longing to belong. To further apply the ideas behind estar norteada/o, I explore my own disorienting family histories as well as the location of the community that surrounds me.

## Norteada/o and Making Sense of Identity and Belonging

Growing up in San Antonio, Texas, I realized my maternal family never spoke Spanish to my brother and me; instead, the family made a deliberate effort to speak to us in English. We cooked in English, watched TV in English, but the two of us were completely lost when everyone else spoke Spanish. We were stripped of a language and an identity that was replaced with conditioned colonial stories of love and valor: the tale where after fighting with her brother, Apolonia hit the ground with a stick and Spanish gold coins popped out of the earth with perfect timing to save the rancho; the legend about Pedro Huizar sailing to San Antonio from the Canary Islands and carving a rose window on a mission in memory of his shipwrecked fiancée; or how my grandmother's father had owned the last piece of what once was six hundred acres of land *from* Spain. These were the family stories told and retold to us every weekend when we visited our grandparents.

We were supposed to be "Spanish," not Mexican, that is white, not brown—a raced and classed form of upwardly mobile identity that assimilates into mainstream America. We were not taught the complex history of Texas—how our family lived in San Antonio when San Antonio was México; how Mexicans like our family fought *for* the Texans in the Battle of the Alamo against other Mexicans. Or how before that, the Spanish side came to San Antonio as part of a larger colonial proselytizing project; that those who came from the Canary Islands were Black and purchased their whiteness as a certificate in order to own land. Or even how that land, the six hundred acres from Spain, was actually part of a land grant that stripped Earth away from the Indigenous people living on it to be redistributed to a "noble" class of people. When everyone died—my grandparents, my parents— and the storytelling stopped, I yearned for the rest of my history, for the history and language I had been deprived of, and the parts that had been silenced and forgotten. I had reached toward institutions for answers; in college I took Spanish and history classes hoping to recover my sense of self and family, but my canonized undergraduate education only further perpetuated colonial myths. In challenging colonial stories from my lived experience, I use food memories to recenter my epistemological understandings of myself and my family in a way that contests colonized notions of power.

For this *autohistoriateoría* (biographical counterstory), my maternal grandfather's agency manifests itself as I begin to recognize ruptures in the processes of decolonizing my own cultural memory. My grandpa walked like the lowercase letter *n*, bowlegged like he had been riding a horse all his life with his feet strapped into Stacy Adams Madison boots. He was informally schooled through second and third grade; he would drop off his own children, my mother and my uncles, at the library so they could have the opportunity to read and learn, and so he could spend time at the corner cantina or local icehouse. By the time I knew him as my grandfather, he was overweight and healing from kidney stone surgery that left him with a front-belt scar. When the doctor told him he couldn't drink beer or soda anymore, he was devastated. Sometimes when my grandmother wasn't looking, he would take me into the kitchen, move away the mess of grocery-store plastic bags stashed under the Formica table pushed against the wall, and show me large pickle jars lined up along the baseboard. "What *is* that?" I once asked. "Tepache" he replied. Then Grandpa would drink the sour-smelling liquid made from fermented watermelon rinds right out of the jar, his Adam's apple pulling down each *trago*, or gulp. My mother told me never to taste the tepache because it would make me go blind. My grandfather made his own moonshine while consciously using every piece of the fruit—nothing went to waste in his house since he understood the hunger of his childhood from the Great Depression.

When my brother and I visited our grandparents, my grandfather would sometimes make us lunch before heading off to the farm—the five-acre parcel of land that meant he had made it in the world. Typically, we had the choice of burnt-tortilla *tostadito* bean-and-cheese tacos or Dagwood-style sandwiches that appeared to clean out the "icebox": thick slices of Texas-Toast white bread topped with scrambled eggs, deli ham, lettuce, tomato, more bread smeared with mayonnaise, sardines, mustard, crackers, and *fideo* (Mexican vermicelli) followed by more bread. We preferred the tacos. My grandfather's creativity and resourcefulness with food extended beyond the kitchen. At the farm, my grandfather knew how to read the land, so when the mesquite tree sprouted leaves in the spring, he could plant certain crops for the summer. My grandfather could not read the Farmer's Almanac, but he knew when and where to plant seeds, how long to wait before harvesting crops, and how to mend the sandy soil. He grew tomatoes, rows of strawberries, okra, radishes, turnips, collard greens, and beans and planted various fruit trees. It was his paradise. I have memories of learning how to scatter oats onto plowed dirt from the tailgate of a pickup truck. My legs dangled, and I tossed fistfuls of oat seeds like a baseball. My grandfather stopped the truck, hoisted himself onto the bed of the pickup, grabbed some seeds, and gracefully broadcasted a handful that distributed like rain in a perfect arc. "Ahora, sí. Like that."

Recognizing my grandfather's resourcefulness about food positions him as having ingenious knowledges that were not recognized in the school system. His denied food literacies served to aid in survival. He was a man of few words speaking a little English, a little Spanish, and a little Caló, but mostly all of them together. Considered textually illiterate, my grandfather could not read numbers or letters and was often lost; later, with the progression of Alzheimer's, he would even get lost in the grocery store. This state of confusion with identity, with space, and with food can be conceptualized in Chicana/o vernacular as being norteada/o. I was a bit norteada in my interpretation of my grandfather's moonshine, thinking these "gaffes" were the acts of a crazy old man. I did not understand that he was performing an Indigenous way of preserving food through fermentation while maintaining literacy through reading and even more so by remembering how one works on the land.

## Reading a Barrio Foodscape: The Thompson Community in San Antonio, Texas

Situating the counterstory of my family to contest colonial histories opens one path to an Indigenous epistemology. I also use estar norteada/o as a way to read the geopolitical space and location of my working-class San Antonio barrio. I base this on a norteada/o visual landscape reading of space and place in the Thompson Community neighborhood. This is one of San Antonio's Westside barrios. I noticed minimal access to fresh food seemingly as a result of the geographic isolation and segregation from the rest of the city. To give a general scope and context to the small neighborhood, the triangular topographic boundaries include these three limits: railroad track, a highway, and a defunct military base. While the military base allowed for working-class Chicana/os to become middle class, its closure in the 1990s shifted the neighborhood from a former thoroughfare for the base to a near ghost town. Just across the railroad tracks and adjacent to the neighborhood boundaries is what is known as the toxic triangle: an area east of the base where higher rates of liver cancer are blamed on groundwater contamination from improperly disposed chemical waste.[29] Arguably, the Thompson Community is not remote enough at a distance from the hazards to be convincingly safe. This neighborhood community is insulated but not in such a way that it functions autonomously from the rest of the city or the rest of the Westside. As far as the local foodscape goes, the closest grocery store is 1.5 miles outside of the residential neighborhood zone, which may be near or far depending on the type of transportation that community members have access to. If traveling by foot, pedestrians must cross under an overpass that is poorly lit in the evenings, which discourages people from walking to this grocery store. This is a classic example of the geography of environmental racism.

Employing an inquiry on the method of the linguistic landscape and reading the environmental signage,[30] I see that the neighborhood's fast-food signs can be assessed to reflect what the US Department of Agriculture terms a "food desert." This is "an area in the United States with limited access to affordable and nutritious food . . . [and] . . . composed of predominantly lower-income neighborhoods and communities."[31] This term is problematic because a desert is a rich biome full of many sustainable organisms, yet the connotations of desert imply barren land with infertile soil.[32] Within a four-block radius of commercial and neighborhood space, twelve (mostly fast-food) restaurants operate, four of which are Mexican restaurants. There are three convenience-store gas stations, and while all of these businesses offer the potential for accessing fresh and frozen fruits and vegetables, the gas station storefronts display wilted cilantro and over-ripened bananas among other overpriced and unappealing fruits and vegetables. The preponderance of fast-food restaurants does not indicate a dearth of food but rather a surplus of processed foods and the inaccessibility of fresh organic produce.

The neighborhood's thriving small-business economy has shifted over the years. Before the base closure in the mid-1990s, when the base had steady flows of traffic, most of restaurants and even mechanic shops were booming. But after the closure, the local businesses suffered, the community's sources of income disappeared, property values decreased (which affected school funds), and the neighborhood's food accessibility worsened. With fewer people maneuvering through the area, in time the area slowed in growth. One exception

Figure 6.1. South San Antonio industrial foodscape. Fast-food chains predominate in the food desert of the Thompson Community. Courtesy of Lee Ann Epstein.

to the area's slowing growth is San Fernando III, one of San Antonio's largest cemeteries. A private school and elementary, middle, and high schools all line the front of the defunct military base, which used to be a busy thoroughfare for a burgeoning population of middle-class Mexican Americans. The foundation of the Mexican American community's upwardly mobile American dream was dependent upon the military-industrial complex, which built on farmland and poisoned the surrounding area's land and water systems. Beyond reading the landscape vernacular, or in this case signage, not front yards, as vernacular landscape,[33] I can employ a theory of estar norteada/o, whose regional meaning of turned aroundness speaks to the confusing food environment of this community.

## Thompson Community's Garden and Mercado

According to the Thompson Community Association, for a short period of time a community garden existed behind one of the elementary schools. Because few community members had the extra time to volunteer the necessary manual labor, the community association and local residents abandoned the garden. Within the past two years, a small *mercado*, or Mexican farmers market, opened up in the parking lot of one of the Mexican restaurants. Pickup trucks stacked with fruit in the truck beds backed into parking spaces, and folks sold their produce for cash. This, too, was short-lived, perhaps because the market was available during the weekdays, when many are working. Without access to fresh produce and healthy food options, many folks are left disconnected from food and rendered culturally invisible.

Furthermore, many Indigenous and Mexican American diets have shifted due in part to the corporate control of processed food production.[34] Reframing the discussion away from a deficit paradigm, which places blame on Mexican American culture and diets and not on the industrialization of food and grocery store redlining, is crucial to recognizing who has power, and the voices of those who are powerful in the margins. Scholars Tuck and Yang challenge what can be defined as decolonial, and view connecting Indigenous peoples to land rights as fundamental. Tuck and Yang assert that with the process of settler colonial land appropriation, "the disruption of Indigenous relationships to land represents a profound epistemic, ontological, cosmological violence."[35] In our context, the neighborhood's isolation from the rest of the city, the presence of toxic and nonoperational militarized zones, and minimal access to food are the forms of structural violence that community members confront daily.

## AUTOTOPOGRAPHICAL SHIFT

Despite my struggles with confronting colonialism in my family and in the community, I am putting decolonial knowledges into practice. I currently volunteer at a Westside San Antonio community garden with my stepmother that is located between our two neighborhoods. The garden is full of roses and other flowers as well as a sizable vegetable garden with a harvest station. Olive, peach, grapefruit, and pomegranate trees separate the shaded flower section from the full-sun vegetable patches. The food here is meant for the community. It is here that I have learned from elders how to dry seeds from collards, companion-plant herbs with vegetables, prune roses, grow okra, and identify many of the "weeds" as medicinal and healing coevals.

In my own backyard, I grow most edibles intentionally and some

Figure 6.2. Author's kitchen garden, San Antonio, Texas, summer 2015. Seen here are nopales growing among fruit trees, medicinal herbs, and decorative plants. Courtesy of Lee Ann Epstein.

unintentionally. The backyard has random pots with collards (some in the ground), Crawford lettuce (resilient and slow to bolt in heat), radishes, nopal cactus, Brussels sprouts, tomatoes, jalapeños, Swiss chard, *verdolagas* (purslane), *quelites* (lamb's-quarters), *hoja santa*, epazote, *estafiate* (artemesia), *ruda* (rue), *hierba anise* (Mexican mint marigold), *chiltepin* (bird pepper), rosemary, thyme, oregano, lemongrass, lemon balm, aloe vera, and spearmint, among some squash volunteers from the compost. Bean and loofah vines climb up the chain-link fence and offer some privacy and flowers as they curl around the rows of barbed wire at the topmost part of the fence. Grapefruit, fig, and pomegranate trees face north, as they can tolerate the mild San Antonio winter, and there are a couple of blueberry and blackberry bushes as well. One sprouted avocado tree stays protected near the house close to a still-potted grapefruit tree grown from seed from one of the sweetest Rio Grande valley grapefruits I have ever tasted. Surely there are other useful plants in the yard that I cannot yet identify, but it continues to be a process.

What I envision for the future of this space is to share this "property" as public land space. I have sprouting seeds of collard greens and propagated sprigs of rosemary, two companion plants that grow well together. My plan is to replace the small green patch of grass and dirt between the sidewalk and the street. Even though this is city property, the community must still tend to it. And if it were full of edibles such as greens, fruits, vegetables, and herbs, then folks on foot, headed to the bus stop, or walking with their children, could help themselves. While it may take some time, I have aspirations of planning youth food workshops as after-school programs, as well as having a community kitchen where elders and youth can cook for, share food with, and learn from each other.

## Nortearse, Only to Find Oneself Again

In unpacking the concept of estar norteada/o, I situate the confusing attributes of colonized familial histories as a counterstory framed as an Anzaldúan *autohistoriateoría* (self-history/theory), and so open vistas on the space/place/geopolitical location of community in relation to food and foodscapes. The observable turned aroundness is important as it challenges *where* in time, space, and consciousness one is located.

Despite the colonial historical connotations to noretearse, the process of finding oneself, or coming to consciousness, includes moments of uncertainty and confusion, and that process is what is working toward contesting systems of power to make meaning. Challenging deficit perspectives recognizes that it is not the people who are lost or turned around but rather the systemic structures of geographic isolation, proximity of food access, economic disadvantages, and racialized districting and food marketing.

With norteada/o situated in moments of decolonial and third-space feminist theory, this methodology situates my own subject position by examining two different but overlapping angles: the yearning and embodiment of hunger and identity, and the geopolitical dis/location where the hungering to belong is pervasive. Using this approach to expand on the concept of being norteada/o allows for oppositional discourse that engages with (mis)representation others have inscribed in our stories. In the case of my personal ruptures of consciousness, I challenge the heroism of my Spanish colonial familial history; within my community, I insist that despite our geographic isolation, we, too, deserve access to fresh produce and other healthy food options, even when our location and the environment prove disorienting. Estar norteada/o is an interstitial space of turned aroundness. It is here in this suspended location where ruptures and re-memberings of cultural identity may come forth to contest structural inequity and systematic discrimination.

CHAPTER 7

# Tortilleras, testimonios, y recetas

## Decolonial Foodways from the México-US Borderlands

### LUZ CALVO AND CATRIONA RUEDA ESQUIBEL

*This land was Mexican once*
*Was Indian always,*
*And is.*
*And will be again.*

Gloria Anzaldúa

In our work and research on "decolonizing diets," we grapple with histories of colonialism, racism, and heteropatriarchy, exploring what they value, what they trivialize, and what they strive to eradicate. Throughout this essay you will hear our two voices: Catriona is a *tortillera*, a queer Chicana, with roots in both New Mexico and Sonora, while Luz identifies as a gender-queer, mixed-race, breast cancer survivor. We met in graduate school in the 1990s, and from the beginning food has played a major role in our relationship. Now, after being together nineteen years, food is central to our life. Food connects us to our ancestors, the land, and our bodies. For the past several years, we have posted healthy, plant-based recipes along with information on ancestral foods to our *Decolonize Your Diet* Facebook group and our website. More recently, we have written a cookbook, *Decolonize Your Diet: Plant-Based Mexican-American Recipes for Health and Healing.*[1] In this essay, we share portions of our "testimonies," our personal journeys with food, in order to explore the meaning of decolonized food in a queer, feminist framework.

## A *Tortillera* Journey: In Catriona's Words

When I was in my twenties, my sister Christine had a friend, Frances, who we felt epitomized gender oppression. Frances cooked fresh tortillas for dinner every night. In fact, she would have nachos ready when her husband came home from work so that he could snack on them while she made the tortillas. Christine and I were both ambitious, career-driven go-getters, so it seemed to us that Frances was accepting the limits imposed on her by a sexist framework that strictly delimited women's roles to the domestic sphere. At the time, (and ironically, like heteropatriarchy) I devalued the "women's work" that Frances was performing while I moved toward "important" academic endeavors. That was thirty years ago, and I am here to tell you that I am now envious of Frances's tortillas, which I am convinced, were masterpieces. In my current work, I think back on tortilla making and try to articulate its changing meanings in relation to gender, sexuality, race, and food.

The image of the Chicana or *mexicana* making tortillas occurs all around us in art and literature. An early representation comes from the sixteenth-century Codex Mendoza, which shows a young girl being taught to grind corn on the metate.

The next image comes from a series of scenes, depicting children from age three to fourteen, in which boys and girls are being taught gender-specific tasks. The boy learns to haul goods and fish from a canoe, and the girl learns to make tortillas and weave. Both are being taught by adults of their same gender. The boy and girl receive equal rations; thus, the codex suggests that their gender-specific labor is equally valued.

In his book *Que vivan los tamales: Food and the Making of Mexican Identity,* Jeffrey Pilcher focuses on what specific foods signify in terms of Mexican national identity.[2] Although his title emphasizes tamales in Mexican culture, the book itself focuses more on the centrality of the tortilla to Mexican life. In fact, cover artwork by Diana Bryer,

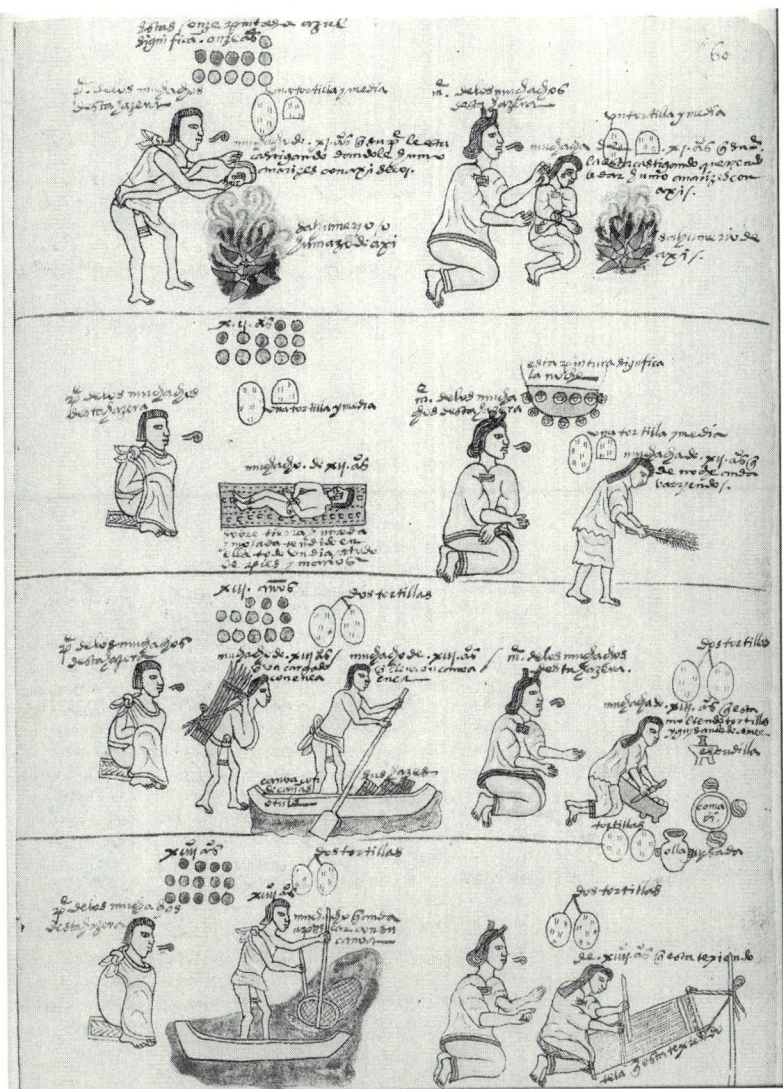

Figure 7.1. Instruction of Tenochca children as depicted in Codex Mendoza.

*The Tortilla Maker*,[3] includes no tamales at all; instead, it depicts a heteronormative *Nuevomexicano* family scene, with the mother making white *flour* tortillas, the daughter watching and learning, and the father and son seated at the table awaiting tortillas.

Pilcher ties Mexican/Indian woman's labor in preparing tortillas with timelessness while echoing Gloria Anzaldúa, "This land

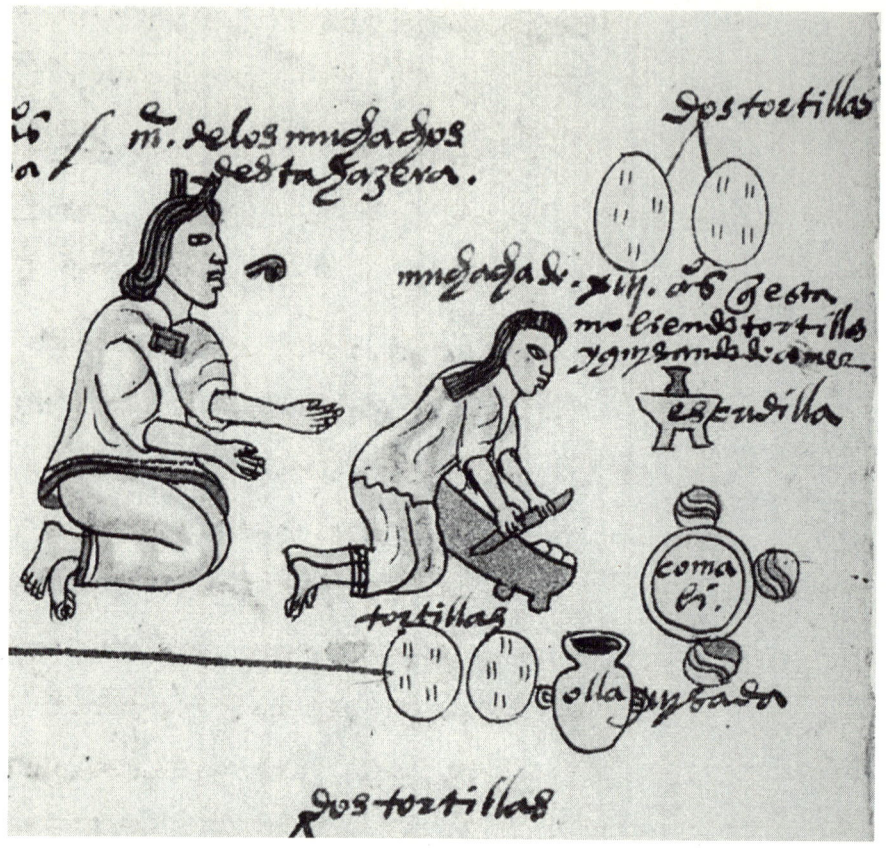

Figure 7.2. Young *tortillera* receiving instruction (detail). Codex Mendoza.

was Mexican once / Was Indian always." But he conflates this labor with Mexican Catholicism, declaring, "For thousands of years Indian women performed penance each morning, kneeling for hours to grind up corn and pat out tortillas."[4] During the colonial era, Pilcher argues, "The staple grains corn and wheat remained for the most part mutually exclusive, with bread feeding a wealthy Creole society and tortillas becoming the province of the poor and Indian communities."[5] Further, he maintains that campaigns to convince Indians to eat wheat bread instead of corn tortillas were present during the colonial era, independence era, Porfiriato, and the period after the Mexican Revolution through the 1940s. Predicated on Europeanizing the Indian, most of these movements were ideologically based on the supposed superiority of European wheat and cultures to Indigenous [Mexican] corn and cultures. Similar ideological projects would take place in the United States through a racialized nationalistic framework.

During the Mexican Revolution, many women served the revolutionary army as soldiers, while others set up camps for military men by carrying bedding and gathering and preparing food. This service is dramatized in Luis Valdez's 1987 production of *Corridos! Tales of Passion and Revolution*. Lizabeta, the heroine of the "Soldadera" segment of the film, has just returned from the battle of La Cadena, in which Juan—the soldier she loved, for whom she left home and followed the army—was killed. Crazed with grief for days, she finally returns to camp following the horse of Captain Felix Romero. After she walks behind him for forty miles, implicitly accepting her new position as his woman, Lizabeta calls a young boy over and orders him to "run and fetch me some fresh water and corn so that I can feed my man." Although the production gestures toward the importance of women's unpaid labor in keeping the army moving, this reference to "fresh water and corn" erases a critical aspect of that labor, for these are not the only ingredients involved in the production of tortillas. The long process of making tortillas involves turning maize into *nixtamal* through the addition of wood ash, slaked lime, or *cal* (cooking lye) to swell or loosen the husks. Moreover, at a transitory location like an army campsite, fresh masa can only be procured from corn if it has already been worked by another woman at her metate.

In these depictions, the tortilla is intimately associated with woman: she is the tortilla maker providing nourishment. Yet the tortilla maker is also invoked to represent the traditional *mujer*, from whom contemporary Chicanas are often at pains to distinguish themselves. The opening scene of Sylvia Morales's 1979 film *Chicana* contrasts today's Chicanas and the challenges they face with the legacy and judgment of their mothers and so-called traditional culture:

> A brief montage of close-ups of *raza* negating the independent Chicana breaking from traditional women's roles in the house follows. There appears . . . an *abuela* making tortillas, a younger "straight" Latina, and finally a Chicano male with a big bigote. . . . The succession of characters, showing their disapproval of the "liberated woman," added with a sound effect of a musical tone

dropping cause laughter for most audiences. As filmmaker Morales puts the *raza familia*'s faces next to each other, we get the picture that traditionally it is the entire culture that does not like *la mujer* to step out of her role of mother, or housekeeper.[6]

Such a politics of representation is not limited to feminist cultural production. In the film *Zoot Suit* (1981), the character of *El Pachuco* (zoot suiter) draws Hank Reyna back to his home, gesturing to "Tu mamá, *carnal*." After this introduction, the audience immediately sees Hank's mother, Dolores Reyna (played by Lupe Ontiveros), making tortillas in the family kitchen. The father returns from work and asks after "the boys." Dolores has been listening to Mexican love songs; in contrast, her sons are listening to boogie-woogie songs as they iron their pants and shine their shoes. The youngest brother, Rudy, tries to skip dinner, instead putting together a taco "*de volada*" (quickly) using his mother's tortillas and fresh beans. Meanwhile daughter Lupe (played by Alma Martinez) is in her room working on her reverse pompadour. She knows her low-cut blouse and short skirt will not pass her parents' inspection, so she climbs out her window in an attempt to avoid a confrontation. Her attempt is in vain: Dolores catches Lupe and brings her back inside, scolding that she looks like a *puta* (whore or loose woman), although, with a glance at her family, Dolores quickly amends her words to a *pachuca*, and Lupe retorts that her short skirt is just the latest fashion like the boy's drapes (zoot suits). The confrontation highlights the tensions between the traditional hard-working immigrant Mexican parents and their pleasure-loving, image-driven Chican@ children.

In a very different kind of representation, Moctesuma Esparza and Esperanza Vasquez's film *Agueda Martinez: Our People, Our Country* (1977) depicts the traditional ranch life of Agueda Martinez, an acclaimed New Mexican weaver. This narrative nonfiction film lovingly portrays Martinez's connection to the land. Martinez is shown to be strong, self-sufficient, and autonomous—at odds with many Chican@ depictions of "traditional" women. Her adobe house has no indoor plumbing, so Martinez has an outhouse and draws water from the well:

> I still don't want anything inside the house because that would hurt my natural habits and the system of my body. If everything

were inside, I wouldn't go outside. I wouldn't feel the vigor and the cleansing of winter.

In one memorable scene, she grinds dried red chiles by hand, toasts blue corn in the oven of her wood-burning stove, pinches off balls of dough from a larger mass, and hand rolls flour tortillas, using a thick dowel as a rolling pin, and with movements so natural they are like steps in a dance. (My mouth always waters during that scene.) She explains later that she sells her rugs to buy the things that she can't supply herself: only coffee, sugar, and shoes:

> I make my tortillas and there is no waste of anything. I use my shoes until they're completely worn, and then, I'll burn them in the stove and make a tortilla.

Martinez's tortillas are "New Mexican style," by which I mean that they are thicker and bread-like. Thus they are very different from thin, Sonoran-style flour tortillas (through which, it is said, you are supposed to be able to either see the moon or read a book). My father, Alfonso, describes the white flour tortillas made by his nuevomexicana mother (Librada Tafoya Esquibel, 1902–1994) as being as thick as pita bread and sometimes made with yeasted dough. *Es que* his Aunt Beatríz's tortillas were even thicker, and she baked them in the oven.

In 1988, Tey Diana Rebolledo, Erlinda Gonzalez-Berry, and Teresa Márquez edited the first published collection of New Mexican women's creative writing: *Las Mujeres Hablan*. When this anthology came out, I was an English major at New Mexico Highlands University. Significantly, the editors made a conscious decision to include works by previously unpublished writers and also to include creative forms of writing that were not considered "literary." In particular, I remember being struck by the fact that they included recipes as examples of women's writing: the first section, "Historia Oral y Memorate," includes Ciria S. Montoya's recipes "Chiles Rellenos" and "Arroz con Leche" (both noted as "estilo nuevomexicano"), and Irene Barraza Sanchez's recipe for *capirotada* (Mexican bread pudding). I remember feeling insulted at the inclusion of these recipes, assuming that the editors were not taking seriously *real* creative writing. As I have with Frances's tortillas, I now look back and appreciate those recipes and the work they performed.

In *Las Mujeres Hablan*, Montoya and Sanchez deliberately carry forward food traditions. Ciria Montoya's recipes come from her mother, Ana María Sánchez. Montoya shares *dichos*, sayings or proverbs, from her father and cookery from her mother in order to represent the ways in which knowledge is gendered. Irene Barraza Sánchez tells a story about her nana:

> Papá Ramón and I chatted while Nana began to mix flour into dough for her delicious tortillas. I watched as she placed a blackened griddle on the wood stove. She let the dough rest a few minutes and then began rhythmically rolling perfectly round circles from each mound of dough. Nana had learned the art of tortilla-making when she was a very young girl. As each tortilla cooked and blistered on the griddle, the smell was enticing.[7]

When I first read these words, I was a twenty-something attempting to ignore my queerness and to live the way I was supposed to in a heteronormative society. Although I was excited about this collection, I was also angry that it was not the fulfillment of all my hopes and dreams, many of which were then as yet unarticulated. You can see here my own tension: between academia and "success" and what were presented as traditional gender roles. As I struggled to live a heterosexual lie/life, I learned to make whole-wheat bread from scratch and bake it in a wood-burning stove. I embraced whole-wheat bread as I had never embraced tortillas. Surely tortillas must be unhealthy, made as they were of white flour and lard.

It was not until several years later, after I had come out as a lesbian and was researching Chicana lesbian fiction, that a particular borderlands poet challenged my assumptions:

> To live in the Borderlands means to
> put chile in the borscht
> eat whole wheat tortillas
> speak Tex-Mex with a Brooklyn accent;
> be stopped by la migra at the border check points

Gloria Anzaldúa, the author of these words, was neither the first nor the last Chicana to play with disparate ideas and ingredients, and her poetry brings these together to speak to the contradictions of the borderlands. In the line "To put chile in the borscht," Anzaldúa references the culinary preparation of a dish outside of one's comfort zone, perhaps borscht from a recipe book, perhaps as a surprise for your lover, or your comrade, or your *comadre*. And so you follow the directions carefully until you get to the last one: "taste and adjust seasonings." The soup is good, sweet and tangy and warm. But it is missing a little something. And when you look in your spice cabinet and refrigerator, you identify clearly what it needs: a little heat from chile colorado. You add it sparingly, tasting it over and over again. Before long you have the best damn borsht that was ever made.

Anzaldúa's poetic reference to eating (and making) whole-wheat tortillas seems the ultimate contradiction because, unlike making whole-wheat bread, the Indigenous whole-grain version of white flour tortillas is not whole-wheat tortillas but corn tortillas from *nixtamal*—the preparation of the maize with wood ash and slaked lime (*cal*) to swell or loosen the hulls—the original street food: part delicacy, part serving utensil. And yet from reading the codices, we know that the *Mexica* of Tenochtitlan were not purists. Indeed, they played food like jazz musicians, improvising and riffing on particular themes. So there were tortillas made from corn: nixtamal, fermented nixtamal, nixtamal and beans, amaranth (*huautli*), *calabazas* (*ayotlaxcalli*), insect larvae (*amoyotlaxcalli*), green corn (*elotlaxcalli* and *xilotlaxcalli*), prickly pear fruit (*nochtlaxcalli*).[8] To paraphrase Aurora Levins Morale's refrain from her book *Remedios*, "Gluten-free breads are not a new thing with us."[9] Given their creativity and imagination in creating tortillas, Indigenous cooks incorporated wheat flour into their many tortilla variations.

In the poem, Anzaldúa's Chicana speaker has clearly been influenced both by Tex-Mex cooking, which is all about the white flour tortilla some claim, and by California organic, hippy, Euro-American culture, where the whole grains referred to were always European grains. One difference I have noticed between my students in New Mexico and my Bay Area students: When teaching the Anzaldúa poem, I usually would ask, "What are tortillas usually made of," and in New Mexico the answer was always "white flour," and in the Bay Area the students usually answered "corn." The whole-wheat flour tortilla thus shows an elaborate genealogy, of the Mexican corn tortilla being supplanted by Sonoran wheat, and then the borderlands Chicana remaking it with whole-wheat flour to reclaim healthy eating.

## A Cancer Journey: In Luz's Words

In 2006, when I was diagnosed with breast cancer, one of my first responses—a futile attempt to gain control of the situation—was to turn to research. I wanted to know: What is the disease? What causes it? Why did I get it? I was especially curious about breast cancer survival rates among Chican@s and Latin@s. I was aware that too many of the queer African American writers, activists, and poets who shaped my identity and politics in important ways had struggled with and eventually died of breast cancer. Of course, I am thinking of Pat Parker, Audre Lorde, and June Jordan. I knew from discussions about their deaths that African American women had low rates of breast cancer overall, yet they died of the disease at higher rates than white women. I also understood (and this is still largely true today) that no one could fully explain this statistic. With regard to Chican@s and Latin@s, I had never heard any discussion whatsoever, so I set out to review the literature.

One of the first studies I found was especially relevant. This population-based study focused on breast cancer rates among Latinas, mostly of Mexican and Central American heritage, in the San Francisco Bay Area, where Catriona and I live. The ground-breaking study, the first of its kind to consider language use, found that breast cancer rates were a full 50 percent higher in US-born Latinas than in foreign-born Latinas. Further, the study found that rates of breast cancer increased along three axes: (1) the longer the immigrants lived in the United States, (2) the lower the age of the immigrants when they arrived in the United States, and (3) the more acculturated they were, as measured by language. The findings for breast cancer risk in relation to language were especially striking: "The difference in risk between women who spoke Spanish only and those who spoke only English was 4-fold."[10] Since speaking a particular language cannot possibly cause breast cancer, language use in this case is a measure for acculturation. Exactly what part of the acculturation process increases breast cancer risk is unclear. Unfortunately, researchers in this study did not collect data on diet. However, they note that "some data suggest that diet early in life may influence breast cancer risk."[11]

As I was searching for research on Latina breast cancer, I found many articles showing that the health "benefit" of being an immigrant extends to health issues beyond breast cancer. This phenomenon, known as "the Latino/a (or Hispanic) immigrant health paradox," is well documented: immigrants from México and Central America (these populations are not always disaggregated in the studies) were found to have lower overall mortality rates (that is, number of deaths per year per thousand), lower infant mortality rates, and lower rates of many diseases. As was found with the breast cancer studies, this health benefit declines with the amount of time spent living in the United States, ultimately declining and virtually disappearing in the next generation. This phenomenon is called a paradox, because most racial groups, including whites, have better overall health with increased education and wealth and decreased overall health the lower their socioeconomic status. This has not been the case for Mexican and Central American immigrants. The poorest recent immigrants have better health, on most measures, than second- and third-generation Latin@s, and many of their statistics are equal to those of middle-class white Americans.

Unfortunately, much of the literature discussing the Latino/a health paradox ignores the importance of diet. Moreover, when researchers do study diet, their methods are not precise enough to trace traditional versus nontraditional diets. For example, the research collapses significant distinctions in the interests of summarizing data: thus, corn tortillas are not distinguished from white bread or flour tortillas. That said, significant evidence in scientific literature confirms the many health benefits of the traditional diets of México and Central America and the Indigenous diets of South America. As Catriona and I discussed this literature, we hypothesized that ancestral food traditions in México and Central America, particularly those of the more rural and more Indigenous areas, may confer a protection on Latin@ immigrants, preventing diseases such as diabetes, heart disease, and some cancers.

In a 1997 study on diet and breast cancer among Latinas in the United States, the researchers report on the amount of fiber in the diet.[12] High-fiber diets are believed to protect against breast cancer because fiber modulates estrogen, which fuels the growth of breast cancer cells. Mexican-Americans in certain regions of Texas eat a high-fiber diet consisting of beans, rice, and corn tortillas, and this is correlated with lower rates of breast cancer among some "Hispanic" populations.[13] The participants, recruited from a senior center in Houston, Texas, tracked their dietary intake for three days. Evaluating the findings, the authors found that the majority of fiber in their diets came from fruits, vegetables, bread, tortillas, cereals, crackers, beans, nuts, and seeds. The authors conclude that breast cancer incidence rates support the hypothesis of the "protective role of the traditional Hispanic diet" (534). Clearly, the use of the word *Hispanic* here is incorrect, given that one of the traditional high-fiber foods in question, the corn tortilla, predates the Hispanic era.

Inside the racist ideology of US American exceptionalism, US Americans (including acculturated Chican@s) are often unaware that being born in the United States itself poses a risk factor for breast, prostate, lung, and colon cancer. It is important to note that age-adjusted rates of breast cancer in México are among the lowest in the world.[14] In the United States, one in two men and one in three women will be diagnosed with cancer in their lifetime. US rates of breast, prostate, lung, and colon cancer are directly related to the "Western lifestyle," which includes the so-called standard American diet (high in fat, carbs, sodium, and low in fiber and phytonutrients), exposure to toxins, and other as yet unidentified factors. Unfortunately, most of us tend to throw up our hands, convinced that nothing can be done to reduce our risk. Yet, an examination of the exceedingly low rates of breast cancer in regions such as México, Central America, Bolivia, and Ecuador should cause us all to stop and think critically: Cancer is not inevitable.

I am particularly struck by the fact that the countries in Latin America with the lowest incidence of breast cancer are those more connected to their Indigenous cultural inheritance, such as Guatemala, Bolivia, and Peru. Again, we return to our hypothesis that Indigenous ways of planting, growing food, and eating continue to protect the health of those who live in the area now known as Latin America.

Mexican healing traditions posit the existence of a condition called *susto,* a fright that startles the spirit from the body. I believe that in the period immediately following my own breast cancer diagnosis and treatment, I suffered from susto. The shock of the diagnosis and the aftereffects of chemotherapy left me weak and dispirited. My recovery was slow and hampered by the fact that I was afraid to eat, convinced as I was that food was my enemy and had somehow caused the cancer. I could not distinguish safe from toxic. Eventually, we settled on the idea of growing our own food and raising chickens for eggs. I spent months in the backyard, clearing land and building raised beds for vegetables. By then, I was convinced that I should be eating ancestral foods, so I planted beans, corn, squash, and chayote. In my research, I had found that herbs of all kinds conferred strong protections against cancer, so I constructed an herb spiral planter out of recycled clay roof tiles. I read about the health benefits of wild greens, like *quelites* (lamb's-quarters) and *verdolagas* (purslane) and was thrilled to find *verdolagas* (considered a weed by those who don't know) growing wild in my front yard. I planted seeds for *quelites*, and they quickly established themselves in my garden.

Through the very act of getting my hands dirty and connecting directly with Mother Earth, my susto at last started to recede. I began to accept the cycle of life as I observed the seasons in my garden. I felt very connected to ancestors I never even knew. I developed a plant-based spirituality. I started to make my own corn tortillas and ate delicious breakfast tacos made with eggs from our backyard chickens and herbs from the spiral. I felt grounded again. I found the simple pleasure of cooking and preparing foods from the garden, and I regained my strength and equilibrium. I find Anzaldúa's discussion of the "Coyolxauhqui imperative" relevant to my own process of reconnecting body to spirit:

Table 7.1. Age-standardized incidence of breast cancer (per 100,000 women)

| United States | 92.9 |
| Argentina | 71.2 |
| Brazil | 59.5 |
| Cuba | 50.5 |
| Costa Rica | 45.4 |
| Paraguay | 43.8 |
| Venezuela | 41.2 |
| Colombia | 35.7 |
| Mexico | 35.4 |
| Ecuador | 32.7 |
| Peru | 28 |
| Bolivia | 19.2 |
| Guatemala | 11.9 |

Source: Ferlay et al. 2015.

Table 7.1. Age-standardized incidence of breast cancer (per 100,000 women). Source: J. Ferlay, I. Soerjomataram, M. Ervik, et al.; GLOBOCAN 2012 v1.0, IARC CancerBase No. 11, "Cancer Incidence and Mortality Worldwide."

> When fragmentations occur you fall apart and feel as though you've been expelled from paradise. Coyolxauhqui is my symbol for the necessary process of dismemberment and fragmentation,

of seeing that self or the situations you're embroiled in differently. It is also my symbol for reconstruction and reframing, one that allows for putting the pieces together in a new way. The Coyolxauhqui imperative is an ongoing process of making and unmaking. There is never any resolution, just the process of healing.[15]

Years earlier, on a trip to México City, I had purchased a small replica of the round stone of Coyolxauhqui. Later, when I was completing my garden, I placed my statue of Coyolxauhqui in the herb spiral, so that she would rise like the moon from the roots of ancestral plants.

In my view, healing must take place not only for the individual but through our culture as well. While my health journey has been deeply personal, I believe that we must act collectively both to resist that which is and to (re)create the world in which we want to live. Our call for a decolonization of food tries to hold many factors in play: individual health and healing; healing our communities from the legacies of colonization, racism, and patriarchy; and the urgent need for fundamental changes in the way we grow our food and exist on land that belongs to Indigenous peoples. This work can and should happen simultaneously.

Moreover, as Chican@s living in the United States, we have a unique obligation to engage in political organizing in solidarity with the rural Indigenous farmers in México, where globalization is having disastrous consequences: forcing Indigenous farmers off their land and leading to dependence on commodity foods imported from the United States. Rural Mexicans are now beginning to exhibit health problems associated with the Western diet. As a result, the immigrant paradox is likely to diminish or disappear in the years ahead, as immigrants will arrive in this country without the protection afforded by being raised on a traditional diet of heritage corn and beans.

## The Sunshine Room: Chican@ Diets and Americanization Programs

In the 1940s and 1950s, the Mexican American generation came of age. These decades witnessed the coming of age of the children of the first large wave of Mexican immigration to the US during the years of the Mexican Revolution. The Mexican American generation was under immense pressure to acculturate to the United States, that is, to re-form themselves according to the language and culture of Anglo America. Through English-only education, where students were paddled for speaking Spanish, to Americanization programs emphasizing "American" foods, this generation was forced to give up its culture.

In *Becoming Mexican American, Ethnicity, Culture, and Identity in Chicano Los Angeles, 1900–1945*, George Sánchez explains the Americanization programs imposed on Mexican families in Los Angeles:

> Reformers encouraged Mexican women to give up their penchant for fried foods, their too frequent consumption of rice and beans, and their custom of serving all members of the family— from infants to grandparents—the same meal. According to the proponents of Americanization, the modern Mexican woman should replace tortillas with bread, serve lettuce instead of beans, and broil instead of fry. Malnourishment in Mexican families was not blamed on lack of food or resources but on "not having the right varieties of foods containing constituents favorable to girth and development." The typical noon lunch of the Mexican child, "thought to consist of a folded tortilla with no filling," could easily be the first step to a lifetime of crime.[16]

Reformers reasoned that the Mexican child, eating only an "empty" tortilla for lunch, would look upon the supposedly more bountiful homemade lunches of their Anglo classmates and be filled with envy, which would lead to theft, which would lead to a lifetime of crime. Rather than envisioning inequality as a problem, these reformers fixated on and obsessed over Mexican culture, as typified by the tortilla, as the cause of larger social ills. Moreover, it is clear that the reformers had little knowledge of the rich health properties of Mexican American cuisine of that time. For example, the reformers did not promote the health benefits of daily consumption of freshly cooked beans, a common practice among Mexican immigrants of that era.

These Americanization programs touched our own families: Luz's father, Tony (Antonio Hector) Calvo, recounted his experience in grade school at San Fernando Elementary, in San Fernando, California, in the early 1940s. For one year, he—along with the other Mexican American children—was taken out of class and put in what was called "the Sunshine Room." Each day in the Sunshine Room, the children were given a glass of orange juice, forced to shower, and then forced to take a nap. He could not remember what else happened in that room, but we cannot help but wonder what the Anglo school children were studying while he and the other Mexican children were pulled out of their regular curriculum. The school clearly viewed Mexican children as deficient in health and hygiene and perceived these lacks as being more important than education.

## Indigenous Models

In the United States, American Indian scholars and activists are challenging the devastation wrought on Indigenous communities by the American industrialized diet. Their work draws links between colonialism and community health and survival, often providing important models for conceptualizing Chican@ food activism. As Chican@s claim a decolonial food politics, we believe that it is critical that we acknowledge and support the work of American Indian food activists.

For example, Winona LaDuke (Anishinaabeg), an author and internationally known activist, brings attention to Indigenous struggles—for sovereignty, for the earth, and for traditional foods—and contextualizes these struggles in the histories of Native peoples in relation to the US government and capital. In *All Our Relations: Native Struggles for Land and Life* (1999), LaDuke provides case studies of environmental challenges faced by nine different Native communities while highlighting the stories of the Native activists who led these struggles for survival. In "Akwesasne: Mother's Milk and PCB's," LaDuke chronicles the industrial contamination of Akwesasne, a Mohawk reservation on the St. Lawrence River in upstate New York. Decades of air and water pollution by Reynolds Aluminum, the Aluminum Company of America, General Motors, and Domtar Paper have contaminated the reservation ecosystem, affecting water, fish, plants, people's bodies, and ultimately women's breast milk.[17] "Women are the first environment," explains Mohawk activist Katsi Cook. "We accumulate toxic chemicals like PCB's, DDT, Mirex, HCB's. . . . They are stored in our body fat and excreted through breast milk. What that means is that through our own breast milk, our sacred natural link to our babies, they stand the chance of getting concentrated dosages."[18] One study of fifty new mothers found "a 200 percent greater concentration of PCBs in the breast milk of those mothers who ate fish from the St. Lawrence river."[19]

The river fish are central to the traditional diet of the Akwesasne Mohawk; however, due to the threat of their breast milk being tainted by PCBs, Mohawk mothers are changing their diet to protect their infants. This change signals more than a trade-off: "Our traditional lifestyle has been completely disrupted, and we have been forced to protect our future generations. We feel anger at not being able to eat the fish."[20] Cook emphasizes that even as the Mohawk women amend their diet to protect the health of their infants, doing so "does not preclude the corporate responsibility of General Motors and other local industries to clean up the site" that they contaminated through industrial pollution.[21]

In *Recovering the Sacred* (2005), LaDuke continues her documentation of Indigenous struggles for land and food sovereignty, including the harvest of wild rice by Ojibwe in the Great Lakes region of North America. Despite this complex of struggles, community members continue the tradition of harvesting wild rice, some for personal consumption, some for school lunch programs or boxed groceries for elders, some for sale. Unfortunately, globalization disrupts these traditions: "The rice produced on Blackbird Lake is being eclipsed by rice production far away—rice grown from patented seeds on diked paddies, nourished with chemical additives and harvested with huge combines—yet still called 'wild' rice."[22] As LaDuke's case history illustrates, wild rice is both the cultural and legal patrimony of the Ojibwe people, whose ancestral lands span the US-Canada border, yet corporations and universities continue their efforts to patent, clone, and modify wild rice. Monoculture farming of sterile strains of the rice—developed through the University of Minnesota and grown close enough to the wild rice beds that these may be affected through pollen drift or duck migration—threatens the future of wild rice beds and the traditional diet of the Ojibwe.

In her chapter "Food as Medicine: The Recovery of Traditional

Foods to Heal the People," LaDuke depicts the actions of Indigenous communities to reclaim native foods as a means of survival. LaDuke's analysis demonstrates how colonization and removal policies have stripped Native peoples of their ancestral diets. Fifty million buffalo were killed by nineteenth-century military policy specifically to destroy the Indians' food source. When Native peoples were removed and confined to reservation systems, their traditional diet of wild rice, corn, maple sugar, wild game, and greens was replaced with commodity foods, namely white flour, white sugar, and lard. LaDuke explains:

> The high starch, sugar, and fat content in commodity foods caused high blood pressure levels which stress the pancreas. If stressed repeatedly, the pancreas essentially becomes poisoned and insulin metabolism becomes permanently impaired. That is where diabetes comes from. Before the 1940s, there is little record of diabetes in most Native communities. In contrast, it is now the second most common diagnosis for Native Americans admitted to the hospital.[23]

LaDuke describes a transformative collaboration between three O'odham nations (Akimel O'odham, Tohono O'odham, and Hia c-ed O'odham) and Native Seeds/SEARCH:

> Seeing that nearly all of the government funding to research O'odham diabetes was focused on their genes and not their diet, the nonprofit Native Seeds/SEARCH asked O'odham families if they might donate native food samples to determine their value in controlling diabetes. These samples turned out to be the first Native American foods that nutritionists ever analyzed to determine their impact on blood sugar levels, insulin, and diabetes.[24]

The findings of this research were groundbreaking: "When a person eats acorns, mesquite pods, and tepary beans or prickly pear cactus, the special dietary fiber in these foods slows down the release of sugars into the blood. . . . In short, these 'slow-release' native foods protect Native Americans from an imbalance of blood sugar and insulin following a meal."[25]

LaDuke further notes how O'odham families, working with Native Seeds/SEARCH cofounder Gary P. Nabhan, a food scholar and activist, have conducted studies reconstructing a nineteenth-century O'odham diet. They found that "the traditional high-fiber complex carbohydrate and low-fat diet resulted in a lower release and uptake of sugars from the intestines."[26] After two weeks on this traditional diet, the subjects spent two weeks on a modern diet based on foods available at nearby convenience stores: "The convenience store diet soon produced higher blood sugar levels severe enough to trigger diabetes if that diet had been maintained" beyond the period of the study.[27] In spring 2000, a group that included youth and diabetic elders walked from the Sea of Cortez to Tucson in a "Desert Walk for Health and Heritage," all the while eating only native foods provided by host communities. The purpose of the Desert Walk was "to emphasize how combining traditional and scientific knowledge of desert foods and medicines can directly benefit contemporary communities of Native Americans, to heighten awareness about the epidemic of diabetes among Native American communities, and to promote inter-generational cultural exchanges among the Seri, Tohono O'odham, and Yaqui people."[28] The walkers and host communities shared food, information, and songs while discussing important questions concerning Native health.

Another partner in the desert walk was the Tohono O'odham Community Action (TOCA), an organization that continues to spread the message of the health benefits of native foods and the dangers of commodity and convenience foods. In 2010, TOCA produced *From I'itoi's Garden: Tohono O'odham Food Traditions*. This self-published volume (available through blurb.com) features history; traditional stories; growing, foraging, and harvesting techniques; and recipes. The recipes focus on ten native foods of the Sonoran Desert—Tohono O'odham squash, acorns, cholla buds, saguaro cactus, mesquite beans, prickly pear fruit, agave, wild greens, sixty-day corn, and tepary beans—along with other domesticated and wild foods. In summer 2013, TOCA launched a new food magazine, *Native Foodways: Celebrating Food, Culture and Community*. The magazine features food news from all over Native America, recipes by Native cooks using traditional foods, and editorials from food activists. Nabhan explains: "For the O'odham, and other recently Westernized Indigenous peoples, a return to a diet similar to their traditional one is no nostalgic notion: it may, in fact, be a nutritional and survival imperative."[29]

Numerous studies of Indigenous peoples have documented both

the health problems directly associated with an industrialized diet and the health improvements subsequent to returning to a traditional diet. One widely cited study by Kerin O'Dea and her colleagues focused on fourteen aboriginal Australians.[30] This study followed five men and five women who were diabetic, and two men and two women who were nondiabetic. The participants spent seven weeks pursuing a traditional hunter-gatherer diet and lifestyle. As part of the study, they walked for ten days from their community to the coast, where they spent two weeks; then, they returned inland to an abandoned homestead area, where they spent an additional two weeks before beginning the ten-day trip back to their community. Many of the participants were skilled hunters who had consumed a heavily processed diet before the study. Their urban diet consisted of "flour, sugar, rice, carbonated drinks, alcoholic beverages (beer and port), powdered milk, cheap fatty meat, potatoes, onions, and variable contributions of other fresh fruit and vegetables."[31] Their diet during the study varied as they traveled: on the outward trip, they initially ate beef, kangaroo, turtle, bream (a freshwater fish), and honey. The coast offered few vegetables, so participants ate mainly fish (80 percent) and supplemented their meals with kangaroo, crocodile, and seabirds. At the inland location, their diet consisted of kangaroo (36 percent), bream (19 percent), and yams (28 percent), supplemented by honey, figs, birds, turtle, and yabby (a freshwater crustacean). At the end of the study, all the participants had lost weight; significantly, the diabetic subjects showed an improvement in glucose tolerance.

Another research group conducted a similar study during the late 1980s and 1990 with Native Hawaiians indicating risk factors for heart disease including obesity, high blood pressure, and high cholesterol.[32] The study was run through the Wainae Comprehensive Health Center as a community-based intervention strategy to reduce chronic disease risk factors in Native Hawaiians. The study diet was modeled on a pre-Western Hawaiian diet, featuring taro, poi, sweet potato, yams, breadfruit, greens, raw or steamed fish, and steamed chicken. Although the study limited the amount of fish or chicken in the diet, subjects were encouraged to eat as much of the other foods as necessary to feel full. At the end of the three-week study, all subjects showed significant weight loss, lower blood pressure, and lowered cholesterol.

In another community-oriented intervention plan to fight diabetes, the Anishinaabe Center at White Earth Reservation put together the Anishinaabe Hunter-Gatherer Traditional Foods Pyramid. This pyramid calls for the largest amount of food to come from "traditional meats, fish, birds, and eggs," followed by "traditional grains, nuts, and beans," then "traditional vegetables," "traditional fruits and berries," and a small amount of "traditional sweets and oils." Each of the categories includes suggestions for "healthy modern additions." For example, the traditional vegetables listed include squash (summer and winter), stems, sprouts, new shoots, wild rhubarb, spring greens, wild mushrooms, Jerusalem artichokes, and wild roots (bitter root, camas, cattail). The healthy modern additions suggested include "spinach, Swiss chard, and other lettuce greens, miscellaneous fresh herbs, onions, peppers, carrots, parsnips, potatoes, yams, string beans, cauliflower, and broccoli."[33] Notes on the pyramid point out that dairy foods do not form part of the traditional diet and recommend that calcium intake come from traditional calcium and mineral sources such as bone soup or broth, fish-head soup, canned fish with bones, vegetables, and greens. Other notes emphasize that traditional foods are local and organic and warn that "processed man-made things like sugar, artificial sweeteners, soda pop, bleached flour, partially hydrogenated vegetable oil, most packaged foods, convenience and fast foods" should be avoided. Finally, the poster highlights seasonal foods, with sections that catalog the seasonal availability of various meats and vegetables.

Contemporary Indigenous food activists are turning away from the culinary heritage of colonization, particularly variations on fried dough such as fry bread, bannock, and *sopapillas*. LaDuke and the communities she studies connect commodity foods with the determined eradication of Native American foodways, whether through the annihilation of the buffalo or the decimation of Indigenous heritage crops. First Nations Canadian health programs urge participants to avoid "the five white gifts"—flour, sugar, lard, milk, and refined salt—to which their ancestors were indoctrinated in residence schools. Apart from their connection to diseases of development, these "gifts" emblematize the ways in which the sustainable, healthy foodways and lifestyles of aboriginal peoples were systematically dismantled and replaced with the commodity foods and sedentary lifestyles inherent to colonization.

As Chican@s in the United States, many of us are not as close to our Indigenous roots as the Native subjects of these health studies. However, we are also not that far away. For decades, Chican@ and Central American community groups have focused on a return to Indigenous traditions by reclaiming Indigenous languages like Nahuatl, Zapotec, Mixtec, and many others; forming circles for ceremony, dance, and prayer; and organizing Indigenous runs across the continent. Throughout colonization and its aftermaths, Indigenous ways of cooking and growing food have been kept alive, most often by women, who hold great, if mostly unrecognized, knowledges about the local, regional, and ancestral foods of México, Central America, and South America.

Colonization has had disastrous effects on the health of Mexican and Native peoples throughout the Americas. The imposition of the Western diet and lifestyle has resulted in skyrocketing rates of diabetes, metabolic syndrome, heart disease, cancer, alcoholism, and asthma. We have reached a critical moment in which the survival of our generation and those that follow depend on our dramatically changing socioeconomic systems that promote toxic monoculture crops and commodity foods laced with, if not entirely comprised of, poisons.

## No se puede descolonizar sin despatriarcalizar

Bolivian feminist María Galindo argues that to decolonize, we must dismantle patriarchy.[34] Similarly, we understand that as Chican@s, we cannot simply return to the plant-based, home-cooked diet of our ancestors without addressing gender oppression. We are not simply calling for a revaluation of Chican@ women's labor: instead, we are calling for a complete reconfiguration of gender. We are not calling for a return to the kitchen: we are calling for the liberation of the kitchen. We challenge gender binaries and call for an expansion of gender identities, including two-spirit, gender queer, transgender, and more. We want to expand cultural spaces and recognition for all—those who identify as male, those who identify as female, and those who exist in the borderlands of gender. We want to see women like

Frances passing on vital cultural knowledge to their queer grandsons by teaching them to make tortillas; to forage for *verdolagas*; and to clean, can, and share nopales. We want to see Chican@ lesbian street activists taking back the land and using it to grow herbs and vegetables to sustain our communities. We want to see gender queer Chican@s painting slogans that call the status quo into question. We want to see queer danza groups that reconfigure the meaning of gender balance. We want to see an explosion of community gardens that function as the people's pharmacy, growing plant-based remedies and functioning as centers where ancestral knowledge is gathered and shared. We want *curanderas* who can help people through the traumas associated with colonization and disease to perform *limpias* to cure us of our *sustos*. We claim food as an important site of pleasure, and decolonial cooking as a creative act of resistance. Our identities as *tortilleras* (a slang term for queer women), now take on a double valence, as we now make our own tortillas blended from native corn, amaranth, mesquite, and whole wheat.

We live in the space of Anzaldúa's Coyolxauhqui imperative, which she describes as "an ongoing process of making and unmaking. There is never any resolution, just the process of healing."[35] We continue the ongoing process of making and unmaking . . . of healing. Whereas Catriona identified her earlier resistance to recipes in Chican@ feminist texts, we end our essay by sharing a recipe from our cookbook. It is inspired by our queer Chican@ feminism, filled with the healing power of nopales, and offered to sustain revolutionary love. Our take on decolonized food is not "purist": we occasionally use ingredients that are healthy but not native, such as garlic, cilantro, and rice. Our overall ethos is to encourage the consumption of healthy native foods, such as beans and nopales.

Chicana feminists have reclaimed Coyolxauhqui, a Mexica goddess figure, as a rebellious daughter who resists patriarchy. We intend this dish to provide sustenance to women, gender nonconforming people, and male allies as we heal ourselves from the violence of colonization and patriarchy.

Figure 7.3. Coyolxauhqui Bowl. Courtesy of Luz Calvo.

## INGREDIENTS

2 tbsp extra virgin coconut oil

2 garlic cloves, chopped

½ white onion, thinly sliced

6 *nopal* paddles, cleaned and diced

1 fresh jalapeño, minced

1 tsp sea salt

¼ cup chopped cilantro

2 cups (500 mL) brown rice, cooked

4 cups (500 mL) beans (pinto, bolita, or other heritage bean),
    cooked and prepared *de olla*

Raw Green Salsa (recipe follows)

Several sprigs cilantro

## INSTRUCTIONS

In a sauté pan with a tight fighting lid, heat oil on medium high heat. Oil should start off hot. Reduce heat to low, add garlic and swirl to release aroma, about 15 seconds. Add onion, *nopal*, chile, and salt. Stir to combine. Cover and cook for about 10 minutes or until *nopales* are tender. At this point, *nopales* should have released their liquid and the mixture should be extremely "juicy." Remove lid and continue cooking until all liquid evaporates. Depending on size, age, and season *nopales* were harvested, this may take 20 minutes. If it is taking longer than 20 minutes, increase heat and stir constantly to prevent burning. Remove from heat when no liquid remains. To put together bowls, warm beans. Put two generous scoops of brown rice in each bowl. Add ½ cup of beans. Top with a generous scoop of *nopales*. Drizzle green salsa in a spiral around bowl. Garnish with sprigs of cilantro.

## Raw Green Salsa

### INGREDIENTS

8 large tomatillos, husks removed
1 small white onion, quartered
1 garlic clove, peeled
1–2 fresh serranos, stem and seeds removed
½ bunch cilantro
1 tsp sea salt, or to taste
1 tbsp lime juice
2 tbsp water
1 avocado, peeled and seeded

### INSTRUCTIONS

Place all ingredients in blender or food processor, and blend until cilantro is blended into even specks distributed throughout salsa.

(The recipe is reprinted here with permission from the authors and the Arsenal Pulp Press.)

# CHAPTER 8

# Chicos del horno
## A Local, Slow, and Deep food

JOSEPH C. GALLEGOS

The chicos-making process in prose. It is quite an achievement to undertake and to become immersed in the transformation of the tender, milky corn kernels into the golden popcorn-hard nuggets that will store for many years: You start with an ear of concho corn, unhusked, with the cob full of shiny, pearl-white flint or dent corn. The earthen oven's (*horno*) fire transforms the corn into a cooked cob filled with robust, tasty, smoky-flavored kernels. When the kernels become dehydrated, they shrivel into teeth-breaking hard corn; the color blends soft golden and brown hues. This tiny dehydrated kernel (*chico* means "small") is a reduced and shriveled form of the plump and milky corncobs that went into the overnight *horno* roast, which is filled with the ashy remains of the piñon, cedar, and aspen wood used to fuel the fire.

The whole process of eating what one produces takes patience, knowledge, and planning. So, you plan for your meals months in advance. From the butchering of a steer to the luck of shooting and dressing an elk, the unwritten rules of planning food for the winter are always on the minds of the many farming families that surround us. Recipes and the timing of meal schedules are never written down; this is just part of the process of seasonal eating. The chicos are one of our versions of a durable food with a deep history. We can save and store chicos corn for the times when the weather is freezing or otherwise inclement, and let me tell you that there is nothing as comfortable after a day of dealing with severe weather than sitting down to a hot, steamy bowl of savory chicos. The comfort of having a wood-warmed home and eating farm-grown stews and soups that fill not only the stomach but the soul is an irreplaceable feeling. We never, ever go hungry or become disgruntled for lack of a delicious meal whose basic recipe has been in the family for uncounted generations. An old farmer that lives down the way told me one day, when I was complaining to him about the short prices I got on some sales calves, "a person will never get rich here but you will never starve to death." Feeding oneself on the rancho is easy as long as you are patient and don't mind cooking. The slow-cooking recipes seem to be tastiest, but when a quick meal is needed, frying food takes less time, and there are some fried recipes that make my mouth water. At this moment, I am thinking about the sweet-tasting fried potatoes we enjoy in the spring, when the potatoes we preserved in our root cellar (*suterrano*) seem to have a sweeter taste than a potato stored in a warehouse far away or bought from a grocery chain. These home-grown potatoes—fried in a well-seasoned iron pan—may be our version of local, slow "fast food."

From the planting of the corn, through the stripping of the roasted kernels from the cobs (a task called *desgranando*), to the cleaning and packaging of the dry hard kernels—the whole process requires family, friends, neighbors, and patience. Without this help, it would be diffi-cult to make enough chicos to sell with a sufficient amount reserved for the family. Spending money on labor for chicos requires that you have a healthy bank account and that you have a good marketing sys-tem to sell enough to get a return on costs. In the old days, there was plentiful family labor to provide the great amount of work invested in an intense and short period of time; without such cooperative labor, the chicos just wouldn't come to fruition.

The family is extremely important to chicos production because siblings sell or barter with one another and are very reliable when the time for laborious work is at hand. My sister Marie and brother Jerry are a great help when it comes to getting the chicos process started. Marie will usually bring friends from the city to celebrate the chicos season. The chicos season is an annual cycle, and every year visitors and newcomers arrive to learn how the traditional white-roasting corn is processed. With the arrival of all the company, the chicos har-vest and roasting season gets kicked off.

The gathering of family is a strong asset that the work of making chicos produces. That's right! The corn makes the people. Without the concho, no chicos roast, no storytelling or sharing of ancient know-how. No sense of place, no community, no seed saving, no family, no food, no life. We are the chicos corn people. The stories and shared histories that the elders and others discuss at the chicos roast are funny, inspiring, tragic, and comic but always interesting. These are the moments that become our memory of place. Just the other day, my daughter Patricia showed great excitement when she told a friend that the ranch was going to work the hornos and the family was going to gather. She showed pride and excitement about the chicos tradition, which in turn made me very happy about my decisions in life.

When chicos are being cooked, friends and neighbors appear out of the woodwork. The friends are always helpful, and they enjoy the fire and the social events that occur for weeks around chicos preparation. The work ends late evening, and then the morning comes with the opening of the adobe ovens, and this cycle also offers great memories and some real fun times.

The sealing of the horno is the last step each night, and after hours of work the evening draws to a close, and the still-hot, bright embers that cook the corn inside disappear beneath the earthen works. The horno walls emit warmth, and a bit of steam escapes through the wet plaster seals, made visible by a moon that is at that time of year usually the brightest light around. This scene has always been a great moment to call the day a day and get much needed rest.

The morning always comes faster than one thinks. When the morning arrives, the most exciting time is about to happen, the opening of the mighty horno. The opening is the moment when we all find out how well the fruits of our sixteen-hour *jornada* (work shift) turned out. The small family crowd starts asking questions:

"Are the chicos cooked?"
"Did we heat the horno enough?"
"Were there enough embers?"
"Is it gonna rain today?"

The opening takes some time because someone has to slowly chip away at the mud plaster and carefully remove the adobe bricks used to seal the doorway. Once the adobes are removed, the ears are revealed, and they are no longer bright green but a dull gold color, and some look nicely singed and toasted. Everyone in the crowd anxiously waits to taste the cooked ears. The person removing the adobes is usually

Figure 8.1. The "chicos corn people." The annual *horneada* at the Corpus A. Gallegos Ranches in San Luis, Colorado, September 2015. Courtesy of Devon G. Peña.

the first to reach into the steaming adobe to quickly grab and toss out a few of the golden-toasted ears. The steaming golden corn is then stripped of the husk. Usually a well-cooked chicos on the cob is tasted by one of the family members, and if that taster approves, the chicos are judged good and ready to husk. Someone always brings salt to the event, so a few sprinkles on the moist and smoky steamed cobs goes a long way.

Most times we throw some beef or a leg of lamb in the horno with the chicos for the overnight roast. This breakfast makes for a great start to the day before the final morning workload begins involving the husking of the oven-roasted cobs before we spread them out across the homemade open-air dehydrating racks. The cooked meat is moist and tender. The taste of the roasted chicos and *carne desebrada* (shredded meat) combo simply cannot be described. Many get excited about cooking meats in the horno because the hot moisture infuses the meat and slowly cooks it. We have also cooked fish in the horno, and the bones can be eaten since they disintegrate like sardine skeletons. The smoky taste is again prevalent, and the scene is always filled with "oohs" and "aahs." We seldom cook chicos without adding the extra surprise of a rare taste, a taste that revives family and place memories.

## The Chávezes

For generations, the harvest was a way of life that brought family and friends together. Over time the gatherings became rare, and I seldom heard of anyone doing chicos roasts. By the 1970s, my neighbors the Chávez clan was the only local family that was still gathering for chicos or potato harvest season. There are many stories about the huge gatherings of the past, and the Chávez festivities seemed especially large to me, but grandma mentioned that much larger events happened all over the valley "*en los dias antiguos*" (in the olden days). I felt saddened by the neighbors' annual celebration. When we were invited to their horno roast, I always had such a great time and did not want to return home to a place without a ranch horno tradition. The horno gathering at the annual Chávez chicos fest was huge, but I do recall that other families, including our own, had long ago gathered for the harvest season. Unfortunately, I did not experience any big gatherings hosted by my own family during my youth. Our horno had melted into the land under decades of cows, rain, neglect, and forgetting.

These good neighbors, the Chávezes, are farmers on the upper part of our acequia, the San Luis Peoples Ditch. They were a grand inspiration to me when they rebuilt their horno and showed the "acquiahood" what is was like in past times when cooking concho corn was an annual event. Often in the fall, when Dad and I would drive to the Chávez farm to join full swing with chicos making and their family celebration, I would ask my father why we were out of the chicos business. He explained that when the exodus to the city by money-seeking *parciantes* (irrigators) began, all of the acequias had major cutbacks in production, and the large harvest celebrations all but disappeared. With that postwar exodus, the chicos culture weakened, and only a handful of families kept planting and cooking the wondrous concho corn; the Chávezes were among those who revived this tradition.

I was tremendously inspired by Mr. Chávez to get the rancho back into row crops. They taught me by example. The road to our home from town is above the Chávez *extensión* (long lot); every time we passed by their home on the way to or from town, I was able to see what they were up to. We could see their beautiful home garden and their concho cornfield. What really caught my eye was when they made their chicos; the large gatherings sure looked like a good time after hard work. I noticed that most of their family would come to help with the process, and they would do all their chicos on a Labor Day weekend. They would celebrate with huge meals of homegrown meats and freshly picked vegetables. I remember dropping by during one of their chicos sessions, and I was given tasty food with all the fixings. The stories, jokes, and occasional songs by the musically inclined family members made for a joyful gathering. I had such a good time that I knew our farming family was going to have to see what it would take to get back into row crops. Was this even a possibility what with my lack of row-crop wisdom and my dad being rusty? He had made the ranch into more of a pastoral operation with sheep, goats, cattle, horses, and even a hog or two. After discussing the possibility of getting into row crops, Dad would look at me in that certain way I understood meant he was wondering if it was a good idea to increase the workload for a product with few economic returns. Dad told me stories of the bygone row-crop eras when planning for the spring was seriously hectic. He described getting the starter plants going and having to care for and prepare the soil beds. I could hardly imagine the intense workload involved. The traditional starters included cauliflower and cabbage, which the old-timers grew to export as cash crops from these lands (through the 1950s and early 1960s). The starters were important because the growing season is too short to grow the head vegetables. The starters were adapted to the cool summer nights of the Culebra, which is a great location for plants that like a diverse temperature range. Dad said there were "hot beds" all over the farm so that when time came to start the cabbage and cauliflower, most of the planting could be done during the short window of warm and rainy weather that farming here faces.

Our family finally arrived at the decision to get the ranch back into row crop production. This shifted our approach to a lot of things, including the purchase of tractors since this involved the selection of models that "pull and lift" but can also pull slowly with ample horse-power to till or cultivate the soils as gently as possible. We bought tractors geared to operate at high rpms for faster speeds for field crops and big jobs, but we also sought models that operate efficiently when geared low at slow speeds for tilling soil and cultivating row crops. The slower speed prevents tilled soils from piling atop the starter plants. I found that modern tractors were seldom what was needed in the organic row-crop business and that the older, more reliable "Pop'n Johns" and Massey Fergusons are perfect for cultivation.

When the time came, we decided to test the row crops, and we definitely started small. The objective was to get enough concho corn to make at least one horno full of chicos—enough *para la muela* (enough to taste or for family use). We estimated this was about ten rows of corn. I really can't remember exactly how much concho we produced that first time, but I do remember the weed problem nearly overwhelmed us. Growing row crops in this part of the watershed requires recognizing and dealing with various types of weeds that love to grow in freshly exposed ground. Organic row crops require annual tilling, and with our rich, sandy loam soils all weeds seemed welcomed. The first years of chicos were dismal, with one or two small hornos. The lessons we learned those years were invaluable; the following seasons offered more lessons, especially relevant to where I am today facing limitations on production due to not being able to find people willing to engage in the intense manual labor required for organic weeding work. If I get too ambitious and plant too many rows of concho, then fields of the crop can go to seed because of the lack of labor to harvest the corn and make chicos. Of course, once you find a good skilled worker, it is understandably in their own interest to immediately want more pay. The window for making chicos at harvest point is about two months of nonstop work: From harvesting to cooking to cleaning and packing the golden kernel, the work is constant, and I cannot blame workers for wanting more cash. I am happy when they agree to accept a portion of the chicos in lieu of full cash payment for their good labor. Our *cultura* makes this possible.

## "Cheater" Chicos and Suterranos: A New Era?

The chicos-making tradition was a big deal in the past, but during the mass exodus of farm families from the collapse of sustainable agriculture in the Culebra River valley that began with the mad city/mad money rush of the 1950s and 1960s, the culture of corn took a dive. My personal experience as a youth growing up was a case of being undereducated in the corn culture. I only recall one time we used the ranch horno to make roasted chicos. My grandma made chicos, and I recall there were *ristras* of chicos corn hanging in our doorways for a long time. I also recall when the cows knocked down the horno by using it as a back scratcher. Usually, a cow will seek objects like trees, houses, vehicles, or posts to scratch their itches, and this time they decided to use our adobe oven.

When the livestock destroyed the rancho's horno, the family went into a sad new chicos era. This was the time of "Cheater Chicos!" Cheater chicos are produced through a process that is different from the horno-roasting method, and it usually involves the use of a "cheater corn," perhaps a Nebraska sweet white corn variety or some other out-of-state commercial hybrid that is not a local heirloom. The cheater is a nonheirloom corn that is boiled in a huge cauldron and then dried under fans. We have "semi-cheaters," which use nonheirloom corn that is cooked in the traditional horno method, or an heirloom corn that is boiled and dried. This semi-cheater is what I call "imposter" chicos. My family made imposter chicos for many years, but the taste difference between the two types is dramatic. The cheater and imposter chicos may pose as authentic "chicos del horno," but a real chicos connoisseur can definitely tell the difference.

Sure, we did not totally reject cheater chicos at the table, but the difference between the two is like night and day. I noticed in the store, just the other day, that the bags of packaged chicos corn they were selling were imposter chicos. The sad part was that the imposter chicos were very pricey, and I know that if the buyers of such chicos are dismayed by their purchase, that is a sad result because all chicos will take the blame and the chicos experience is compromised. I told my cousin, the store manager, that they were selling "imposters," and he immediately wanted an explanation.

I explained what a cheater is, and then he looked at me in dismay and asked, "Well, what do I do?"

I suggested that he get rid of the imposters and get some local chicos del horno. So, I went to my house and got into my supply and replaced the cheaters. My cousin was very happy with that solution.

This is more than food or nutrition for the long winters. We try to store cold-resistant tubers, potatoes, carrots, onions, and so on in our *suterranos* (root cellars), but I personally find it easier to dig a large hole in the ground and toss the veggies, like rutabaga, into the hole; I cover the "cornucopia hole" with an old blanket and about two feet of dirt. The root cellars require much maintenance. The holes are easy to dig, and the stored foodstuffs are easily extracted. A root cellar is primarily for larger production yields; now that there are fewer garden fresh crops to store, the cornucopia hole method seems much more practical. I never cleaned my grama's cellar, because she was in a transition when the large gardens were becoming smaller. I did work on and clean the large cellars of a potato grower when I was young, and that is where I learned how important cellars are as necessary structures if we truly want and need to be sustainable. I have a root cellar that Devon G. Peña's students built for the farm, and it was used for a few years (I will talk about Dr. Peña later), but again, the maintenance needs and lack of use mean the structure is now in dire need of repair after some twenty years. The changing times with less family labor on the farm mean smaller production of staples, and the maintenance of a cellar makes it impractical for our current operations. Across our valley, the suterrano has been replaced by steel buildings that are insulated and heat controlled by a thermostat.

Money considerations have totally overtaken the desire to work with native resources when it comes to the topic of the suterranos. The surrounding materials like the dry soil and logs that came from *La Sierra* (common lands) that were used to make the ever-important structure were replaced by the mined and processed steel and iron. I am ever impressed by the steel buildings and how they have revolutionized the storage of farm products, and I would someday like to own a smaller steel building for storage of farm products, but there is a cost associated with keeping the building at prescribed temperatures. If I were to own a steel building for product storage, the propane truck would soon be backing up to the new steel structure, and the farm would need even more money to pay for energy use. With suterranos there was little cash needed to pay to keep the root cellar at a constant cool temperature. The steel building guy will definitely argue that the money saved by not maintaining a labor-intensive cellar will offset fuel costs, but I can't imagine how this would occur. From my experience with repairing a root cellar roof and fixing the interior, I can see the positive benefits of not having to leave the area for materials to repair or build a new cellar. The logs were close by, and the straw was from the same farm or a neighbor's farm. Back when there was plenty of family and available local skilled labor, the suterrano was extremely easy and inexpensive to operate. I still think that even today if we paid expensive labor to repair and maintain a root cellar, the expenses for a suterrano would not exceed those of a steel building.

As far as time savings and easy access to and around the facility, the square steel storage is simply more efficient. Efficient is a dangerous word around here with all the factors and differences of perspective that apply to the case of suterranos. As I said before, getting into and out of a steel building is easy; with an earthen structure, a single doorway coming in from an incline driveway provides more limited access to the suterrano and definitely takes more time. I recall many trucks getting stuck in the cellar because the incline was too difficult to negotiate, especially when it was wet and slippery. When I think of efficiency, the topic becomes debatable because its meaning is contested. The ever-popular discussion of defining efficiency often even results in a court case, and I believe it is better to define efficiency in a way that is based on how one perceives life. Complicated, yes. So, how did a simple topic like suterranos become so muddled? My view is that using efficiency as the major basis to make decisions as to how we will do our work is fine, but only if we consider if the farmer is able to use personal and local logic. With all the different definitions of efficiency come all kinds of ideas flying through the air, and the debate starts; the work comes to a halt.

Consider the case of chicos: Are they an efficient value-added crop, or are they a labor nightmare? I know that if labor was inexpensive, the chicos would be a food that could make substantial money. But it is difficult to meet payroll and find laborers for work that was once thought of as nonskilled and is now seen as highly skilled. Still, is living a sustainable life with chicos that keep the *panza llena* and having a stockpile of chicos better than having to make money? I feel more secure about the tangible chicos in my pantry, ready to cook and satisfy those who have winter hunger, than I do with money in my pocket when I look into the wintery blowing night, wondering if the rural store will have any fresh food in stock so I can convert my money into food. It just seems more "efficient" to work to make chicos than to work for money to buy the chicos, which would likely be imposters. I wonder if valuing efficiency takes into account the security of someone's food supply and the comfort that knowing how to grow, preserve, and serve food will contribute to the positive health of the earth and its inhabitants. I still believe in the efficiency of the suterrano and what I perceive to be important factors for keeping this farm sustainable. The suterrano provides a different way to preserve food and takes the place of spending cash on modern steel buildings. My belief is the suterrano has the least impact on the earth and is much more sustainable.

All this reflects on our place-based family values, which include an appreciation for hard work; the labor time to be an acequia farmer is tremendous, and the profits are close to nil. The benefits of self-reliance and self-satisfaction are priceless. You can't get too rational about it. We have our own values, and money is just another asset. I believe that my brothers and sister may disagree with me about my view of money, but the history of our area has proven that money is not reliable.

I say this because my father had issues with my grama about her saving everything, "Para los tiempos duros." My dad once criticized

Grama for saving all kinds of plastic sacks "for the hard times." So I asked Dad about the "hard times." He said that he was only a child when the *tiempos duros* really hit the county and the country. He recalled how he never knew that he was in hard times because he never felt hunger; he never suffered from the terror of not having work to do. What he remembered from the Depression era—and this was a stark reminder that the country was in turmoil—was when the immigrants would travel from state to state looking for work and end up at our farm. Grama and Grampa would make the hungry immigrants food packages of in-season crops and vegetables, and they would be off to their destination of hope. Dad said that he had never seen poverty like the immigrants suffered and that the poverty they faced was a product of what seemed far away from our farm. The security of the rancho relied on the family, and the constant work kept the area prosperous and secure. We had great food, water for agriculture, fine neighbors, ample opportunities for celebration, religion, adventure, culture, and all the necessities for a comfortable life. This is why Dad never felt the hunger that victims of the dust bowl suffered. Grama would pray her heart out that hunger and poverty would end for those less fortunate than those on the rancho and that the *gente* would hold together to endure the "hard times." I am sure there were other signs of the economic depression of the 1930s in the watershed, but when Dad would tell us the story of those times, I would let my imagination run free and try to envision what San Luis was like: a utopia or fairy-tale place where the ugly wolf of economic depression was at bay? It all made sense to me how the *familia* endured the "hard times," so every once in a while when there is a chance to save a little more hay or grain for the winter, the adage "Para los tiempos duros" will come out, and we will talk about our past.

The economists have come and gone, creating theories that basically portray the end of the chicos-making way of life. They keep saying the lack of economic efficiencies will make the farms here disappear. I agree somewhat, but I have a lot of hope in our ability to adapt and persevere through dynamic times. During my lifetime in the Culebra watershed, cattle replaced sheep, and row crops declined, adaptations so we could survive the changing economics of farming. With little labor or costly labor, the cattle market can be managed in this age with fewer people. Mechanization has made haying a one-person job; I am amazed at the equipment that is out there today. Cutting twenty acres of grass expended the energy of three or four people in the past, and that does not include the people at the homestead supporting the field-workers. A modern diesel windrow mower will cut the twenty-acre field in a few hours; it does cause fatigue as the farmer tires of the repetitive motion and loud engine drones.

With all the troubles and changes taking place in agriculture, we are still here. Living working, irrigating, farming. Right here in place. We were always here; the water was waiting and moving.

## Help on the Acequia

Today, we are trying to revive our chicos-making culture. A person that has pushed and assisted the farmers in reviving the culture is my close friend Professor Devon G. Peña. The professor, or Peña, as we refer to him, has been a real educator about where the chicos culture came from and why it is important to save this culture.

Through Peña's skills as a professor and educator, I was encouraged to keep carrying out my return to row crops as another method of becoming diverse in the sustainable agriculture arena. Peña and I met in the late 1980s when we had to endure some environmental battles with large corporations. Through our common values about the environment we became good friends. With his assistance and connections, the ranch was able to increase the chicos production to a larger capacity when he brought some of his Colorado College students to build a larger horno than the one we were using.

I wish to recall the discussions we have had regarding the benefits of chicos and the possibility of making big on commercial production. This is where Peña and I have had disagreements about where our future lies with concho corn and ultimately the chicos culture. He believes that we can make chicos more commercially viable as an organic niche product; I argue that it will take many years to develop this capacity.

When I first met Peña in 1987, my immediate response was to be suspicious and ask a lot of questions. I still have questions, but my doubts have quelled considerably since then. Upon self-reflection, I see that when I went from the business world back into the acequia farming community, my naïveté was overwhelming. I sometimes wonder how I survived being mainly in the cattle business during the 1990s fall in cattle prices. When I would discuss my economic situation with the professor, I felt that many times he would mix up the real economics of not producing enough product to pay the bills with the natural asset economy that saved me money and kept my life on the rancho sustainable. Peña has now purchased a rancho in the acequiahood, and we are all happy about having him as a neighbor. I am exceptionally happy because he can now live the "happy poverty" that natural assets bring. I have noticed lately that he and his partner, José, are more concerned about producing enough hay to help with some of the ranch expenses. This tells me that he is living some of the hardships of making a small farm pay off.

When newcomers get into the small-farm ag business, my first thought is: I hope they have a revenue stream. Basically, keep your day job, and try to make the farm a practical part of your life until the economy changes or the volatile stock market crashes; always try to keep a consistent amount of cash coming into the ranch. It does not have to be a great amount of revenue, but it must be consistent. I learned this the hard way because I seldom had a revenue stream, and when I would sell farm commodities, I would get cash in lump sums and have many months of lean times before the next sale. The time between sales is when the rancho gets into trouble with bills, and many business entities do not anticipate or respond to this. For example, the necessity of feeding livestock is continuous, and expenditures are necessary. The nonrevenue months for the required work will suck up the sales money, and when that is used up, the rancher will get desperate and start borrowing money, and that is a vicious trap for any small farm operation. With a small revenue stream, bills can be partially paid until the next sale allows you to pay bills in full without having to go to the short-term, quasi solution of bank borrowing. Walking through a small store in Pueblo, I noticed a decorative dish with the adage "Behind every successful farm is a wife working in town." I immediately thought about the revenue stream. Many of the small farms here in the Culebra lack a revenue stream, and that is why they face economic issues, but those that have a revenue stream are still chugging along enjoying the natural assets.

Figure 8.2. Joseph C. Gallegos irrigating crops in heirloom milpa. Gallegos had just finished planting the annual mix of maíz de concho, Bolita beans, *calabaza*, *calabacita*, and *habas*. June 2014. Courtesy of Devon G. Peña.

Now that Peña has a different perspective on small-farm ag, I can have practical conversations about how chicos and other value-added crops will not be the way to keep our farms from falling prey to land speculation, water grabs, and all the enemies of a successful small farm. There have been several pushes to get farmers into value-added crops, like chicos or bolita beans, but the organizations and cooperatives seem to fall apart when true money economics are applied and ignore the value of these natural assets. My neighbor blames subsidies for the problems blocking the realization of a small-farm, fair-trade society. He says that a government subsidy makes the playing field uneven and competing with huge subsidy users cannot be done. I may believe some of that, but until change occurs, I need to find safe strategies to make enough cash to pay the daily farm bill.

Because of dominant economic pressures and organizational issues, the acequiahood needs more cash flow to stay afloat in this day and age. I have managed in the past to keep the farm liquid, but I was in a special situation with a strong land base and little farm debt and expenses. Most newcomers to the business will not have the same ideal situation as I have had, so I advocate for farmers to secure their revenue stream. The basis of all this is that, in the short term, the natural assets of concho corn are not easily converted from their natural asset worth to money; the two do not match, and chicos have the lower value. Selling chicos for $10 a pound can never replace the value of the fun and celebration and family ties the chicos experience brings.

No matter what happens with the chicos culture, Peña should be recognized for inspiring and assisting a growing number of acequia farmers to build their own hornos while supporting community-based food traditions and local jobs. More families started to caravan from Denver and other points east and north to help the matriarchs and patriarchs make chicos. A culture was revived, and much thanks goes to the professor.

## El espiritu de los chicos

After their first visit to the Corpus A. Gallegos chicos del horno roast, a couple went back to Denver to rest and recuperate from the long weekend of hard labor. As they settled in, they felt a type of spiritual uplifting. A strange but strong feeling of accomplishment brought out inner feelings that can only be felt and shared by those who attend the horno celebration.

I believe that this spirit, a sense of place, which overcomes us, needs to be felt by more people, and that is why I encourage more involvement by anyone who wants to participate in the magic of chicos. So, when you get invited to a chicos roast, don't fear the unknown. Enjoy the ancient culture that started the chicos. Like my friends from Denver and other places, I too have felt the chicos spirit, and I have tried to understand why it is such a strong feeling.

My only answer to such an abstract question is that the hornos and the corn came from ancient civilizations thousands of years ago, and the millions of people of the past, present, and future generations that have had to depend or will someday come to depend on the corn for survival arouse in our spirit a sense of direction from within so we can carry on these ancient traditions that honor the earth. Our return to the earth, the corn, and the fire awakens the energy of our pasts, and the ambiance turns into celebration by revived souls. I have at times thought that maybe these forces are vicariously celebrating through us.

# CHAPTER 9

# Travels of a Diaspora Community

## From La Sierra Madre y Tierra Caliente to the Pacific Northwest

### MARÍA GUILLEN VALDOVINOS

As an economic refugee displaced to occupied Duwamish land in Seattle and alongside *mi gente* in Eastern Washington, I am constantly engaging in transnational activism that crosses physical and imaginary borders. As I continue the resistance of my ancestors, I share this narrative of the intergenerational heritage and knowledge of my people. *Mi familia* has carried vestiges of traditional environmental knowledge (TEK) with them across divisive borders from Guerrero and Michoacán, México (Tierra Caliente).[1] We have utilized this TEK as sources of our resistance for survival and continued cultural flourishing in a small farm-working community in the Pacific Northwest. Our struggles and resistance are transnational and extend across the United States and México.

My parents grew up in La Loma Valle (The Hill Valley) near Petatlán and Xihuatanejo, Guerrero, in one of the most economically "poor" yet environmentally rich regions of México. My maternal *abuelita* is originally from Agua Fria, Michoacán, and my maternal *abuelito* is from Los Talleres, Michoacán. During the 1920s and 1930s, *mis abuelitos* migrated to the coasts of Guerrero with their families at a young age, leaving behind their homes to find land and continue planting and harvesting. My paternal grandparents and their families have always lived alongside the coasts of Guerrero and have been fishers and farmers. Therefore, I reclaim my ancestral homeland alongside

Figure 9.1. Author's parents and great-grandparents, Mexico City, 1976. Luis (father) and Felipa (mother) with Amador (great-grandfather) and Emilia (great-grandmother). Courtesy of María Guillen Valdovinos.

the coasts of Guerrero and Michoacán, which can also be referred to as Tierra Caliente or La Costa Grande and Costa Chica.

Michoacán y Guerrero are located in the mountainous region of Sierra Madre del Sur, which is known for its incredible biodiversity and a large number of endemic species of butterflies, orchids, and birds. It is one of the original locations where the wild turkey was domesticated. The South Pacific dry pin-oak forest is situated on the southern slopes, and the Balsas dry forest, located near the Balsas River, is known for its diversity of mammal species, which include the collared peccary, ocelot, coati, and jaguarundi. In the lagoons along the coasts of Guerrero, Michoacán, and Oaxaca are found Mexican South Pacific mangroves. The diverse ecological biodiversity is interlinked with cultural biodiversity because for thousands of years our people have developed deep connections to the land, sea, plants, and animals.

*Mis familias* of P'urupecha and Afro-Indigenous descent have been moving back and forth from Guerrero and Michoacán for centuries. We have retained our traditional practices yet also incorporated Spanish, Middle Eastern, and West African traditions.[2] My families' narratives begin prior to colonialism and imperialism in both the Americas and West Africa under the Spanish Empire and carry deep memories and stories of repression, oppression, and exploitation for the past four hundred plus years. The enslavement and struggles of my African ancestors must be acknowledged; therefore, I declare that I am of mixed descent and my African ancestors were brought against their will to the Americas.

The displacement of West Africans into "New Spain" began during the slave trade, 1521–1821.[3] The Spanish enslaved West African Indigenous peoples for a slave labor force and utilized their bodies as tools for economic profit for royal and mercantile elites. African slavery is deeply associated with the mistreatment of Indigenous people across the Americas because we were all viewed as inferior and we had to be exterminated. It is important to acknowledge that Indigenous peoples were aware of the injustices committed toward people of African descent, for our struggles are all very similar. In 1829, slavery was banned in México. Additionally, it is important to recognize that

> Blacks accounted for the majority of the troops in México's 1810–1821 War of Independence with Spain. As a group that had known subjugation and oppression, they gave full support to the war in an attempt to end racial inequality and oppression. In fact, one of the principal leaders of México's independence movement, later president of México, Vicente Guerrero, was known as México's "first Black President."[4]

We must acknowledge and recognize the contributions and efforts of both Afro-Mestiza/os and Indigenous peoples during the independence of México from Spain yet at the same time remain critically aware that this has not necessarily brought about a racially harmonious Mexican society and equality among diverse ethnic peoples. After independence, mestiza/os and people of Spanish descent in México still controlled most of the country, and social and racial inequalities played a key role in the construction of the young Mexican nation.

Throughout the 1920s, José Vasconcelos's theory of *la Raza Cosmica*,[5] also referred to as "racial homogenization" or Mestizaje, aimed at creating the "perfect" race or "fifth" race in México after the end of the Mexican revolutionary war. His theory was implemented by those in power in institutions and the education system and attached to all people in México despite the cultural differences among us. By homogenizing (and "whitening") the Mexican population under a mestizo racial identity, the contributions of Indigenous, African, Middle Eastern, Pilipino, and other communities were erased from the history of México in order to create a sense of a nationalist Mexican society. The invisibility or "disappearance" process is a strategy that ignores the contributions of ethnic groups in many Latin American countries and is a method that has long been accepted as means of solving ethnic, racial, and social issues through restoring whiteness by "bleaching" out people of African descent and other ethnic groups. Vasconcelos developed the pedagogical theory to implement this program and was encouraged by those in power to enforce it in the education system. The Mexican government campaigned to create one country through education, the arts, and the media to demonstrate that México was "modernized." According to this logic, Indigenous people and people of African descent kept México from becoming modern.

Therefore, the goal was to "civilize" México by imposing the Spanish language upon the entire population. The majority of Mexicans spoke Spanish as a second language or did not speak Spanish at all. This was seen as backwards, and for progress to occur, Indigenous people had to be either assimilated into Mexican society or disappeared. Vasconcelos believed that Indigenous dialects could not be used as education tools. Therefore, "education" programs were implemented to supposedly decrease illiteracy and provide education for the poor, yet it is important to recognize that this was a tool to assimilate Indigenous people into one Mestiza/o culture despite most of the Mexican people being of Indigenous descent.

In addition, there is a danger in glorifying only one Indigenous past (Mexica or Maya) without acknowledging Indigenous peoples' resistance to assimilation into mainstream "mestiza/o" society, because it neglects the struggles of Indigenous peoples and communities today. I argue that instead of México being a Mestiza/o society, we are a multiplural, multiethnic society, and we should distinguish the different people of national and regional origin from what it means to be a nationalist. Through an intense process of assimilating Indigenous peoples into one homogenized national identity, Spanish was forced upon my great-grandmother during the 1920s, and I was stripped nearly bare of my language and culture. Yet we retained our traditional knowledge and traditional medicines and ways of healing. The resistance of our ancestors has kept our people alive.

My sisters and I were born in Zihuatanejo (Cihuatlán, Place of Womyn; in Nahuatl). This is a P'urhépecha women's sacred land where for thousands of years our ancestors buried our placentas and honored the goddess of warriors and women that died in labor, Cihuatéotl.[6] Our umbilical cords are buried alongside the umbilical cords of our ancestors on our sacred land, which is now tragically the third-ranking tourist destination in México. We carry the legacy of my maternal ancestors as we continue to pay honor to the struggles of *mujeres* in my *comunidades*. Our *mujeres* have led our people, therefore I was mentored to be a *mujerista* (womanist) and a carrier of culture for the survival of my *gente*. Our land and resources are now used for profit and consumption, and our cultures, peoples, and histories are a source of entertainment and recreation for the world's rich who come to México for a vacation yet deny us a right for survival outside of our lands. Those of us who are displaced often do not have the opportunity to return to our homelands, because of documentation and immigration status and class inequalities in the Global North. As a result of the social inequalities in México, we were forced out of our homeland and to migrate north into occupied land. The struggles for home and land are common to all Indigenous peoples in the Americas. Therefore, the struggles of the liberation of my people are interconnected with the liberation of Native people in the Northwest and throughout Abya Yala (Turtle Island, North America).

*Mi familia* was displaced to a small farm-working town in Eastern Washington in 1993 after México signed the North American Free Trade Agreement (NAFTA), which forced many Indigenous peoples and campesina/os from México and Central America to migrate to *el Norte* for economic survival. Don Luis, my father, had been emigrating and "crossing" the border back and forth from México and the United States since the 1970s and obtained permanent residency during the Reagan administration in 1986. Since his arrival *en el Norte,* he has been a farmworker working *en los campos* (in rural areas). Doña Felipa "salío de su querida tierra con niñas y niños en 1993" (departed from her beloved land with children in 1993). This was done for the good of the future prospects of *la familia,* and she painfully left the rest of her family behind in México. We arrived in Eastern Washington and lived with other farmworkers and diaspora *familias* and communities that also traveled far from the Tierra Caliente y Costa Chica/Grande zones of México. To survive, our web of connections and acquaintances back in our homelands allowed us all to find each other and create a community and *familia* in a predominantly conservative small town. We recognized that our homeland is in the Tierra Caliente region, yet we are here because we cannot survive in our homelands.

In the predominantly white town where we grew up, the history of my people and peoples of Mesoamerica was never mentioned, and we were given only distorted fragments of information about the Mexica and Mayan peoples as "uncivilized," bloodthirsty, ritual killers who only sought to kill others. So, of course, we were taught to internalize racism, have self-hatred toward our people, and question our definition of home and place. We were denied access to an education system that acknowledged the struggles of Indigenous peoples, immigrants, people of color, queer/trans folk, and other marginalized communities in the United States and, in particular, the legacies of the imperialism of the United States in the Global South. To recover my family "her-stories" and "his-stories,"[7] I relied heavily on *los cuentos y leyendas* (stories and legends) my parents told us to connect to our land, traditions, and ways of life back in the homeland. My parents told us over and over again that "nuestra tierra y gente siempre nos han levantado" (our people and our land have always lifted us). As resilient and self-sufficient displaced peoples, we brought a bit of our sacred homeland in our heirloom seeds. Mi mami dijo que, "Cada semilla tiene una historia, cada planta tiene vida; lo que le hacemos a los demas nos regresa mucho mas fuerte" (Each seed has a story; each plant has a life; what we do to others will return to us with greater force). Honoring

Figure 9.2. Author's *abuelita* Basilia during a community land reclamation project. The family was very active in the *tequio* (community labor) centered on caring for the land. Courtesy of María Guillen Valdovinos.

what the plants provide for us and how we treat those around us generated a sense of political consciousness and autonomy among my community. We relied on each other for our well-being and survival.

As I engaged in autoethnographical research on a specific *huerto familiar*, I came to learn a lot about life in the small farm-working town of displaced peoples from the Tierra Caliente region of México and the historical migration patterns of displaced people from the south to the north. I realized that we live in a society that is obsessed with fear. This fear is mobilized by shrewd politicians to exclude Indigenous people and people of color from public spaces and impose environmental racism and disproportionate risks on our bodies and communities. This led me to recognize and appreciate resilience as a living labor capacity to create spaces that reaffirm a sense of place and a sense of belonging in places far from our homelands.

It is important that we recognize the knowledge that elders in my community, including my parents and grandparents, have acquired through intergenerational experiences. For example, we have a *dicho* (aphorism) in the family: "Nosotros no somos mejores que las plantas y animales; somos todos parte del mismo ambiente" (We are not above the animals and plants; we are all part of the same environment). For my community, this meant that to have a better sense of the world around us, we must recognize that there are no exact eternal hierarchies in ecosystems. We are all interconnected, and each organism plays a role in making the world around us our mutual life-support system. This is an essential ethical principle that Indigenous peoples use as a basis for socially constructing coeval relations with the environment and to develop methods of healing through the ancient traditions of our place-based millenary civilizations. The tools for survival of economically and environmentally displaced peoples are not limited to creating autotopographic spaces in community gardens and *huertos familiares* in both urban and rural settings. This involves retaining traditional ecological knowledge; thinking local, slow, and deep in a globalized society; and looking at our deep food as medicine for community health, resistance against environmental racism and sexism, and engagement with food sovereignty for the survival of displaced and communities of color. Our resistance and resilience will determine the future of our *gente y pueblos*.

## Transnational Mesoamerican Identities: Displaced Peoples, Resistance, and Autotopographical Spaces

The experiences and struggles of *mi familia* to create a space to plant and harvest our traditional foods and crops are part of a larger struggle by displaced peoples and communities to create a sense of place outside of our original homelands. By mainly focusing on the experiences of one family's diaspora and displacement, I hope to further expand the idea that home kitchen gardens are not particular to one region of Mesoamerica. Rather, they are significant to understanding the importance of traditional foods and kitchen gardens for many Indigenous peoples and displaced communities around the Americas and the world.

Mesoamerican peoples and diaspora communities from the Global South use autotopographical spaces to create a sense of belonging and home and to adapt to the ever-changing cultural dynamics of the world. Spaces of autonomy that are dedicated to establishing and building local food sovereignty are opening in thousands of local places across the world. These spaces break down the separation between consumer and producer. Devon G. Peña and Teresa Mares state that

> *el jardín* (garden) is a space for the charting of individual "autotopographies"—self-telling through place-shaping. This is certainly true of the classic home-based kitchen gardens . . . that continue to spread across the urban US. These *jardincitos* are spiritual and political symbols of a process involving nothing less than the re-territorialization of places as a home by transnational communities.[8]

In addition, community gardens function as spaces for community organizing and resistance. However, community and kitchen gardens are not the only ways to create autotopographical spaces. Rather, there are alternative ways to self-tell by place shaping, including taquerías and food stands. Here, people create spaces to enunciate or tell the stories of our experiences as displaced peoples. Our cultural identities are at the core of our placemaking and place-based subjectivities.

In a prescient article addressing the theory of autonomy and environmental justice, Devon G. Peña affirms that "local knowledges are constantly mobilized by communities engaged in struggles to restore and protect local spaces in rural and urban spaces and sustain right livelihoods against the intrusive vagaries of the globalizing market and . . . 'meta-state' of the World Trade Organization."[9] For this reason, by engaging in transnational struggles to preserve our heritage foods, we are creating community-based spaces to reclaim our identities and retain the knowledge of our plant relatives for the well-being of our communities. Transnational activism allows us to organize with other Indigenous and displaced communities to address the inequalities driven by the global market and the privatization of our environment and food systems. We engage in addressing the lack of access to our resources and how it is related to the disproportional distribution of wealth, commodities, and foods between the Global South and Global North.

In a time when neoliberal globalization creates mass displacement of rural communities that migrate to the Global North, Mesoamerican farmers and peoples are moving to metropolitan centers to find means of survival. They are vulnerable to social and racial injustices as they seek to improve their economic conditions. This process of structural violence experienced by displaced communities forced out of our homelands and ways of life is linked with neoliberal economic policies that further disfranchise those that are already marginalized in our homelands.

These struggles derive from the effects of international trade agreements, such as NAFTA, which have caused a huge increase in food prices, devalued the local economy and currency, and opened the doors to dumping genetically modified and subsidized food commodities in México, drastically devastating our communities and lands, especially those of Indigenous peoples and campesina/os. It is estimated that approximately 1.3 to 2 million rural farmers from Mesoamerica were forced out of their lands and have been displaced to the Global North between 1994 and 2004. We did not chose to migrate; rather we were forced to abandon our ways of living for the survival and well-being of our people. As a result, we seek only to go where our food is going because we have relatively no access to our own traditional foods within our homelands. We are in the United States not by choice, but because staying home meant something less than a bare-life option.

## *Huertos Familares*: Homegrown Kitchen Gardens and Community Gardens

Our ancient seeds have a deep connection and history with the land and people. These seeds and heirlooms can be traced back approximately five thousand years in the Mesoamerica region. There is an immeasurable responsibility and relationship with growing and nurturing our sacred heirlooms that have fed our people for thousands of years. Each seed holds cultural value and memory that connect us to our ancestors and our sacred land. Our *huerto familiar* is located

Figure 9.3. "Each seed has its story." The author's family heirloom maize. These eight landrace varieties were brought from the Tierra Caliente region of Guerrero, México, and are now cultivated in Eastern Washington. Courtesy of María Guillen Valdovinos.

in an arid region in Eastern Washington with severe seasonal weather including heavy snowfalls and hot, droughty summers. Our family had to readjust, get acquainted with the land, and learn new ways of planting and harvesting, such as how to utilize sprinklers and tractors and how to work around the weather, while incorporating our ancient ways of working with *la tierra*, such as making fertilizer out of cornstalks and husks and planting certain types of maize and peppers that could survive the climate. These all have played a key role in the cultivation and process of harvesting in the community gardens of the region as well.

The work at the garden usually begins in late March or early April and includes clearing out the plot and getting the soil ready for planting the Tierra Caliente seeds and plants in early May. One of the first plants my father recalls bringing with him in the late 1970s was epazote. He brought that specific plant with him across geographical borders because he believes that epazote is vital for healing both the spirit and body. He is a curandero (traditional healer) and sobadero (a person that uses body work including massage as a healing practice). Plants are linked with his identity and his connection to the land and his people. He recalls knowing about thirty different types of maize and over sixty other types of plants, including plants used for medicinal purposes, which he harvested from a young age. Due to the displacement from his homeland and lack of access to the plants, he has lost knowledge of their names and how they were utilized in the past. He only has memories of them. For that reason, I have a limited knowledge of the plants my people have utilized for thousands of years.

My father states that he did what he could manage and brought some of the variations of maize to the Pacific Northwest, as he was forced to migrate and learn how to harvest in an arid region that was very different from the soil he was used to back in Guerrero. He says, "No se me hizo difícil aprender a trabajar en este país porque siempre hemos trabajado con la tierra desde que nací" (It wasn't difficult for me to learn how to work in this country because we've always worked with the earth since I was born).

So in a sense our people have always been resilient even as we engage in waged farmwork that disrespects our rights and dignity. I realize that to maintain our connection with the plants and tierra, we need to learn how to work with different methods of planting, harvesting, and seed saving. Forced into these jobs, we learn how to incorporate our traditional knowledge and build community among ourselves in order to establish a sense of place and belonging.

In the midst of animosity toward displaced and economic refugees in the United States, mi familia was able to collaborate with other families from Guerrero and Michoacán to look for a piece of land to create a small community garden to feed our families. They felt displaced and needed to bring a little bit of home with them. Through organizing in the campos, they asked their patron to provide them with a small piece of land to harvest maize and other plants. I consider this another form of unrecognized community organizing that was essential for our community to retain our traditional knowledge and cultural identity that links us to the land.

By appearing passive and complicit, they were able to convince the rancher to give them the space. I argue that this success gave many of the workers and families the courage to demand more rights and an increase in salaries from the rancher. This space created a sense of community and an opportunity to mobilize at a small scale by empowering the workers and families to establish a center for the community that allowed us to practice our ancient ways of harvesting and planting. The community garden was influential in empowering our community to demand more rights. I argue that we must also incorporate community gardens as vessels to organize for social change.

The garden is a community project where during harvest season any of the families can go and harvest the food they need for the week. Harvesting usually occurs August through late October, and a few times during this period all the families will come together and harvest as a unit and distribute the food among the families. This is the way of our people; we cannot let anyone of us go hungry. It is important to note that no profit is made out of this community garden. The main objective is to create a space where we can have access to our traditional foods and provide healthy ways of eating to our families.

Families will also come together to make tamales, mole, and other traditional foods to consume as a community and to share stories and knowledge with each other. In a sense we are retaining our community and developing new ways of creating community and a sense of a large *familia* to preserve our cultural identities. This support system also works as a collective to raise funds for any emergencies occurring within one of the families.

Clearly, my *familia's* knowledge of medicinal plants, planting, and harvesting was a key component to adjusting to living in the United States as farmworkers and laborers. To create a sense of identity and home in the United States, we began planting and harvesting our traditional foods and ways of healing. Eventually, my parents were able to bring approximately eight variations of maize and other plant species to the United States. As others do in *huertos familiares, mi familia* cultivates and harvests a mixture of maize, frijoles, *calabacitas*, chiles, tomatoes, and *pepinos* (cucumbers), alongside traditional medicines such as different *yerba buenas* and *manzanilla* (chamomile).

Maíz and *calabacita* (Mexican squash) were some of the first native crops coaxed from wild relatives and domesticated and diversified by Mesoamericans approximately five thousand years ago. This is important because "community gardens around the world provide significant ecological, social, and economic benefits, which are consistently underestimated in the face of urban development."[10] People of Mesoamerica have a complex history with *huertos familiares*, and our ancestors have played a key role in the domestication of wild plants and conservation of plant species. The plants domesticated in Mesoamerica were and are still major contributors to the global economy. Numerous plants and traditional medicines are still produced today in kitchen gardens, and a variety of maize, frijoles (beans), *calabacita* (squash), and chiles have transformed global diets. Mayan agricultural systems and milpas have multiple layers of crops that imitate the rain forest and prevent surface erosion, and today many of those methods of harvesting are still practiced. Kitchen gardens have historical links to our communities and are an essential component to our identity, in particular in creating autonomous spaces.

## Traditional Ecological/Environmental Knowledge

Respect for human, animal, and plant life has been an important component of how my *familia* conceptualizes the world around us. Each plant, animal, seed, and other organism has a purpose and a story. Thus, we cannot disregard the contribution each provides to our lives. One of the most important pieces of knowledge passed down by elders, our grandparents, and our parents was to acknowledge the feelings and emotions of the plants and animals that surround us. Dualism and understanding that you cannot separate animals from their community and habitat allowed me to understand what it feels like to be displaced from your homeland and community. I learned that we cannot continue cycles of exclusion and separation. Animals and plants have the same rights as we do to have a healthy environment. We must take care of our lands and environments because we cannot destroy the homes of others. How my *familia* conceptualizes our lands and environments is an ancient worldview of living in an interconnected way where we are all part of the same ecosystem. Thus, we must not destroy what is linked to us.

It is vital for activist scholars to return to our communities and do research on the TEK that our Indigenous ancestors and relatives have preserved for generations. My ancestors remained resilient and continued planting, harvesting, and coexisting with our traditional foods for generations. TEK is a particular form of place-based knowledge of the plant, animal species, landforms, watercourses, and other biophysical qualities in a given place.[11] As we continue to document and protect this knowledge, we will do well to follow the advice of ethnoecologists Fikret Berkes and Serge LaRochelle, who clarify that the protection of TEK means "research projects need to be participatory in nature, with the community [as] partner in the cooperative process of knowledge creation and sharing, rather than being the object of research."[12]

## Thinking Local, Slow, and Deep

The decolonial analysis of globalization and the displacement of communities is essential to clear strategic thinking in the food justice movement. In particular, it is important to understand what it means to act locally while thinking globally and deep. Mares and Peña describe going deep as follows:

> Engaging communities that have been historically excluded from the mainstream alternative foods movement is critical in the movement for food justice. Within food justice, it is simply not enough to examine the ethics of going slow to go local. One has to go deep, and this means respecting local knowledge, wherever and whenever it is found. As discussed in our opening vignette, there is a wealth of multi-generational place-based agroecological, ethnobotanical, and gastronomical knowledge within Native communities in the United States. However, there is also a wealth of this knowledge in diasporic and immigrant communities that have faced parallel histories of colonization, displacement, and environmental racism.[13]

We have to make connections between where the food is produced, who produces food, and whether local communities and peoples even have access to consuming local food. At the same time, we must value and respect local Indigenous peoples' knowledge and their deep histories within their origin and adopted lands. We must recognize that long-distance systems of food production involving processing and shipping cause enormous environmental stresses on both local and global ecosystems, including the pollution of the land and rivers, and limit local people's access to their lands and traditional foods. Consequently, Indigenous peoples, immigrants, farmworkers, and other producers suffer from tremendous health problems caused by pesticides and other environmentally hazardous chemicals, and from other stressors, including starvation wages.

## Food as Medicine: Healthy Environment, Healthy Bodies, and Healthy Communities

Lost access to our traditional foods has led to the devastation of the environment and health of displaced and Indigenous peoples. Planting and nurturing plants and seeds are a way of honoring life and are vital for the well-being of our communities. Winona LaDuke states that "as colonizers drove Indigenous peoples from our territories, we were cut off from access to traditional foods."[14] Traditional practices of gardening and harvesting are essential for both the nutritional and physical well-being of Indigenous peoples because they keep our communities engaged in physical activities.

Social, economic, spiritual, and physical factors contribute to the loss of our physical health. Robert Gottlieb notes that "the connections between what people eat, what kind of food is produced, and how it is accessed is very important in low income communities where lack of access to fresh, affordable, healthy food has direct health and nutrition consequences."[15] Inequalities threaten the well-being of farmworkers, working-class people, and the poor, who are repeatedly being denied natural resources, access to healthy food, and the means to retain cultural traditions and ways of living. Waste and other environmentally hazardous materials are often buried near low-income neighborhoods and communities of color and drastically take a toll on our health and living conditions, in particular on the bodies of *mujeres*.

## Environmental Racism and Sexism

Ecofeminism, proposed by French social theorist Françoise d'Eaubonne in 1974, is the idea that the domination and exploitation of women and of the environment are interconnected.[16] Feminists of color argue that class, gender, race, nation, citizenship, nation, and sexual domination are all interconnected and are deeply linked to legacies of colonialism and domination, and we cannot separate the liberation of our bodies from the liberation of our lands and peoples. There are significant lived experiences of poverty, discrimination, and violence directed at our women and unleashed by dominant white heteronormative structures that seek to embed and control our dissident subjectivity.

Direct experiences with structural violence and ecological degradation lie beyond the social spaces usually occupied by white middle-class women. Not all women live and are working in polluted places. Not all women experience racial profiling and polite brutality. Not all women are subject to the poverty of deprivation and hunger. These differences in the material conditions of life that white women versus women of color experience are associated with divergent ideologies of nature and variations in perceptions of environmental risk, the ability to take action, and the forms of plausible organization.[17]

Decolonial scholar Andrea Smith argues that "environmental racism is another form of sexual violence, as it violates the bodies of Native and other marginalized peoples. . . . women of color are suffering not only from environmental racism but environmental sexism."[18] Often mainstream environmentalists place the blame for ecological degradation on the Global South, immigrants, people of color, and primarily women of color. In other words, the "poor" are to blame for the destruction and "overpopulation" of the planet. The assumption is that women of color, economic women refugees, and women from the Global South become the perpetrators of the environmental degradation of our homelands. This view fails to acknowledge how food produced in the Global South is often exported to pay off debts to the World Bank and Global North finance capitalists rather than utilized to feed local communities and peoples. Worse, people cannot feed themselves, because transnational corporations have displaced them from the land to produce luxury crops for export to the First World. Corporations and political systems remain unaccountable when the blame is placed upon the displaced. For that reason, we must realize that

> any damage done by indigenous peoples, peasants, and Global
> South farmers cannot be compared to the damage done by the

multinationals and the World Bank, so the claim that stopping "overpopulation" of peasants and Indigenous peoples in the Global South countries will "save the environment" is baseless.[19]

Indigenous peoples and communities of color, especially women of color, are seen as threatening because we have the ability to reproduce the next generations of our people, and that stands in the way of the governmental and corporate control of our lands and bodies. Consequently, we are seen as pollutants that stand in the way of the well-being of the colonial body and empire. There is no doubt that our bodies are political and our ability to reproduce stands in the way of continuing the conquest of native people's lands and bodies. Control of our bodies is seen as a mechanism to control our communities. The aim of systematically targeting women of color, immigrant women, and Indigenous women and our children is to destroy our people, so in a sense we are not guaranteed bodily integrity and respect. The bodies of the *mujeres* in my *familia* are extremely political and carry legacies of resistance that bind us to our lands. Our bodies are interlinked with the cultural survival of our people. Therefore, I recognize that I have been culturally raped.

Smith reminds us that "by extension, because colonizers viewed Indian identity as inextricably linked to animal and plant life, Native people have been seen as rapable, and deserving of destruction and mutilation."[20] The attempt to transform Indigenous peoples into souvenirs and objects of consumption by white people, a creation of what is rapable and deserving of destruction, is therefore also part of a Western ideology of conquering the other, in particular stripping Indigenous peoples of our identities and humanity. Being culturally raped has long-lasting emotional effects on our people. We must acknowledge that we live in a gendered, violent, racist, imperialist society that will not recognize our right for self-determination, and we must recognize our strength in resilience, for our struggle is far from over.

## Conclusion: Food Sovereignty Is Knowledge for the Survival of Our Communities

Our survival is linked with retaining knowledge of traditional foods and foodways that has kept our people alive for thousands of years.

In the essay "A Flower in the Hands of the People," Gustavo Esteva suggests that "instead of losing our roots, as globalization encourages, we have opened up to broad coalitions of the discontented across national borders, while always asserting ourselves in our own places. That's how we have moved from resistance to liberation."[21] Our resistance and retention of our cultural identities are central to creating spaces, in particular community gardens, for the well-being of our families and communities. Most importantly, environmental and food justice organizations need to account for our struggles as displaced peoples and collaborate with us to create just and socially conscious movements that respect human rights for all, Indigenous sovereignty, and struggles for self-determination. Echoing La Via Campesina, Mares and Peña define food sovereignty as the right of peoples to healthy and culturally appropriate foods, produced through ecologically sound and sustainable methods, and under conditions where the people can define their own food and agriculture systems.[22]

Our struggles form a vital multinodal stream in the environmental and food justice movements. The survival of Indigenous and other communities of resistance will depend on cultivating the well-being of the people in our own locales through the retention and practice of deep ecological knowledge of food and foodways. This essay provided but one example of a resilient community that has created a sense of place and home through the autotopography of the *huerto familiar*. As my ancestral homeland calls me home, I reclaim being deeply rooted in the struggles of my ancestors and transborder communities, aware that my personal political liberation is linked with the liberation of my people and the land.

# CHAPTER 10

# Food, Class, Ethnicity, and Race in the Classroom
## A Teacher's Testimony

### JULIA CURRY RODRÍGUEZ

Food, food preparation, and sharing food with others have always been important in my family. My mother's food became particularly important after she, my sister, and I came to the United States, relocating permanently out of México in 1965. My mother is originally from Guadalajara, Jalisco, but as a family we never lived there. We lived in various places in México because of our parent's employment in Cuernavaca, Morelos; Colima and Manzanillo, Colima; and Salina Cruz, Oaxaca.

As long as I can remember, my mother has been famous for her cooking and food presentation. In one of her first jobs in the United States she became the most sought-after cook in the restaurant where she worked, such that when she was not working, some of the clientele refused to order. North of the border, I have encountered odd comments about my mother's cooking as being somehow inauthentic as to what is expected of Mexican cuisine. My mother's cooking, her food selections, and aesthetic presentations have been central to my sense of self throughout my life, so to me those comments were odd. I wondered how people could say my mother's cooking was not authentic! Who determines what is authentic in my family and in my home? Indeed, once we found ourselves settling in the United States, I am pretty certain that my mother made our settlement smoother and kept our heritage alive in the labor of her food.

I intentionally use the word *labor* to talk about my mother's cooking because I wish to make evident that preparing food and feeding people involve actions and behaviors influenced by economics, culinary skills, and decision making, none of which ought to be taken for granted. As a newcomer to the United States, my mother made decisions about how and where to procure groceries, how to use her wages to pay for our food, and also how to prepare foods that were familiar, healthy, and savory to us. Many years later, as I redesigned my class at University of California, Berkeley, I drew on my observations of my mother's quotidian actions throughout my life. I am fully aware that without this opportunity to reflect on how I came to use food as a means of teaching a course on the family, I would not have come to the conclusion that I have carried my mother's lessons unconsciously throughout my journey in higher education and also as an educator.

I honor my mother's teachings in my food habits. Like her, I make community by sharing food. I also use her wisdom in my food selections—my mother made me a healthy eater long before I encountered health-food markets and self-proclaimed vegetarians. I have used my mother's recipes for my own food preparation, modifying them as I have learned new ways and expanded my palate. I am not the magnificent cook my mother is, but I am her daughter, and my gift of observation helped me to learn from her even when neither of us was aware of the pedagogical element being enunciated through her family role. My mother's food habits inform and sustain my work, my pedagogy, and myself.

## Message for a Future Class

> Prepara la comida siguiendo estos metodos, pero a tu gusto.
> (Make the food following these methods, but use your taste.)

In the early 1990s at UC Berkeley, I was assigned to teach a course on the "family." I began to use food as part of my teaching when I found many dilemmas with the major paradigm of the time, which was literature that depicted Mexican families as dysfunctional, maladaptive, and culturally deficient. In class I wanted to debunk these cultural determinist approaches and myths about Mexican and Chicana/o families.

A quarter of a century has passed since that time in my teaching experience. By writing about this class now, I have been inadvertently forced to reflect on my teaching. I began my academic career as a teacher of Chicana and Chicano Studies in the late 1980s. My teaching assignments have always encouraged me to focus on consciousness and recognition of the labor of women, the impact of immigration, and the uses of theory and methodology in research. When I joined the Chicano Studies faculty at Berkeley in 1990, I had already developed some ideas about what worked in my classrooms. In the family class I wanted students to critically examine their family histories from the point of view of structures of inequality. I knew that some of the topics they would encounter would make them uncomfortable, but I wanted them to reflect on the meaning and sources of their discomforts. I had already learned that teaching gender and sexuality critically always encounters resistance because of the normative assumptions people hold about gender and sex roles.

At first, being assigned to teach a family class made me squeamish. I did not like the conventional frameworks and much of the mainstream literature, which I found uninspiring—dry, absurd, and pedantic. Scholars often focused on marriage and social problems, lineage, sex roles, and machismo. I had no interest in teaching about so-called dysfunctional families and the presumed lack of support such families gave children in their endeavor for higher education or any other efforts seen to be affecting their prospects for upward mobility. These approaches seemed like a kind of remote social science involving the reductionism of a "mathematization" of complex social mores, relationships, and cultural frames. Thus, I initially designed the class using oral history approaches. I also drew on literature from critical feminist scholarship that allowed me to develop an analysis that made the changing nature of women's and men's roles more salient, exciting, and complicated.[1] After teaching the class a few times, I determined that asking students about families created a different kind of resistance even when the families we learned about were resilient, creative, dynamic, and ambitious as opposed to simply "dysfunctional."

Following several failed or at least frustrating efforts, I hit on the idea of using food as a way to encourage students to examine stories of class, culture, and familial relations. After I ran into several problems while trying to use conventional life-history approaches to develop sociological studies of families, I learned that many of my students knew little about the origins of their own families, some came from blended families, and in some cases several students were problematizing the heteronormative cultural impositions that seemed to them to be desirable in Mexican American families. Prior to joining UC Berkeley, I had never encountered so much self-questioning about sexuality and sexual identity among my students. I am convinced that the time, the campus, and the unique group of students were a significant factor in my observation, and this influenced my reflection on how and what to teach in those years in a class dedicated to the study of the family. In a sense, some of these students were searching for ways to hold onto their myths about family, while simultaneously deconstructing families into those they wished to form themselves.

In this essay, I reflect on this class to unpack elements of a decolonial pedagogy and highlight some of the transformative learning outcomes of using food as a means to decolonize cultural reflection. My aim in the class was not just to have students talk about the food practices they observed in their families, but rather to learn about themselves and their families as members of class and racial groups that conformed to gender traditions both imposed and invented by their particular social conditions. Through a focus on their favorite foods and their most esteemed family cooks, students reflected upon familial social-class standing, gender relations, and power dynamics as tangible changing forces. In their analyses they repeatedly demonstrated an understanding of the structure of the family as something to which they have a responsibility to consciously adhere or to transform.

As preparation for the research assignment, I provided students tools for observation, notes on methodological ethics, and the admonition to engage actively in observation when they went to their family homes to gather their evidence. As the class progressed, I often listened to students comment on how their mothers made the best "[fill in anything from enchiladas to tamales]." I also heard discussions about intergenerational aspects of food culture and how grandmothers were essential to the creation and sharing of memories of food as well as to developing the skills to bring family traditions from the past into the present. And finally, I also discerned a very subconscious understanding of class position and gender roles that greatly affected how students thought about food and how they interpreted the role of food in their lives prior to attending university.

Speaking about family food memories gave students a way to reflect on their cultural and relational assets and knowledge in ways they had previously taken for granted. The discussion of favorite foods brought about nuanced discussions of class position. Some students talked about their favorite foods while coming to the realization that many of the foods they remembered as favorite were inaccessible to their families because of their financial limitations. Therefore, some students developed menus depicting foods they desired, and juxtaposed these with the foods their families actually ate. I found these discussions particularly interesting because they illustrated that the family economy shapes not only the health outcomes of food consumption but also the imagination of how to address the awareness of limited financial means.

Students developed their assignments to challenge concepts such as assimilation, poverty, and inequality and explored traditions of communal practices of sharing and resilience. Asking students to do research on their family food habits and to name their favorite family cooks was an intimate entry into the study of families, memory, and culture, as well as gender roles.

The reality, at least in 1990s Berkeley, was that many students shied away from discussing their families. Some of the reasons they gave me were that they came from "blended families"; that is, biological families and families resulting from a second marriage, biracial families, bilingual/bicultural families, and/or families of varied legal status. Some students were deeply conflicted about whom to speak to in their families, and what to disclose to people outside of their kinship group. In these discussions, they uncovered issues of familial loyalty, political marginality, and family secrets that I could not really address in the class.

For other students, the problem was that they came to class with the preconception that the stereotypes about Mexican families were inescapable. As students beginning their quest for social mobility, they recoiled at the idea of having to deal with their families who were being portrayed as living marginal lives, as immigrants, as poor and uneducated, or as bigoted and heteronormative, but always as though they were coming from defective backgrounds, which they were forced to abandon in order to succeed.

I struggled to help them find ways to cope with rethinking the analytical approaches to their personal experiences, explaining that their families were not just marginal; they were also ingenious in their survival strategies, resilient in their endeavors, and critical in their understanding of inequality. Moreover, I focused on finding ways to teach students to not accept the societal definitions of their lives—they had to figure out how to challenge from within their own understandings and analyses of their living conditions. I asked students to conduct oral histories with one family member to learn about their lives as students and not as family members. I really just wanted the students to talk candidly with a parent or sibling and ask about their lives, their struggles, and their successes—but not in a direct manner. This assignment produced affirmations that their critiques would result in more fulfilling explanations about their racial, gender, and class positions. But for some students, their assignments highlighted despair over family problems that they had escaped by leaving for college.

After reflecting on years of frustrated outcomes and mixed experiences, I decided that what I was encountering in the classroom had more to do with how students were transforming their consciousness through their studies and the process of reflecting on their own lives. The truth is they were experiencing changes as students going through transitions in the hotly politicized spaces of UC Berkeley, where identity politics, challenges to traditional cultural practices, and encounters with diverse sexualities made it difficult for them to want to study something as "conservative" and "boring" as "the family" from the conventional approach of dysfunction. I had to figure out a way to reclaim the space of family studies through a decolonial methodology in order to teach these students that the study of the family was not inherently a conservative or mundane matter. The field of study was filled with challenging and relevant perspectives on power, knowledge, politics, inequality, oppression, resistance, resilience, and transformation.

I realized that I had to move beyond the conventional sociological study of the family to something that focused on what families do without explicitly asking students to think about their family roles. I decided that talking about food patterns might allow us to discuss class, culture, and values without stirring personal conflicts and emotional upheaval. Thus, I explored teaching family by teaching food.

## Teaching "Family" through Food

I drew first from Manuel Gamio's classic ethnography *The Mexican Immigrant: His Life Story*, Laura Esquivel's novel *Like Water for Chocolate*, and Patricia Preciado Martin's oral histories of octogenarian women in Arizona, *Songs My Mother Sang to Me*.[2] I imagined my class doing oral histories of families through the favorite foods they prepared at special occasions to document memory and possibly reconciliation.

The assignment asked students to observe and talk with their family members, but I asked them to focus on their favorite foods, giving special attention to who prepared the food, how the food was prepared, and what role those foods played in their family. I also wanted them to think about the importance of familial bonds and traditions through shared meals—how their families held on to the cultural knowledge in the kitchen. Focusing on favorite family cooks made students make choices about their ethnographic assignments. But thinking about food also made them consider labor, economics, and social membership in racial and ethnic communities.

I intended that students find a way to ask about the food, its preparation, transformation, and presentation. This approach moved them away from simply announcing that their mothers or grandmothers made yummy food, toward instead thinking about what it took to make those meals and why they were made in the accustomed manner in their families. I asked students to choose one favorite cook and three of their favorite recipes to develop their research papers.

Their task was to ask about the place of food in their family's social practices and how these practices affirmed a heritage and tradition that was mostly different from the mythologized views of Mexican and Chicana/o families. Students were required to observe their favorite cook prepare the designated meal or entrée, ask for recipes, and inquire about what the food meant to the family and their traditions. I also asked students to reflect on their observations and whether they had ever paid attention to the labor involved in preparing the foods they loved. Lastly, I asked them to think about how their family financial status shaped their food choices.

Students prepared an oral history that included images and recipes of favorite foods. Their analysis focused on issues of marginality, inaccessible ingredients, and the uses of time in the organization of reproductive labor in the household. They also focused on themes of affection and memories that they took for granted in their family food strategies. Using their observations, students conducted library research to help them situate themes, understand the process of food preparation and presentation, and interrogate the role of their designated "ethnic" foods. For the final assignment, students brought one of the favorite dishes to share with the class and to celebrate their rite of passage as researchers collaborating with their families.

Manuel Gamio long ago proposed that it was important to determine the way that Mexican immigrants who came to the United States preserved and transformed their culture. He and other anthropologists were supported in doing research that focused on how Mexicans preserved their culture in the United States. One of the cultural practices they observed was centered on food habits. In particular, these scholars were concerned with the effects of the forcible "Americanization" policies of the early 1900s and how these affected immigrant families who were encouraged to adopt US habits, including so-called American food and other consumption patterns.

One of the findings was that, indeed, Mexican immigrants preserved their food habits but they also adapted food preparation techniques by using available foodstuffs to replace items more suited to their familiar taste and dietary palate. Thus, new combinations of spices, herbs, and white flour, rather than corn masa for their tortillas, were documented. What this information shows us is that Mexican Americans and immigrant Mexicans preserve their eating patterns by adaptation and innovation. Gamio documented changing consumer patterns that might have been used by the food industry to seduce immigrants to accept and use American foodstuffs, but the oral histories of my students demonstrated that the Mexican people made important decisions about their food consumption not only in regard to taste but also due to concerns related to health outcomes and cultural preservation.

## Ingredients for Making Memories

When students use food as a window to the study of the family, they produce meaningful themes and unanticipated perspectives that allow them to renegotiate an understanding of cultural resilience in a society demanding conformity to Anglo assimilation models that devalue the self and grounding in family and community. This section presents a series of menus, recipes, and reflections from my students that illustrate the journey from cultural ambiguity to a sense of critical consciousness of the right to be different and deeply appreciative of one's own ancestral traditions and knowledge.

## 1. A MENU THAT SHOWS DESIRE AND FINANCIAL INACCESSIBILITY (GABRIEL, SACRAMENTO, CALIFORNIA)

Gabriel showed up for the final with a box of cheerios, a gallon of milk, and Styrofoam cups. He also had a poster board on which he had an elaborate menu. In his presentation he told us that as children he and his sister would often play at what they wanted to eat. Each one would yell out a favorite delicious food, and they would write it down on paper. They hung on to their paper knowing that they would not be able to have that food because their parents, who earned farmworker wages, could not afford most of those items.

Gabe described how they usually would write down their menu items and hold them up on a board to pretend as they ate their cereal that they were eating one of the coveted entrées. In his analysis, Gabriel spoke about how many farmworker families have insufficient money to purchase essentials for their families because of the low wages they are paid for their labor. Some students challenged him to become a more conscientious eater and demanded that he reject animal products (the milk) and eat only organic food. They confronted him in ways that made him think of the issue of class privilege and how his family would have to overcome innumerable barriers to comply with these demands. He thought about his family and how his parents loved him and his sister. They gave them both the best they could, given the condition of poverty suffered by the family.

Moreover, in response to his peers, Gabriel asked them to imagine how their critique might have affected him and his sister given that they were in fact poor. By the challenges his classmates posed, he realized that they were clueless about the price of food and the disparities of class relations among workers in their communities and perhaps among various groups of Chicana/os. As for the issue of animal products and organic foods, Gabriel asked them if they had ever thought that when parents buy their children milk, even though they cannot afford it, they are making decisions to provide sources of nutrition for wholesome development despite their limited income.

## 2. CHILES RELLENOS, FOR THE ONE WHO ASPIRES TO BE A DOCTOR (ROBERT, LOS ANGELES)

Robert was a transfer student and a bit older than the average students at UC Berkeley. He aspired to becoming a doctor and eventually returning to his community in Los Angeles to practice family medicine. For his assignment, he decided to observe his mother prepare his favorite all-time favorite food: *chiles rellenos de picadillo dulce* (roasted poblano peppers stuffed with sweet ground meat).

Robert was a US-born child of immigrants. His family came from Mexico City, and he was accustomed to eating black beans, *bolillos* (Mexican baguettes), and many varieties of *sopas* (a pasta and tomato-based soup). For Robert, the challenge of his assignment was learning to make perfect chiles rellenos. His mother helped him pick the *chiles poblanos* (poblano peppers), the ones with the least wrinkles so that they could be peeled easily after the required toasting. The chiles were washed, dried, and held over an open fire to toast until the seeds stopped cracking. After the toasting, he was told to place the chiles in a paper bag (or tortilla warming cloth) to allow them to sweat, which would make peeling easier work. Meanwhile he was to prepare his *picadillo* (sautéed ground beef).

> *Picadillo Dulce* (sweet sautéed ground beef)
> 1 pound of the leanest ground meat fresh from the local
>     butcher (not the supermarket)
> ¼ cup of raisins
> ¼ of a white onion
> ½ cup of blanched and slivered almonds
> a small handful of green Sevilla olives
> the freshest sweet paprika
> salt to taste
> *Salsa de Jitomate*
> 4 ripe red tomatoes to be toasted for the salsa
> ¼ of a white onion
> a *manojo* (handful) of *tomillo* (thyme)
> salt to taste

Robert told us about preparing the *picadillo* in his mother's favorite *sartén* (cast-iron pan) with very little oil because the meat would release the natural oils—simply add the paprika. Then add the onion, and cook it until it becomes translucent. Add the raisins and the almonds; then the olives. Lastly, add the salt. Taste to ensure the balance of each ingredient. Set aside.

The salsa is sprinkled over the chiles, creating a beautiful color palette of deep green from the poblanos and bright red from the tomatoes. The preparation: Grill or boil the tomatoes. Peel; add to blender with onion, *tomillo*, and salt. Puree, and pour into a glass container. Set aside.

Peeling the chiles is the task in this recipe that requires the most skill because you must be careful to take all of the peeling without tearing the roasted flesh of the chile. Sweating the toasted chiles after roasting them over the open flame creates a beautiful aroma of toasted fresh and hot, earthy, green delight. Each chile must be carefully handled while peeling and then removing the seeds and veins to reduce the possibility of spicy heat. After each chile is peeled and prepared, it is laid out to be filled with the *picadillo*.

After filling each chile, place it on a plate, add the salsa on top, and call the fortunate bystanders to share in a delicious burst of taste. Tortillas and beans are good side dishes but not necessary. The chile alone is like a gift from the gods, or in Robert's case, from his mother.

Robert explained that he chose this dish because of its difficulty and because the process made him think about the kind of student he wanted to be. He strived to emulate his mother, someone who is patient and careful to take time with each step to ensure the final perfection. His goal was to apply to medical school, and to succeed, he relied on the valuable lessons from his mother in the kitchen! Another important point is that this recipe is not the common chiles rellenos coated with egg batter and fried. Mexican cuisine presents many regional variations to food preparation, and this version is unique for its careful measure of fat content. Whether Robert's mother's cooking influenced his desire to pursue medicine is unclear. But in his choice of recipe he demonstrated attention to detail and consideration for food that gave him pleasure and demonstrated skill.

## 3. FOOD IN THE FIELDS FOR *CONVIVENCIA*: *CAFECITO* AND GOOD *CHISME* (VINCE, MADERA, CALIFORNIA)

Vince, a junior, came to my class thinking he would learn about his culture and history. Like other students, his goal was to read, pass his tests, and continue on with his studies. From the beginning, Vince relished speaking about his parents, who worked in the fields. He spoke about the moments of talk in *la madrugada* (dawn; daybreak), when his mother prepared the daily *lonches* and the morning *desayuno* (breakfast) for his father, him, and his siblings. He spoke about the smell of *canela* (cinnamon) and boiled beans and how just the thought of those things together took him back to the time of his childhood and relationship with his grandmother. She introduced him to *salsa bruta* (brute, raw salsa), which he described as having only two ingredients, serrano chiles smashed in the *molcajete* (pestle and mortar) with a bit of salt. Vince swore that this salsa was the best way to get people going in the morning. For Vince, *salsa bruta* was the most important part of a day when work was to be carried out under the sun and often under harsh conditions. He also stated that when he had papers to write and assignments to finish at school, he often wished for *tacos de salsa bruta* as a help for getting on with his task.

Vince offered three recipes for his assignment: coffee, *ensalada de fruta* (fruit salad), and *tacos de sesos de cabrito* (goat brain tacos). For Vince, the assignment demonstrated his wonderful sense of humor. Having already established that he was a child in a farmworker family, he also made it clear that he was never poor despite the family's low income. His entire clan enjoyed talking, sharing a generous sense of humor, and offering the gift of art.

Vince started his presentation by stating that coffee was a misunderstood drink that did not receive enough recognition. It is important to note that Vince was talking about this beverage before the United States had seen the commercialization of coffee unleashed by Starbucks, and was not speaking about drinks with exotic names and various sizes. Vince focused on coffee for its simplicity and was in fact referring to *instant* coffee. His recipe was hot water, sugar to taste, two spoons of coffee, and a guest who likes to *comadrear* (talk) with lots of *chisme* (gossip). Fundamentally describing a convivial social engagement, Vince continued with the following instructions:

Allow water to boil until guest is ready (to spill the *chisme*).
Add sugar and coffee while listening to petty formalities.
Sip while hot, preparing for *chisme* of the day.

Clearly, Vince understood that food is not merely about nourishment of the body but also serves to strengthen kinship and social relations. The recipe for the *ensalada* (salad) includes the following description:

This ensalada "crujiente" (crunchy salad) goes perfect with a hot day. Like the coffee, the salad must be enjoyed with company— especially with someone who has a good sense of humor. The setting is of great importance for full effect.

The ingredients are:

*coco* (coconut)
*jícama*
*pepino* (cucumber)
*limón* (lemon)
*salecita* (salt)
*hielos* (ice cubes)
*calor* (a hot day)
*arbolito con sombrita* (a shady tree)
*alguien con buen humor* (someone with a good sense of humor)

Vince reveals something fascinating: the idea that a recipe is as much about *where* it is prepared and eaten as it is about the ingredients and methods. He demonstrates that a sense of humor in the presence of company makes any condition enjoyable. He also speaks about the person whom he chose as his favorite cook, his mother. In his last recipe he indicates that all food is to be "taken with company":

Nothing tastes better than brain tacos in the fields when you are extremely hungry. Like the other two recipes, the setting and company are crucial. This dish tastes best when the one who prepared them is present, i.e., Mom.

Vicente describes his food by integrating the elements of working in the cotton fields: hot sun, dirt, and (from working hard) being in a state of *hambre de perro* (hungry as a dog). In the recipe for *tacos de sesos* that Vince presented, what struck me the most was how he listed, among the tips on preparation, references to place and people. He is careful to provide directions about the food in terms of ingredients, texture, and the inevitability that if you work in the fields, dirt is bound to land on your food, which will "naturally . . . [add] extra flavor."

## A Food-Based Pedagogy from My Home: The Curry Rodríguez Family

To conclude, I offer some ruminations on my mother's kitchen. But I want to reiterate that food preparation for us began with the consideration of the economic means by which we are defined as consumers in the dominant food economy. I wish to note that my mother's work history in the United States was primarily in the garment manufacturing business and briefly as a cook for two Mexican restaurants in Southern California. Her first job, in 1965, was for a Mexican American who owned a restaurant called Nena's. My mother worked from 6:00 a.m. to 10:00 p.m., primarily in the kitchen. For this work, her boss paid her the minimum wage of $1.65 per hour with no overtime or benefits.

Before leaving for work every morning, my mother made breakfast for my sister and me, which varied only by season. In the winter she left us a pot of oatmeal; in the hot months, a *licuado de plátano* (banana smoothie). We always heard noises and smelled the results of her cooking long before we were nourished by the products of her skills, evident in the way she used her hands. Her *avena* (oatmeal) was made in a small pot of hot water, to which she added *canela* (cinnamon) and milk. I have never thought of the gruel I first saw in the dorms as oatmeal, because my mother's was so delicious and beautifully smooth, whereas the dormitory oatmeal was dry and often lumpy and offered absolutely no delight to my senses.

I accompanied my mother to do the shopping for our household on many occasions. I always admired the conscious, patient, and delicate way she searched for the best deals not in terms of cost but rather nutritional value. Even with her limited resources my mother shopped by preparing her list of goods and then by searching the supermarket or asking for the groceries she wanted.

An especially memorable shopping adventure was when we went to the San Pedro Pier in Los Angeles. She walked up to the fishermen who were selling their catch of the day. The prices were extremely high. But my mother asked them what they were doing with the fish

Figure 10.1. Author sitting left of her mother, Elsa, and sister María. Courtesy of Julia Curry-Rodríguez.

heads. One man asked her if she wanted them. She told him she did, and he wrapped them up and handed them over. She also bought $3 worth of a white thick fish, perhaps halibut or cod? We went home with a huge package of fish heads and our white fish. At home, my mother turned the fish into a scrumptious soup, which she shared with a Yucateca family she often gave food to. She explained to my sister and me that they had greater need than we did and they had more children, so she helped them eat. This is how I learned generosity from my mother. But I also learned how to buy premium goods within limited financial means. She taught me the art of frugality. Here is her recipe for *Caldo de Pezcado con Verduras* (fish soup with vegetables):

> *pezcado en trosos* (chunks of fish)
> *cabezas de pezcado* (fish heads)
> *cebolla blanca picada* (diced white onion)
> *jitomate grande* (a large tomato)
> *manojito de yerbabuena* (a small bunch of mint)
> *sal al gusto* (salt to taste)

Allow the fish heads to make a broth with all the ingredients except the fish chunks. When the broth is dark, taste for seasoning and flavor. If the broth has a good strong taste of fish and not water, the broth is ready to turn into the soup. Remove the heads, and pour the broth through a sieve so that you are left only with the liquid.

> Ingredients (adjust your quantities on the number of people you wish to serve):

> peeled potatoes diced into large chunks
> peeled carrots cut into large chunks
> washed *calabacitas* (organic zucchini is fine) cut in large chunks
> fresh oregano

> Preparation and presentation:

> Add the vegetables and the oregano to the broth. When they are nearly cooked (just before the vegetables are soft), add the chunks of fish. Finish cooking, but don't let the fish get tough. Turn off heat as soon as the fish changes into a solid color (i.e., white if using something like halibut).
> Serve piping hot in deep bowls with a generous supply of lemon. This is a very nutritious food that is especially good for women who are pregnant and/or breastfeeding. I also learned to make nutrition part of my eating decision-making from my mother.

A food that teaches patience was my mother's *consommé*. She loves a perfect *consommé*, which she describes as needing the most meticulous attention. Her ingredients: a whole chicken cut up in quarters and with all the skin and fat removed, a whole white onion, three cloves of good fresh garlic, and salt to taste. If you have made salsa verde, save and wash the husks of the tomatillos and add them to the water since they help to reduce the foam.

The process that leads to perfection in a simple *consommé* requires a careful eye on the chicken for each time it boils feverishly. Each time the chicken boils, remove all the *espuma* (fatty foam), which includes the particles of bones and blood that could stain the broth and reduce its clarity. This process requires a careful and patient cook who must watch and gently remove the foam each time it comes up until the broth is clear and pure. The chicken cooks for close to an hour. The onion and garlic nearly dissolve and disappear. She used the chicken for other recipes, such as enchiladas, *taquitos*, and chicken salad. For me, heaven was the hot cup of clear broth my mother and I shared on a cold day. She made it especially beautiful with a relish of fresh cilantro leaves, diced onion, sliced chiles, and lots of lemon. This was one of my mother's culinary gifts and the lesson I used in eventually thinking through the best methods for handling that family class when the literature and the students could not share common outcomes.

## Conclusion: Food as Living Praxis of Labor, Skill, and Heritage

I began this chapter thinking I would share the experience of asking students to observe food preparation among their families. This way they could examine how class, gender, power, and racial classification shaped their lives and their families. My hope was to help students realize the importance of food preparation as a living labor capacity involving various elements, often performed by women, which was almost always overlooked by the students and society. This valuable, mostly women's labor is part of the family household economy and reproduces not just our sustenance but also the maintenance of cultural pride and traditions. It has the potential to become a source of our daily lived resistance against assimilation pressures.

Reflecting on the students and their assignments brought forth memories of my mother's own negotiation and decision making around food. My mother taught me lessons in her kitchen, while shopping for groceries, and by revealing how she made decisions about sharing food she purchased with her hard-earned wages. The lessons that were nurtured by my mother's food habits continue to inform and sustain my work, pedagogy, and personal well-being.

To tie this essay to the struggle to decolonize or "re-indigenize" Mexican-origin foodways, I wish to return to the idea of "authenticity" albeit by responding to the view posed by some of my friends who declared that my mother's cooking is "not authentically Mexican." How exactly can we determine the divergent forms of maintenance and resurgence of Indigenous culinary practices that millions of Chicana/os and Mexicans adopt? What if, in our everyday routines, we use our shared lived experiences within our families and communities as the ultimate authority on authenticity? We have come a long way from the dogmatic years when we insisted that focusing on class and race were the only approaches to our scholarship. In many respects we continue to reinvent ourselves while keeping essential the intersectional analyses that examine race, class, gender, and sexuality because these shed light on who we are and where we are headed, especially given the context of capitalist sociopolitical structures and power/knowledge relations that constantly assault our bodies and communities. I did not intend my class as a decolonizing effort. I did require critical, if uncomfortable, tension in the ideas the students and I rejected or kept. I was particularly concerned about those unchallenged ideas rooted in assumptions about class, race, and gender norms, which we often heard reduced to the adage, "In our tradition . . ." We learned that we have myriad traditions and many of these come from resistance and the need to reinvent our foods in the face of displacement, marginality, exploitation, and oppression.

Food is not just the lens through which the students and I examined "the family." We also used food to reflect the "living praxis" of labor and skill that helps all of us willing to partake in what Esteva and Prakash call *la comida*—as distinguished from food as eating without the conviviality dimension.[3] Given the experiences of personal and historical trauma that we have all faced, teaching through food then relates to food as a daily lived practice that helps us to heal ourselves, our family and friends, and community by becoming the means to mediate and negotiate over the conflicts and depravations experienced by our own families and communities.

# PART III · Organizing

## Decolonial Movements for Food Autonomy

# When Corn Silk Withers

TEZOZOMOC

When corn silk withers
I will have waited 100 years
for your brown feet
to cross these rows
your hands to reach
and turn the soil.

When corn silk withers
I will have waited 200 years
for manure to become phosphorous;
that we may see each other
in the dusk of day.

When corn silk withers
I will have waited 300 years
for the beekeeper to free
trichogramma wasps
to eat the earworms
embedded in the cranium.

When corn silk withers
I will have waited 400 years
for your eyes to see the splendor
of emerald green
for the rustle of corn wings

to pass in your dreams and deliver
the sweet milk of tender corn nubs.

When corn silk withers
and ears hang full
it will have been 500 years.
and if we wake up
the food of our souls
will be the spread
of rainbow Inca,
Black Aztec,
True Gold,
Apache Red,
Hopi Pink,
Hopi Turquoise,
Hopi White,
Taos Blue,
Oaxaca Green Teosinte.

But if we miss the shelling,
Corn silk will wither,
full ears will hang low,
skin will turn yellow-tan
to white and corn
billbugs will eat our poriferous brain.

When corn silk withers.

# CHAPTER 11

# Fragmentary Food Flows

## Autonomy in the "Un-signified" Food Deserts of the Real

### TEZOZOMOC AND THE SOUTH CENTRAL FARMERS

## Prelude

*It is not so simple to flee from a system of control that is built to absorb escapes.*

—MARK PURCELL, *The Down-Deep Delight of Democracy*

*When the world ends, it will be like when the names of things are changed during the peyote hunt. All will be different, the opposite of what it is now. Now there are two eyes in the heavens, Dios Sol and Dios Fuego. Then, the moon will open his eye and become brighter. The sun will become dimmer. There will be no more differences. No more men and women. No child and no adult. All will change places.*

—Huichol proverb, as quoted by Ina Woolcott (2015)

# Post–South Los Angeles Uprising

After the 1992 South L.A. uprising, in a warehouse area south of downtown Los Angeles, a group of more than 350 poverty-stricken local residents started transforming a rat-infested, garbage-strewn fourteen-acre vacant lot into a world-famous and widely celebrated oasis of biological and cultural diversity. Over the next seventeen years the mostly Indigenous and mestiza/o people working garden plots at South Central Farm (SCF) coalesced into a unified food justice movement while establishing the largest self-organized urban community farm in the United States. The farm, the only green space in a low-income inner-city neighborhood surrounded by warehouses and junkyards, hosted farming methods, crops, and wild relatives among a diverse mixture of people, many of them part of the post–green revolution and post-NAFTA diaspora of displaced Mexican and Central American Indigenous farmers. This iconic episode of the environmental and food justice movement has profound implications for our understanding of the politics of space, place, and radical subjectivities in the genealogy of alterNative agency.[1]

On the surface, the South Central Farm was a place where people could grow cherished family heirloom crops, medicinal herbs, and sacred trees. Look a little deeper and it becomes evident that the farm was also a place where local residents could actively reclaim the right to urban space as a "common"—an *ejido* or *un terruño* (patch of native land). Entire families, including the elderly, youth, and children, could escape the concrete straitjacket of the surrounding city simply by sitting and visiting under a fruit tree as they sipped homemade cold herbal teas and listened to relatives and friends share stories about the family, the besieged origin villages, and the grandmothers' herbal remedies during the heat of the summer day. Through such everyday lived experiences, the farm became a space where multiple generations participated in the passing of knowledge of all our relations, of plants, land, water, and sun (climate), from parents to children. A vacant, rubbish-filled grid space produced by neoliberal enclosures was remade into a situated place filled with material and symbolic culture artifacts, practices, and mythological placemaking. This heterotopia emerged from the recollecting and sharing of ancient memories and the transmission of knowledge. This was accomplished in a man-

ORGANIZING

ner first established by ancestral Azteca (Mexica, Nahua), Chontal, Hñähñu (Otomí), Maya, Mixteca, Triqui, Zapoteca, and other heirs of the originary Mesoamerican civilizations that were present in our diaspora urban *ejido*. The farm was our *calmecac*, a Native institution of higher learning. It was also our zone of escape from the neoliberal capitalist regime that is the latest iteration of white settler colonialism. Such accomplishments are seldom left to their own autonomous development and usually become transient heterotopias, so our decade-long persistence was unusual.

It is hardly coincidental that a survey of ethnobotanical varieties by Devon G. Peña between 2003 and 2006 found that the South Central Farmers hosted a minimum of 150 different species of fruit, vegetable, grain, medicinal plant, and tree varieties, most of them indigenous to the greater Mesoamerican "center of origin," as described by Nikolai Vavilov.[2] The bounty ranged from avocado and banana to the "Three Sisters"—corn, bean, and squash. People were growing multiple varieties of these as well as cactus, chile, chayote, guava, mango, sugarcane, walnut, and a bewildering variety of herbs and decorative floral plants. This agrobiodiversity was realized through a specific spatial organization of each garden plot with one recurring pattern consisting of multicropping and "stacking" (the use of vertical, not just horizontal, space). Companion planting was evident in the patterns Peña documented, and this led to his characterization of the farm as a world-class ethnobotanical and agroecological landscape mosaic (see figure 11.1). The original study, conducted in 2003–6, explained how this particular group of Mesoamerican diaspora farmers were reproducing the home kitchen gardens of their native lands. These farming spaces were products of "mobile place makers," Peña argued, expressing a view more recently confirmed by other scholars, like Leslie M. Johnson and Eugene S. Hunn and Pierrette Hondagneu-Sotelo.[3] Peña first introduced the term *autotopography* and invoked Michel Foucault on heterotopias and gardens in a lecture at University of California, Berkeley, in 2006 to define this as "self-telling through place-making,"[4] but he left mostly unsettled the matter of how this subversive land-shaping activity was linked to institutions of collective action that emerged from the farmers' efforts to organize and govern themselves and to defend their land in robustly ever-shifting forms of reterritorializing subjectivity.

Figure 11.1. *Huerto familiar* with nopal fencing at South Central Farm, June 2003. The SCF plots followed patterns of perennial and annual intercropping evident in Indigenous origin communities. Courtesy of Devon G. Peña.

In a subsequent study, Teresa Mares and Peña noted the presence of a sacred tree, the *pochote* (*Chorisia speciosa*, silk floss tree) and described it as being representative of the farm's agroecological diversity and its distinctive "spiritual ecology."[5] In this interpretation, the mobility of a capacity to deploy a deeper sense of place among displaced land-based peoples is verified by the fact that the pochotes growing at South Central Farm carried the exact same signification they possess in the Mixteca Alta, México; at Chalma the tree's spiky green trunk is celebrated in myth and legend. The Nahuatl name for the pochote considers the dragon-serpent deity brought forth by the four forces of Tezcatlipoca ("The Mirror That Exhales") and names her *Cipactli*, which translates as the "Tail of the Mother Earth Lizard," and this is the name of the pochote in Mixtec communities. It was also the name assigned to the tree by the displaced Indigenous farmers in L.A. and was part of their will to power through the symbolic politics of "naming" to reterritorialize space.

The pochote is in this semiotic sense more than a mobile biological artifact and signifier. Instead, it accompanies the movement north of an entire Indigenous cosmogony understood as a "world in constant motion," which is the quality brought forth through placemaking narratives that unfold in particular "out of place" locations. In this manner, the agroecology and spiritual ecology of the South Central Farm represents an alterNative epistemology and ontology—evolving different ways of knowing and being in the world yet rooted in the ancestral indigenous histories of displaced peoples. This epistemological capacity for *re-emplacement*, indicative of a peoples' cultural resilience, figures prominently later when we consider the problem of neoliberal sovereignty and Indigenous autonomy in the context of a discussion of the politics of de- and reterritorialization.

Between 2003 and 2006, the farmers' struggle to save the farm from being bulldozed became a cause célèbre among the media, the broader public, and people of note in the United States and around the world.[6] The 2009 Academy Award–nominated film *The Garden* documented one vision of the farmers' story and the tragic loss of the farm to bulldozers in 2006. Outspoken supporters on our behalf included actors Leonardo DiCaprio, Danny Glover, Daryl Hannah, and Alicia Silverstone; singer and musician Ben Harper; singer Willie Nelson; elected officials, like Maxine Waters; and numerous distinguished research scholars and dedicated graduate students from a wide range of universities and colleges, including California State University, University of Southern California, and University of Washington in Seattle.

## A Farm in the "Republic of Property"

We argue in this essay that the struggle for control of the fourteen-acre common known as South Central Farm was essentially a conflict over the administration of public planning policies that define and establish a hierarchy of privilege in property rights in the context of environmental and food justice struggles resisting neoliberal enclosure and privatization of urban spaces. *Los Angeles Times* journalist

Al Martínez noted in an editorial opinion piece: "The value of property outweighs the needs of the people. It is the philosophy of kings to amass and armies to protect their possessions."[7] This tension between property as relation and property as possession was at the heart of the SCF saga.

On June 13, 2006, more than forty-four activists and farmers were arrested as the authorities moved to seize and close the farm with an army of 385 sheriff deputies and LAPD storm troopers. All 350 farm plots were bulldozed on July 5. This brutal act of structural violence targeted families and the land. The character and tactics of the LAPD, L.A. County sheriff, and antiterror units that were deployed against the South Central farmers anticipated the patterns of increased militarization of police that has been rendered more visible as a major public policy problem and noted with the rise of the Black Lives Matter, Standing Rock, and related social protest movements against police violence and mass incarceration of black and brown bodies.

In this essay we seek to understand why and how such acts of state violence are fomented and tolerated under the current neoliberal regime for the administration of property rights in urban planning. The eviction occurred despite the fact that a group of us had worked hard to secure a pledge of $16 million to acquire the land as a community urban farm trust.[8] The titled owner, Ralph Horowitz, declined our offer to purchase the parcel through a land trust. In the film documentary *The Garden*, Horowitz claimed that "these people were not the kind of people for America. How dare they demand that I sell the property." The threat narrative that emerged from Horowitz and his defenders completely racialized the discourse, and this resignification was deployed to foment a flippant dismissal of the farmers, who were portrayed by right-wing pundits as "illegal squatters."[9] These pundits failed to recognize the irony of attacking the farmers for displaying what they misrecognized as excessive arrogance (chutzpah?) for simply demanding the exact same opportunities so easily granted Horowitz because of his "white privilege"—which was evident in an earlier backroom deal for the land with the city for a mere $5 million price tag, which he then quickly revalued at $16 million right before the subprime mortgage crisis of 2007–8. What is clear is that the farmers experienced the eviction as violent dispossession and a direct consequence of an asymmetrical juridical and discursive process that privileged private enclosure over community public health and welfare.

The June 2006 eviction was not our end but the beginning of a new stage of political reorganization. The story of our continued existence, resilience, and development remains largely untold. To keep farming, we adapted and seized an opportunity to work in Fresno when land to lease was offered. For a year, starting in the spring of 2008, we commuted four hours round-trip from L.A. to Fresno every Friday night after our full-time urban jobs. We piled into an old school bus and never missed a weekend for a year. To reduce travel time, we found other land to lease closer by, in Bakersfield, and have been sharecropping there ever since. Water for irrigation is of course a major challenge, and for us one of the first key moments in our redeployment came in 2009 when we secured our first irrigation pump to irrigate farmland in Buttonwillow, California. The effects of the current drought present a set of major challenges for us that are beyond the scope of this essay. The drought is directly affecting the cooperative farm as is evident from the increased acreage we have left fallow. We are also implementing the use of drip irrigation technology and plan to increase the varieties of drought-resistant heirloom non-GMO crops.

Today, in our new home some two hours north-northeast of the L.A. Basin, we farm on eighty acres; also, in Arvin, California, we lease a one-acre modern greenhouse surrounded by fifteen acres, and in Lake Hughes we have leased an additional forty acres of land with proper drip irrigation to save precious water. We also collaborate on research and education programs on organic cooperative farming with a wide range of groups, ranging from Harvard researchers to local minority groups. We have created a vibrant Community Supported Agriculture (CSA) program with thirty-three drop-off points and over three thousand customers in Los Angeles and Kern Counties; our food distribution reaches more than forty thousand Angelinos a year. Most important among our program work is our "Food for the 'Hood" project through which we provide fresh organic produce to low-income inner-city residents of Los Angeles where it cannot otherwise be found. We are donating organic produce every month to community organizations, like the Revolutionary Autonomous Communities (RAC), that are directly working with the homeless and underserved families. As of 2014, SCF had donated over twenty thousand pounds of fresh vegetables.

## Bringing Food to the 'Hood

The plan for the Buttonwillow farm and our cooperative is unique. We are following deep-rooted models used to create self-sustaining, small, organic farm cooperatives around the Central Valley. The project goal is to replicate the South Central Farmers' business model to educate local residents in low-income areas and provide employment to under- or unemployed farmworkers. Co-op members are not just given jobs. We are all given the opportunity to work at a multifaceted career that provides paths to an improved quality of life and a successful future as a member of an extended community and its multileveled exchange and action networks. There is evidence of "strategic essentialism" in what we do: Our model can be misread as a case of neoliberal self-care because our approach can permanently shorten lines at unemployment and public assistance offices and presumably create productive tax-paying members of the community. However, this is not individual action aiming to fatten the tax coffers of the state or reduce the obligations of a social safety net. It is instead a form of self-organization that helps farmers help themselves through their own traditional knowledge and autonomous institutions of collective action and mutual aid, which is something their ancestors have been doing since well before the advent of settler colonial invasions.

Our underlying goal is to support a localized and autonomous food economy in order to eliminate hunger and malnourishment in California. This is multifaceted and involves building a market for locally grown food; creating food equity across California, especially by focusing on underserved communities; growing and processing food in our communities; and educating, innovating, and organizing "food champions." Organizationally our current focus is on strategic growth by creating critical mass to support the goals of the cooperative and our nonprofit's mission, and a refocused energy directed toward participating in the growth of local urban agriculture networks.

Lessons learned have shown us that if we don't have our own marketing and distribution channels and control of the means of production, we will always fall into the "anarchic" market conditions throttled by corporate distributors who create a poverty of deprivation and coerce producers to sell low because they cannot hold out any longer. Our commercial kitchen will include dehydration services in a certified wholesale commercial facility. There is a high-speed bagging line available to other affiliated producers to take the products to market as consumer-packaged goods. A certified cannery will allow for co-op producers to process their products into consumer-packaged goods that will be organically shelf-stable for distribution. These vertically integrated services will focus on creating and focusing on healthy products, natural products, and organic processing. Our efforts have been recognized by organizations like the Natural Resources Defense Council (NRDC). They awarded our group the 2013 "Growing Green Award for Social Justice."

## On Being Grown out of Place

Noting the diasporic quality of the agroecosystem at South Central Farm, Peña once observed that the farmers were growing many varieties of landrace crops that were no longer being cultivated in the origin villages in México due to the displacement of farmers by the violence of NAFTA and the intensification of the neoliberal narco-state complex.[10] Many of the landrace varieties of Vavilov's Mesoamerican center of origin—corn, bean, and squash among them—were now being preserved through the farming practices and seed saving of Indigenous people working small home kitchen garden plots in urban and peripheral urban spaces up and down the US West Coast from the L.A. Basin to Anchorage, Alaska. These Indigenous gardens are a transborder decolonial phenomenon. The act of planting Mexican heirloom crops in the United States revitalizes the capacity for self-expression and self-governance and opens the possibility that such a simple act can be activated in sustained political struggle generated by the embodied daily life experiences of these displaced farmers who seek only to remake place out of newly encountered spaces, or as Johnson and Hunn phrase it:

throughout human history, the phenomenon of displacement, delocalization, and relocalization have been pervasive and important, and migration of individuals and populations, voluntary or not, is a hallmark of our times. Thus we must consider the effects and consequences of removal from Place, of becoming delocalized, and of (perhaps), relocalization as migrant populations settle, set down roots so to speak, and develop a new relationship to place.[11]

This insight is more politically significant than has hitherto been discussed. The destruction of heirloom crop lines in México is a direct result of the dispossession of Indigenous farmers by the triplex forces of neoliberal capitalist enclosures including NAFTA, the associated wave of narco-state violence, and the spread of GMOs. This destruction of heirloom seed lines proceeds in mostly invisible form (to the mass public) but remains a slowly advancing ecological and sociocultural catastrophe that is almost unparalleled in scale in recent Mesoamerican history and is certainly on par with the threats posed by GMO crops (see chapter 15, this volume). In 2006, Peña proposed that the rise of transborder seed savers is illustrative of the resilience of Indigenous farmers who are learning to grow their native crops *out of place*. Resilience is the ability to bounce back from disturbance and displacement. It is the ability to reground Indigenous cultures as a whole way of life through our conscious political projects and collective actions. Notably, the ability to exercise this capacity for resilience is not an automatic driver of our existence, and our agency is constantly undermined and blocked by the mechanisms of neoliberal governmentality and its regimes of resignification of value *in and of property*.

This is good but so far leaves unexamined the organizational adaptations that had to be part of the process of innovation that we engaged in order to *re-emplace* our community-based food system in a milieu that imposes a variety of structural barriers related to race, class, and citizenship distinctions that are deployed to constantly thwart and block our autonomy. To grasp this problem, we need to understand how it was that the South Central Farm served as a major focal point of the widening conflict between neoliberal privatization and enclosure of space and the regrounding of the common through the praxis of *reterritorialization* from the bottom up.

## De/re/territorialization: Heterotopias and
## Subjectivity of and without Place

The term *de-territorialization* first occurs in French psychoanalytic theory to refer, broadly, to the fluid and dissipated nature of human subjectivity in contemporary capitalist cultures.[12] The most common usage has been in relation to the process of cultural and economic globalization. While different inflections are involved, a common theme and general implication are that globalization needs to be understood in cultural-spatial as much as institutional and political-economic terms. In this broad sense, de-territorialization has affinities with the idea of the "disembedding" of social relations, as is argued, for example, by Anthony Giddens in his analysis of the globalizing properties of modernity.[13] This misleadingly implies the end of things that are purely "local" or "localized" and that the process of the spatial inscription of globalization removes the ground for a subject to position herself in place as an oppositional force.

Misusing Marx, the so-called Third Way argument presumes all that is solid in relation to place dissolves and is replaced by the constant recodification of space as a globalized postmodern neoliberal pastiche always subject to the rules of the commodity form, or if you will, to the logic of what Antonio Negri, challenging Giorgio Agamben, calls the state of *economic* exception.[14] There is no real agency (or subject) in Agamben's reduction of bare life to bodies deprived of access to the use values required for life itself. For us, agency is a latent property of the subject and is perceived and perhaps enacted only by virtue of the multiple shifting and evolving subjectivity of the self in relation to others in a field of power/knowledge relations. For capital, the only form of agency considered a naturalized quality of what being human means is that no one, including government, can interrupt or prevent the individual from realizing self-interest through the pursuit of profit. The neoliberal narrowing of possible ontological horizons presents the specter of the de-subjectivation of alterNative rationalities because it denies us access to our search for multiplex subjectivity as a deliberative and relational practice. Subjectivity is valued as a reiterative and reflexive *process of becoming* rather than an inflexible and fundamentalist commitment to only one way of being human. It seems as if as soon as we start enunciating alterNative principles and logics for due process and equal justice, the philosophers and jurists declare an end to the matter of the underlying epistemology of law. Under the dominant construct, only the subject as lone rational actor remains; only selfish egos may exercise freedom in the market protected by inviolable contractual obligations enforced by the juridical order.

## HETEROTOPIA INTERRUPTED

The hegemonic organization of space requires a regime that renders all places into grid spaces that can be rendered legible to the locus of power in the administration of property. Space must be legible to the juridical eye in order to police boundaries and enact and enforce zones of exception and enclosure. Violent acts of enclosure continue to impose the rule of law based on the commodity measure of value, and this creates the conditions for the universal imposition of the demands of a rather small but peculiarly coercive, possessive, and acquisitive social class.

The many forms of heterotopia present today are perhaps fleeting counterhegemonic moments as conceptualized by so-called human geography in departure from the original elaboration by Foucault to describe places and spaces that function under nonhegemonic conditions. These are sometimes spaces of "Otherness" or "liminality" and can also be read as "interstitial" or "betwixt and between," places neither here nor there yet everywhere as a possibility all at once. Heterotopias can erupt in unusual places and perhaps most subversively in the form undertaken by diaspora Indigenous communities. These are subversive because they are not always rendered legible to the administration of space as enclosed property.

An interesting passage for us in Foucault's essay on heterotopias is the last sentence: "In civilizations without boats, dreams dry up, espionage takes the place of adventure, and the police take the place of pirates."[15] When we think back to that moment of eviction in the hot summer of 2006, we see that the police deployed an overly militarized force that we have now come to associate with the post-Ferguson state of emergency. After Foucault, the irony is that at South Central Farm there actually *was a boat*, our little lifeboat, the farm itself. And indeed it also came to pass that pirates besieged this lifeboat. Except here the police were actually fulfilling the roles of master gunners and powder monkeys while the privateers behind the raid hid, relatively unscathed, to enjoy their loot in secure high-rise office lairs.

This is a good point at which to bring in the work of Maurizio Lazzarato, a theorist concerned with subjectivity, who notes, "Today's 'police' operate through both the division and distribution of roles and the allocation of functions and through the injunction to conform to certain modes of life."[16] Lazzarato reasserts our earlier claim of the current neoliberal crisis as a process creating the types of conditions that led to the eviction of the farmers under a juridical order that will *always* seek to protect the owners of property. In doing so, it reveals the raw unfiltered reality of the regime presupposition of property as possession as the only meaningful form of value orientation under the law: "The current crisis now produces only negative and regressive subjections (the indebted man), and capitalism is unable to articulate production and the production of subjectivity other than by reasserting the need to protect the owners of capital."[17]

According to David Maidan, Lazzarato is merely amplifying Gilles Deleuze and Félix Guattari's differentiation of two *dispositifs* (apparatuses): social subjection and machinic enslavement:

> "Social subjection" is the way in which we are fitted with an identity (sex, body, profession, nationality, etc.). This is what makes us individual subjects. We also experience a process of de-subjectivation—which acts both at the pre-individual and supra-individual levels—that dismantles the individual subject. Lazzarato calls this process "machinic enslavement," a concept that generally refers to a configuration in which two or more devices are set up in such a way that one has unidirectional control over the others. Lazzarato understands "machinic enslavement" as a relationship in which a subject becomes "the slave" of a machine. He writes: "Capitalism reveals a twofold cynicism: the 'humanist' cynicism of assigning us individuality and pre-established roles . . . in which individuals are necessarily alienated; and the 'dehumanizing' cynicism of including us in an assemblage that no longer distinguishes between humans and non-humans, subject or object, or words and things." "Social subjection" and "machinic enslavement" are objective processes, not ideological distortions of reality.[18]

Given this "machinic" destabilization of subjectivity, the usual post-structuralist approach to the concept of the heterotopia has resulted in the celebration of rather fleeting ontological experiments that are simultaneously physical and mental, "such as the space of a phone call" or "the moment when you see yourself in the mirror." We see beyond such silly notions how the heterotopic agency we exercised was politically mobilized *as a more permanent and permeable occupation of space*, one that can be a single real place juxtaposing diverse elements from multiple spaces. An urban garden or farm is a heterotopia because it is a real space and a microcosm of different environments with plants from around the world in association with the sets of knowledges, beliefs, and practices that sustain such a configuration. This describes our ontological project. It had a nemesis in the form of the dominant regime, which privileges property as possession and dismisses the demands of the multitude for the restoration of the common.

## PROPERTY AND POWER

In one dominant Western tradition, power is routinely presented as the grounding concept and property as its derivative. S/he with power owns things. An alternative tradition was undertaken to reverse this order of logical and generative priority. The "eco-nomic" tradition has consistently rejected an allegedly groundless or abstract notion of power in order to locate this absent ground in property, albeit in the last instance, whereas the "political" tradition has stood matters on its head, arguing that property is itself in need of grounding, since it ultimately constitutes a form or function of power. This is the argument put forth by Dick Pels in his treatise *Property and Power in Social Theory: A Study in Intellectual Rivalry*.[19] According to Pels, by the time of the eleventh-century rediscovery of Roman jurisprudence and the subsequent revival of Aristotelian political philosophy at the newly founded universities and clerical schools, the West had created the "liberal fissure."[20] In the course of an extended process of semantic decomposition, the inclusive feudal conception of dominion or domain gradually split up into a political and an economic compartment (see figure 11.2). As rights of property were defined in a more absolute and exclusive manner, they were ever more clearly

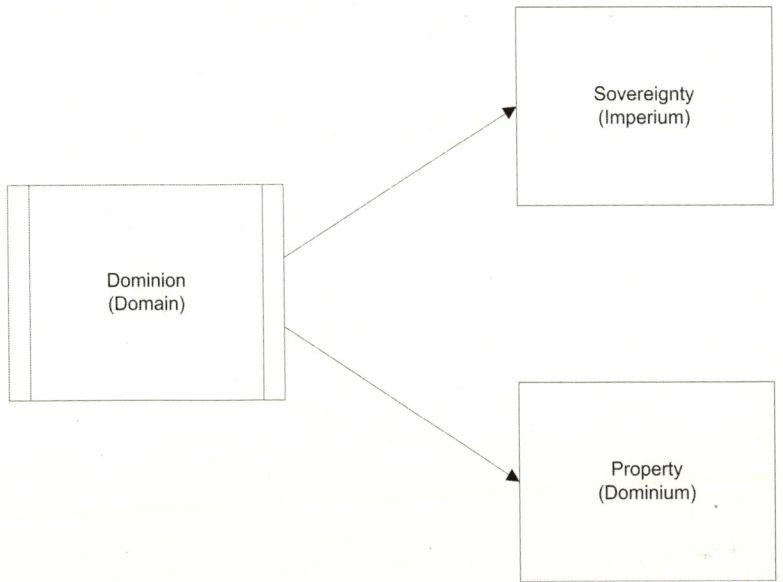

Figure 11.2. The Liberal Dichotomy and Its Dissolution. This diagram illustrates the fracturing of the medieval concept of "Dominion" into Sovereignty (politics) and Property (economics). The binary is false and serves to obscure a regime of signification in which sovereignty is subsumed to the exercise of private property rights. Adapted from figure 2 in Pels 1998, 23.

demarcated from and profiled against rights of sovereignty, which were subjected to a process of concentration and "substantialization" mirroring essential features of the parallel institutional fortification of property rights.

The consequences of this underlying process are discussed by Michael Hardt and Antonio Negri in their book *Commonwealth*, in which they designate the resulting structure as a "Republic of Property." In this social formation, the exercise of power is wed to the juridical-shaping agency of the owners or possessors of property. As they note:

> Said differently, the political is not an autonomous domain but one completely immersed in economic and legal structures. There is nothing extraordinary or exceptional about this form of power. . . . Property, which is taken to be intrinsic to human thought and action, serves as the regulative idea of the constitutional state and the rule of law. This is not really a historical

foundation but rather an ethical obligation, a constitutive form of the moral order. The concept of the individual is defined by not being but having.[21]

This illusive quality of the division of constituted power—between the economic and the political—is a ruse, which leads Hardt and Negri to surmise that the republican form of constitutional government is but

> one specific definition of modern republicanism that eventually won out over the others: a republicanism based on the rule of property and the inviolability of the rights of private property, which excludes or subordinates those without property. The propertyless are merely, according to Abbe Sieyes, "an immense crowd of bi-ped instruments, possessing only their miserably paid hands and an absorbed soul."[22]

Out of this fissure is spawned the contemporary neoliberal fundamentalist tenet that individual greed is good and that this peculiar form of freedom alone can maximize social and ecological welfare. The privileging of property rights over social and biological needs (the excess of life) is posited as an inviolable right and the wellspring for the exercise of *power over things*—that is, property or, more broadly, the all-encompassing commodity form including commoditized bodies, genes, rDNA sequences, RNAi insecticide formulas, financial derivatives, and their algorithms, etc., etc.

## PROPERTYLESS/POWERLESS

Following Pels, we can further note the three elements of an illusive disjunction between "propertyless power" and "powerless property" that inscribe this signifying binary: (1) power exercise is (active) doing, while property exercise is (static) having; (2) power is about persons, while property is about things; (3) power is shared, multiple, and limited, while property is an absolute and a zero-sum construct (see figure 11.3).

The final concept to take from the model proposed by Pels alludes to the idea of dominion abstracted into this binary fracture, but this also creates a recursive regime of abstraction. We propose that this process of abstraction continues all the way through current capitalist signifiers, like collateralized debt obligations (CDOs) and credit

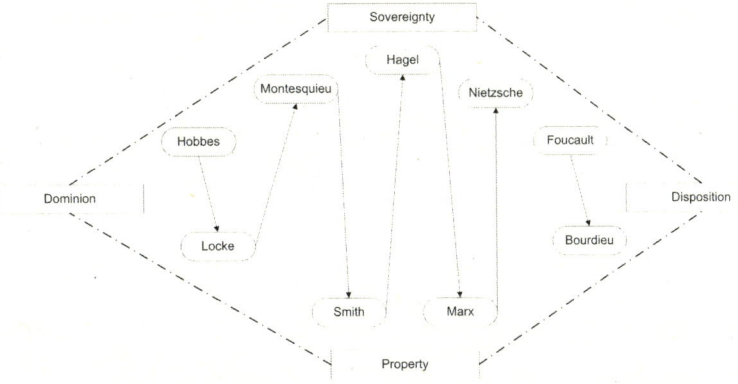

Figure 11.3. Power as Relation or Power as Property. This diagram illustrates the progression of intellectual rivalries charting the course of the liberal binary and its dissolution, which for us finds its most thorough and clear enunciation in Marx, who is first to insist that the political and the economic spheres are actually one and the same since all production is social and subject to conflicts over the disposition of rights to property. Adapted from figure 4 in Pels 1998, 48.

default swaps (CDSs)—that is to say, a regime in which everything is subject to the floating signifier of the commodity relation since "surplus value can be manufactured out of thin air, at near the speed of light" without a connection to real productive activity, as Peña recently quipped in reference to Marazzi's theory of the violence of finance capital.[23] This also reveals how the redeployment of property as possession, which currently forms the basis for the juridical interpretation of actual contract disputes in the disposition of property rights, is really also necessarily a regime of signification.

REGIMES OF SIGNIFICATION

The idea of regimes of signification comes from the political economists of the so-called Regulation School. Their notion of "regimes of accumulation in regimes of signification" involves study of the meaning of the cultural objects produced by capitalism. For Scott Lash and Jonathan Friedman, all regimes of signification comprise two main components: a specific "cultural economy" and a specific mode of signification, the latter being the mode in which cultural objects are defined and located in a particular relationship between signifier, signified, and referent.[24] If the configuration of property rights is the exhaustive signifier defining the limits of the domain of the law, then this itself seems a case in which the regime of accumulation expresses political dominium through the gross privileging of a peculiar cultural object that is erroneously assigned standing as the universal standard of due process before the law in the settlement of property (space, place) disputes.

## UN-SIGNIFIED "FOOD DESERTS"

Under this regime, the existence of "food deserts" presents a dilemma because the food autonomy movement refuses to submit to such nonsensical semiotic closure. It is not so easy to reify structurally induced hunger just because it is embedded in the finance nexus of governmentality and its regimes of accumulation/signification. Through a decolonial lens, the spaces of systemic hunger reveal how the bare life is an everyday form of structural violence against the propertyless—those "miserably paid hands" and presumably "absorbed souls" rather than some unusual state of exception.

Finance capital thrives from reification, and everything must appear to be other than what it really is. This constitutes the deceptive machine of signification deployed to sustain the reign of the commodity form. Reification involves the process of getting people to believe facades and veiling simulations instead of the actual performance of any given act of violent expropriation and exploitation. Capital wishes to render food deserts "re-signified" by presenting these as random occurrences "innocently" reflecting the freedom of consumer choice in free markets. To us, this constitutes a hegemonic discursive strategy involving what Terry Eagleton terms "systematically distorted communication,"[25] which can be deployed to conceal underlying power relationships and intentional class violence. In the undistorted discourses of decoloniality, food deserts are "un-signified" because these enunciate hunger as nothing less than a dietary form of class warfare—these are our food deserts of the real.[26] As Peña states in the aforementioned blog post, "Vampire capital sucks the blood out of the material world and the poor get hungrier." In this context, hegemonic resignification requires reification, but our refusal to submit to hunger in spaces literally stripped of meaningful access to healthy foodstuffs reveals that you cannot hide systemic injustice behind the veil of free-market behavior. Capital faces the challenge of our epistemic refusal and strategic "delinking" from the machinery of capture. You gave us food deserts? We refuse to eat there. Instead we are creating autonomous spaces that lie beyond the reach of this commodity relation to food.

In this manner, decolonial subjectivities perform a wide range of subversive acts that can evade capture and resist the tactics of de-subjectivation, as in the case of the ruse that food deserts are merely reflecting consumer preferences; this is an ontological position that tries to reduce the multiplex human being to one dimension, captured by the naturalized disconnection of individuals free to exercise "rational" choices in a marketplace of similarly reified objects. What is the human, then, in the un-signified food deserts of the real?

Decolonial scholars Jarrett Martineau and Eric Ritskes provide a way for us to appreciate the alterNative epistemic refusal and ontological project of the South Central Farmers.[27] Let us recall the Mares and Peña's discussion of the revelatory episode at SCF that replaced wire-mesh fences with edible cactus rows. The authors note that this was done in direct defiance of the regional food bank seeking to impose a regime of spatial governmentality over the farmers (see figure 11.1).[28] The regime of accumulation segregated us into food deserts; then the liberal do-gooders tried to segregate us from each other inside the farm. What we take as relevant to our rebuke of such repeated enclosures is from Martineau and Ritskes, who note the connection between mind, land, and place in revolutionary indigeneity:

> While we see so-called epistemic disobedience and de-linking as necessary and important aspects of decolonizing art and struggle, we choose to anchor our use of aesthetics within the artistic praxis of artists and communities engaged *in material struggle* for decolonization—those that connect to Fanon's (1963) call to liberate "land and bread" (p. 44). As anticolonial activist-intellectuals like Amilcar Cabral have warned, if all we seek is decolonization of the mind, then we will have already conceded the loss of the most precious and transformative foundation of decolonization: land and place.[29]

Being resistant to gross reification, our brutal experience with un-signified food deserts de-legitimizes the narrative regimes of progress, modernity, and so-called postcoloniality. Food deserts stubbornly remain the ungrounded spaces of hunger inside the material reproduction of marginality and the bare life. However, our alterNative epistemic refusal to accept the regime of accumulation as signified by the enclosure of the farm led to a different sort of escape from the violence of what was essentially an act of second-motion colonial dispossession. In this manner, we challenged the food deserts of the real that surrounded our bodies and communities.

## Emerging Shifts in the Regimes of Re-Signification

We continue to build a food autonomy movement by relying on sets of knowledge, belief, and practice rooted in our constituents' revolutionary indigeneity. We continue this work fully expecting that the regime of neoliberal capitalist governmentality will throw up new barriers and challenges as it sets about the task of reconfiguring and resignifying dominant-subordinate subject positions to prevent our escapes. The organizational forms we deploy must remain *fluid* by being in all places and no one place in particular simultaneously if we are to enact sustainable (permanent) alternatives now rather than always merely reacting to the shifting terrains of struggle issuing from neoliberal capitalist control strategies.

One recent development that challenges our fluidity involves legislation in California known as AB 1871, a bill initially sponsored by Senator Roger Dickinson and approved by a vote of seventy-five to two on August 25, 2014. The bill involves a set of amendments to existing laws on "Agricultural products: direct marketing: certified farmers' markets." This bill would make it unlawful for any person or entity, or employee or agent of that person or entity, to make any statement, representation, or assertion relating to the sale or availability of agricultural products that is false, deceptive, or misleading, as specified, and would make a violation of those provisions a misdemeanor punishable by imprisonment in the county jail not exceeding six months, by a fine not exceeding $2,500, or by both the fine and imprisonment.

By creating a new crime—literally of daily activities that many of us in the food justice movement emulate—the neoliberal governmentality regime is using legislation to impose an unfunded mandate on municipal localities. Even more troubling is that this creates a perverse incentive for additional policing and surveillance of food autonomy networks that could easily devolve into new attacks on and racial and class profiling of food justice activists and the multitude organized in formal and informal networks of food exchange, sale, and barter, all of which are often conducted in public spaces. Neoliberalism in effect seeks to regulate public space via a regime of signification that privileges the owners of private property over the propertyless, and the documented over the undocumented by feigning concern about consumers' right to know and fraud and misrepresentations of agricultural produce. If the politicians really want to work against fraud and misrepresentation in public spaces where food is bought, sold, and shared, they should require labeling of GMO ingredients.

## CAN AN URBAN FARM BILL HELP TRANSFORM SOUTH L.A.'S EMPTY LOTS?

A recent analysis for the Annenberg Media Center focuses on the 2014 "Urban Agriculture Incentive Zones Act" (AB 551) and addresses how the City and County of Los Angeles plan to implement this state legislation. Curren Price and Felipe Fuentes introduced a council proposal in 2014.[30] The ordinance would provide property tax adjustments for private landowners who convert vacant plots into "urban farms." The Los Angeles City Council's proposal follows AB 551 by defining these as "commercial ventures that sell food." The authors of the council bill explain that they see the property tax adjustment as a way to encourage landowners who are not using their property. Parcels of land between 0.10 and 3 acres in size would be eligible for the tax breaks. The definition of urban farms as "commercial ventures" is itself likely to become a contested concept from the vantage points of food justice activists and analysts.

From our perspective, the proposed ordinance basically acknowledges that the South Central Farmers were on the right track when we claimed, from 2003 forward, that Los Angeles should support acquisition of urban space for agriculture. Our basic mission integrated a unique land trust model with participatory governance and agroecology. We argued this could be emulated on a larger scale. One challenge to the current proposal, which basically follows the conservation easement tax credit model, is that we know little about the composition of eligible landowners. Landownership is becoming more diverse. Since the post–War World War II period, ownership among Latina/os and the people of the Mesoamerican diaspora has gradually increased. Years ago Mike Davis noted that a half million or so Zapotecas living in the L.A. Basin were acquiring apartment buildings and investing collective resources to rehabilitate and renovate communal living quarters.[31] These are the L.A. version of ancestral *vecindad* urban forms.[32] Are there vacant lands associated with these same diaspora communities? We don't know.

Latina/os and Mesoamerican diaspora arrivants are acquiring property instead of remaining confined to the rental sector, but we really do not know how much of this property is vacant or amenable for use under the L.A. tax credits-for-urban-farms schema. The "propertyless" are slowly becoming "propertied," and this seems significant since the epistemic notion of the ownership of property among members of the diaspora communities is defined as an act of relation rather than possession and proceeds on the basis of communal asset-building values rather than individual acquisitiveness.

We can face the challenge of documenting the varied forms of race/ethnic and national-origin ownership of vacant land in the area, but the ordinance leaves unclear the methods for determining where and how the most vulnerable and needy communities will find the substantial monetary investment required to allow vacant land to become farmland while complying with the regulatory apparatus that will screen applicants for eligibility for the tax credits. As it currently stands, the well-intended tax credit program may only reward those property owners in the more class- and race-privileged areas of L.A., the very places that already have access to open space for urban agriculture. Our goal is to advocate and insist that the geography of justice must compel local governments to redirect resources toward low-income and vulnerable communities because these continue to suffer the disparate impacts of environmental racism in the un-signified food deserts of the real.

## Coda: Deer Dancing in the Permanent Heterotopias of *Autonomía*

Clearly, we can come back to the opening epigram from the Huicholes and ask ourselves: What are the qualities that make neoliberal governmentality weak and vulnerable? If regimes of signification are arbitrary and pure socially coercive constructs, then they are vulnerable because they are, as we said earlier, phenomenologically weak.[33] This is perhaps also explained as an observed quality that Tom Lundborg and Nick Vaughan-Williams characterize as being based on

> open, vulnerable, and often absurd systems that continually falter, backfire, and often undermine themselves according to their own logic. By developing what we call a "molecular security" approach, we draw attention to the way in which life constantly evades capture. In this sense, we suggest, there is always an excess of "life" in biopolitics.[34]

Lundborg and Vaughan-Williams are international security policy experts, so their expressed interests diverge from what we have in mind, but they do inspire us to think about developing a "molecular autonomy" approach to defend and sustain the "opacity" of the living networks that life moves through in order to evade capture and allow freely associated subjects to engage in a "joy in excess." In this sense, we (SCF) also offer an "excess of strategic invisibility" to protect right livelihoods and the cultural fabric that embraces and produces life in our spaces of *autonomía*.

In the same way the Huicholes follow the blue tracks of the "Kayumari" (deer spirit), we embrace our condition as openness to change and contradiction. We abide by the ethics of the faltering sacrifice at the "Cerro Quemado" (Burnt Mountain) and the molecular resignification of our resiliency. If the fragmentary flows of the food deserts exist, it is purely from the vantage point of a phenomenologically weak (because empty and totalizing) desire in the latest neoliberal iteration of settler colonial-capitalist regimes of signification with a ravenous appetite and apparatus for capture and containment. Our movement comprises a permanent state of economic refusal escaping spatial enclosures. This involves a return, through daily

Figure 11.4. Huichol Deer Dancer at the Cerro Quemado. Original illustration by David Vasquez. Courtesy of The Acequia Institute.

lived practice, to our institutions of collective action. These continuously erupt and flow across urban and rural landscapes in iterative escapes of placemaking that lie beyond the functional dominium of the apparatuses of governmentality. We do not believe this is all there is. We practice the politics of the impossible and take counsel from the Huicholes, who refuse to formalize or document their inventive and ever-shifting social and ceremonial order. "All will be different, the opposite of what it is now."

CHAPTER 12

# Growing Justice in the Fields
## Farmworker Autonomy and Food Sovereignty

ROSALINDA GUILLEN AND C2C

## Why a Sovereign/Autonomous Food System?

An autonomous and decolonized food system rejects capitalism and forgoes human domination. It is instead centered on the well-being of the Earth—the waters and soils that provide the food, and the people and communities that produce, transport, distribute, consume, and dispose of this bounty. La Via Campesina has stated that food sovereignty is "rooted in the ongoing global struggles over control of food, land, water, and livelihoods."[1] The struggle to recenter the planet's well-being in a manner clearly interwoven with our own prosperity in health and sociality involves a search for just and sustainable livelihoods across all sectors of the food chain. This is reflected in the paths that our food takes from the Earth through our bodies and back to the Earth. Food sovereignty is about building up and caring for the soil that nurses all seeds to sprout. It is about protecting the waters that make life possible. It is about maintaining the purity of the air. It is about lessons from and care for the plants that are our medicine and the base of all food chains. It is about respecting and restoring the place-based cultures that spent generations learning their place's soil, lungs, seeds, and waters before food was mass-produced for profit as it is today.

Each part of the food chain is a site of food sovereignty or autonomy struggles. An ecologically sovereign food system is not measured by the profit it generates but by the health of all Earth's communities.

This is the core vision of a small team at Community to Community (a.k.a. C2C) in Bellingham, Washington, a farmworker- and women of color–led organization committed to growing a local solidarity economy that prioritizes the well-being of everyone in the community.

Community to Community works to create a local solidarity economy center by fostering political movements that define their own agendas toward the creation of such an alternative. We do this work because we believe that another world is possible, and we are active participants in other people's movements. We strive to reclaim our humanity by redefining power in order to end settler colonialism, capitalism, and patriarchy in their external and internalized forms. Food sovereignty is important as a liberation practice because the ability to decide what and how we eat is at the very center of our relationship with ourselves, with each other, and with our planet.

The corporate model that dominates the production of food for enormous profit is evident across the entire agri-food system. Take the distribution component for example. Whole Foods is coming to Bellingham soon. It will set up right down from a freeway exit, less than a mile from the downtown locations of the Bellingham Community Food Co-op, and from Terra—both go-to stores for "foodies" looking for artisan products, and for consumers looking for verifiable ethical standards in the foods they buy. The Community Food Co-op and Terra both source (though not exclusively) local organic foods year-round, along with the Bellingham Farmers Market, which operates for nine months out of the year. Downtown Bellingham is dotted with bistros, eateries, small local chains of restaurants, and microbreweries featuring exotic and local organic meats, vegetables, and beverages. All these local businesses are considered indicators of a thriving local food system.

The availability of local organic meats, dairy, and produce is only one component of a healthy local food system. But the arrival of a national entity like Whole Foods into the local food distribution system could be a threat to the little gains that organic and local food producers and distributers have built in Whatcom County. Big-business players like Whole Foods could drive wages down and worsen working conditions not only for the workers at the Co-op and Terra, but also for the local farmers, who will now be competing with cheaper prices from nonlocal producers who supply Whole Foods. The Co-op and Terra could try to compete using the model of corporate agriculture, or they could commit to building a different model that supports the struggles of self-organized workers throughout the food chain, and by engaging in the fair trade movement with the Domestic Fair Trade Association (DFTA). The DFTA not only develops standards for sustainability and labor justice but also evaluates other labels that claim to be "fair trade." Both Terra and the Co-op have already begun to explore a different model by supporting *local farmworkers*. The agricultural industry is no different. Small local farms could try to compete with or join the big agricultural corporations, or they could build a different model on the local level.

Organic industry standards are a good beginning place to change the food system. The movement for organic foods emerged out of concerns about the health impacts of pesticides on consumers, and the environmental contamination caused by pesticides. Organic foods are generally seen as healthy and ethical alternatives—both for consumers and for the planet. But buying organic is not enough, as this only addresses watered-down pesticide policies and superficially confronts the growing concerns about GMOs (genetically modified organisms) in the US food system. Buying organic does not change this damaging food system enough, because organic labels do not include any consideration for the labor conditions of the farmer, farmworkers, or any other hands that packed and transported the food. Another concern that was part of the organic food movement was the construction of a sustainable and socially just food system, but the organic food industry has already compromised the industry by not providing fair labor practices and sustainable living standards for farmworkers and other workers down the food chain. In a 2005 report, there was a general consensus that organic growers did not have responsibilities to honor workers' collective bargaining rights and provide living wages, health insurance, paid sick leave, or paid vacation. But the organic food industry continues to benefit from the image of sustainability and ethical practices even though the reality of food-system workers does not support that image.[2]

At C2C, we recognize that the corporate and industrial models for growing, processing, selling, transporting, and distributing food are unjust and hurting farmworkers and the planet. These models are based on speed and maximizing possibilities of profits at the expense of workers and the Earth. These models encourage competition not only between small farms and big agricultural corporations but also between different sectors within the labor force. The competition also pits young workers against older ones, male workers against female and gender nonconforming workers, documented against undocumented workers, workers of different races against others, so that they will not act in unison and struggle together for higher wages and better working conditions.

In addition to this, there are many other struggles for justice that are not usually considered but tightly intersect with the mobilization of a sovereign local food system. One example is the privatization of all commons. Enormous waste is generated throughout the corporate food chain, but to protect profits and the notion that food is a privately owned good, the wasted food is protected by laws against redistribution and dumpster diving. Privatizing water, like food, is considered to be necessary to protect profits. When water is polluted though, the public is expected to foot the bill if the pollution is ever mitigated. Extreme fossil fuel extraction not only affects the land and water that grow food, but it also subsidizes the agricultural industry's transportation of foods around the world so that every grocery store can stock nonlocal foods like bananas, chocolate, and coffee. Food sovereignty struggles are connected to the gentrification and displacement of urban communities because while real estate developers claim that urban farming will solve the food desert crises in many cities, these farms cannot adequately meet the food needs of low-wage working families. In the meantime, rural farming communities are simultaneously degraded by large agricultural corporations using conventional farming practices. These communities are then forced to host prisons and detention centers to replace their small and self-sustaining farming economies. Building a sovereign food system means learning about and respecting the food that exists naturally in each biome, and the people and cultures that have cared for each biome over generations. Food sovereignty is more than making sure everyone has enough to eat. We assert the autonomy of communities when we bequeath culturally appropriate foodways to future generations.

This is what C2C strives for with Latina/o farmworkers in

Whatcom County. Food sovereignty "puts the aspirations and needs of those who produce, distribute and consume food at the heart of food systems and policies rather than the demands of markets and corporations."[3] Through the Cocinas Sanas/Healthy Kitchens workshops and the Raíces Culturales youth-mentoring projects, we seek to include farmworkers at the local level as dignified stakeholders in the local food system and therefore in the local economy of Whatcom County. Cocinas Sanas workshops integrate locally sourced organic foods into traditional Latino recipes. When groups of farmworker women, young Latina women, and young Latino men cook and eat a healthy and traditional meal together, they build spaces to discuss health issues they find in their communities—at home, at school, and at work. Recognizing that organic foods are often inaccessible to farmworker families, Community to Community has leased and also been hosted at some garden spaces in Bellingham and in rural Whatcom County to allow farmworkers to have access to land to grow foods that are a part of their traditional kitchens (see figure 12.1).

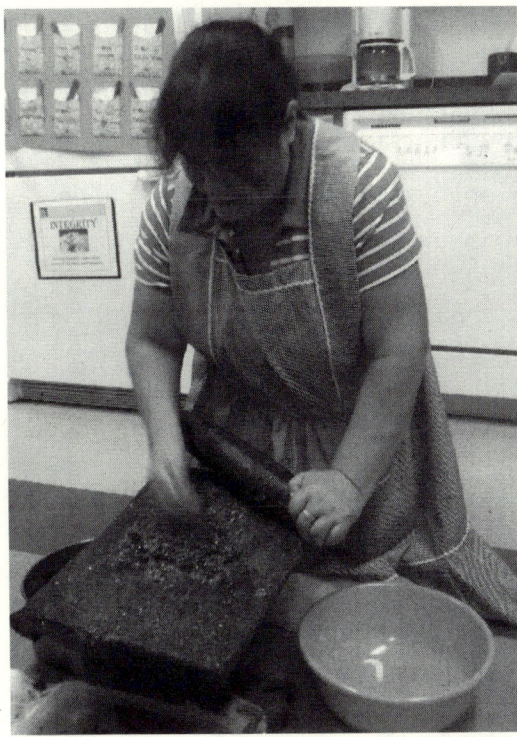

Figure 12.1. Member of Cocinas Sanas using a metate to grind chile. Courtesy of Community to Community.

Contradicting the widespread belief that to succeed, farmworker children must escape from farmwork, at C2C we seek to build spaces where youth can practice growing foods without the industrial, sped-up models of corporate farms. Instead, youth can explore and experiment, observe and learn best practices for growing foods without the use of pesticides and chemicals and with the help of integrated pest management. Together we try to remember how curiosity and natural rhythms are valuable sources of knowledge in our traditions that have been devalued as unscientific and primitive. Since we are not encouraged to follow these rhythms in today's educational systems, we ask community elders to help us remember these knowledges, including seed saving, sprouting, and nixtamalizing corn. By respecting traditional ways of learning and growing foods, we hope to remember that growing food was not always the shameful and devalued work that we know today as farmwork. We try to remember that growing food is a dignified, necessary, and valued skill that our parents and ancestors learned and passed down. The oppressive realities of farmworkers are the result of corruption in today's food system and are not inherent in the act of growing, harvesting, and processing food.

For the first ten years of C2C's work, farmworkers and allies worked to build relationships within the local Whatcom County community. Cocinas Sanas workshops were held at the Bellingham Community Food Co-op, at local church kitchens, in people's homes, and at local restaurant kitchens. Farmworker leaders participated in advisory boards, made public presentations, ran a radio show, gave interviews, and hosted community events and potlucks to demonstrate that farmworkers are indeed a part of the local food community. Through Dignity Dialogues, farmworker communities told about how their daily lives were disrupted and strained by local and federal law enforcement agencies that cooperated to fill detention and deportation centers with undocumented workers, including farmworkers. As a structural support, they identified that ending racial profiling would improve their lives by eliminating the fear they felt was stopping them from fully engaging in civic processes. Thus C2C's work toward comprehensive immigration reform grew out of the grassroots call to protect the farmworker community while supporting the capacity for civic engagement as integral partners in the local food system. In these concrete ways, C2C has worked for ten years to create a local climate that supports independent farmworkers organizing locally.

## Farmworker Autonomy

Outside the United States, farmworkers are generally understood to be landless peasant farmers. Most landless farmworkers in the United States have, most commonly, themselves been from multigenerational farming families that were displaced from multiple Indigenous and *ejidetario* communities outside the United States because of the reach and depth of control that corporate food regimes have on many sovereign governments—which is one reason we must suspend engagement with neoliberal state formations and practice *autonomía* instead of blindly emulating someone else's principles.

NAFTA (the North American Free Trade Agreement) and GMO corn's devastation of landraces are prime examples of this reach and depth of control that has hit particularly close to home for many farmworkers, who no longer enjoy access to traditional and unmodified corn, a staple of their traditional diets. Today most of the farmworkers working in Whatcom County are Indigenous Oaxaqueños who initially followed the "migrant agricultural circuit" from México's berry farms (supplying large distributors like Driscoll's), through California, and then to Oregon and Washington. Many have settled in Whatcom County with their families and children. As displaced Indigenous farmworkers, they are removed from their place-based self-governance structures that framed their autonomy and protections as community members. At the same time, they are excluded from the albeit constrained protective regulations for workers in the United States.

In some significant ways, Indigenous farmworkers are caught in a particularly vulnerable existence as socially stigmatized and degraded communities targeted by racist hiring and supervising policies for the most difficult and worst-paid agricultural work. At the same time, they are attractive to anthropological researchers who feel entitled to research the rapidly changing lives of Indigenous and migrating Oaxacan farmworkers without the accountability provided by traditional governance structures. Even the binational structures that Indigenous Oaxacan communities have built, such as the FIOB (Frente Indigena Oaxaqueña Binacional), are inadequate to the challenge of effectively limiting the vulnerability of Indigenous farmworkers to the ceaseless onslaught of researchers who feign to be looking to produce "balanced" findings, which are then easily used by agricultural corporations to defend the exploitation of farmworkers.

The protection of the dignity of farmworker livelihoods is premised on the autonomy of Indigenous farmworkers to build their own protective self-governance institutions that also protect and assert their traditional consultation and decision-making and knowledge-producing practices. As Rosalinda Guillen, a member of the core leadership circle of C2C, explains, in the 1950s and 1960s farmworkers realized that the way forward for them would be to build their own solidarity systems because what society was currently providing was not working for them. Therefore, they built their own cooperatives of mechanics, a health clinic, a gas station, and a credit union. The current capitalist paradigm does not serve and actively hurts farmworkers, so farmworkers build another system where they can be happy and healthy while allowing for the flourishing of their highest human capacities. In building their independent farmworker union, Indigenous farmworkers have seamlessly integrated their autonomy practices into decision-making and consultation processes. This autonomy is also the basic building block of their health as families and as a community because they are freely associated and can form their own strategies to win equity in workplaces unbound from the limitations of larger and affiliated trade unions.

Another practice that C2C uses to structurally support farmworker autonomy is growing farmworker-owned and farmworker-led cooperatives. A worker-run cooperative is a democratically structured entity that seeks to meet a need in the local community. In 2007, a farmworker family led by María Guzmán and Roberto Bermúdez formed Cooperativa Jacal to grow and sell organic produce at the Bellingham Farmers Market. One of the reasons they both cited for forming the cooperative was to have consumers recognize who grew their chemical-free vegetables locally. Instead of the usual scenario where the farmer and landowner receive all the credit for the beautiful produce, in Cooperativa Jacal the workers themselves are recognized and valued. It also provided an opportunity to build a workplace where they could make living wages, especially since no one who has approached C2C has earned living wages in farmwork.

Although in the end Cooperativa Jacal decided not to continue as a cooperative and instead became a family-owned organic food grower, C2C derived many lessons from the experience. One important lesson was learning to define what counts as success or failure for farmworkers practicing different, more equitable economic relationships with each other and with the larger local community. Cooperativa Jacal took to heart the co-op principle of voluntary and open membership in the cooperative and together decided that the lengthy process needed to build consensus was not the decision-making mechanism for their team at that time. A related lesson was that a necessary and important shift must happen for co-op builders to have the mental clarity they need to make decisions in cooperative spirit.

It is a difficult and an ongoing struggle to make that necessary shift inside a co-op from the sped-up conditions of the industrial factory-style corporate farms to the mind-set of persons in a cooperative, making joint, democratic, and transparent decisions that impact their daily working lives. The factory-style conditions at corporate farms no longer existed at the co-op workplace: There was a machine that could set the speed of the work, but the place was without a supervisor who could make the negotiation of working conditions inherently hostile. It was surprising how deeply internalized dominant workplace norms were, and it took daily practice to avoid reproducing these conditions at the co-op farm. To undo the deeply damaging isolation drilled into farmworkers from a very young age, we learned that it takes time, constant practice, and compassion for ourselves and for others. It takes an active effort to switch to the rhythms of nature and the growing season from the regimented, artificially exhausting time schedules of the industrialized farm. It takes constant remembering not to extract every bit of life energy and not to overwork our bodies to the point of exhaustion each day. Deciding to reset what counts as successful farmwork was also a useful lesson.

## Ecofeminism and Farmworker Women's Economic Autonomy

Community to Community also supported the formation of farm-worker women-led cooperatives. Apart from the autonomy of farm-workers in the larger labor force, we also believe that women's leader-ship is key to growing a local solidarity economy. In the ten years that farmworker women have been using the C2C space to organize, they have expressed—in different ways but in common—that their economic independence is important in their self-determination and their leadership of their families. Indeed, we found that it is hardest for women to develop cooperatives when they are with men who thwart women's freedom to pursue development. It is as if women forming a cooperative and being powerful together as women is too often per-ceived as a threat to the men in their lives—whether they are brothers, fathers, sons, or husbands. Consequently, men often place obstacles in the way of the women, preventing them from moving forward with their quest for economic autonomy.

Deeply affected by the ecofeminism of the Brazilian activists at the World Social Forums in Porto Alegre, Brazil, C2C is intention-ally a women-led organization because we are trying to develop the movement processes and spaces for women to be able to say what they think and then even more importantly to act on it, without fear and with dignity and respect. As women we are on the front lines facing the impacts of every social justice issue—whether it is climate change, workers' protections, or wage gaps. Because we face these impacts first, we can address movement struggles in different ways. Our colleagues in Brazil who have been building and practicing their local solidarity economy shared with us that something shifted when at least half of the leadership of the solidarity economy were women. This sugges-tion strengthens our resolve and supports what we have already seen: that impacted women leading efforts to change their own conditions qualitatively changes the politics and the direction of the movement.

As women, we see the hurt that patriarchal systems perpetuate in ourselves, in children, in gender nonconforming companer@s, and in men. In low-wage immigrant communities these wounds are rooted in the exploitation of our productive capacities in the labor force to extract seemingly endless profits and in the process hurt our spirits and our bodies. But these harms are felt and borne by women disproportionately through the exploitation of our reproductive capacities as well—both in the workplace, where women and gender nonconforming workers are subjected to lower wages and are more vulnerable to harassment and violence; and at home, where we bear the majority of housework and child care responsibilities, which shift to our daughters, not our sons. Our liberation from these systemic vulnerabilities is contingent on our economic autonomy, the same as it is for our male colleagues. C2C is committed to ending sexism and patriarchy because this liberation requires control of our reproductive *and* productive capacities. We are also committed to this because men and masculine folks cannot be their full human selves when they are oppressed as men, with worth defined in a limited manner based only on their productive capacity and their social control over women and children. We all deserve the full expression of our humanity beyond the narrow confines of rigid and controlling gender roles and hetero-normative sexualities.

Worker-run cooperatives are important components for building a resilient ecofeminist local movement toward a solidarity economy. For us, a local solidarity economy is another economy that we are building that *does* work for us, the workers. A solidarity economy is one where we can all be well—healthy, happy, and able to live up to our highest human capacities. It is an economy that is in solidarity with the Earth, which means that we relearn our skills and listen for the logic, rhythm, reasoning, and methods that the Earth communicates to us. It also means that despite our histories, which colonization tried to erase, our legacy should include how to be in a just relationship— that is, in solidarity—with the place and peoples where we are from and with the place and peoples where we are now coinhabiting.

In a solidarity economy we make decisions together through democratic and participatory processes using culturally appropriate methods to discuss and make decisions. These would include making budget and financial decisions, developing accountability processes that build and repair relationships, maintaining our commons, and together charting political directions for ourselves as a community. We would expect that a solidarity economy would look different in each place because it would respect the unique characteristics of that place and that community.

We believe that the solidarity economy should be led by those most directly affected by damaging policies and degraded environments. We are the ones with the clearest thinking about how to end the damage and to heal by making a transition toward policies that create and protect wellness. This means that people who make up the global majority need to be the base of movement building. Because we think that the solidarity economy is a political, not just an economic, system, participation and decision-making processes must begin at the roots. Workers, young people, women, gender nonconforming folks, elders, and those who have been impoverished by this capitalist economy are the base and must lead the movements we need to be well.

Because a solidarity economy uses democratic and participatory processes, there is a critical intersection between autonomous self-governance and political education that engages everyone in the political process, including collective and participatory budgeting. We do our best to create democratic and culturally engaged strategies for learning and engaging political consciousness. We then always manifest that learning and consciousness in action, effecting the change as we are learning what that change is. We learned that to build trust with each other throughout the cooperative learning process, we have to securely know what brings all of us to the process, and what our goals, commitments, and beliefs are about oppression, liberation, and working together. Without this base, we have a shaky foundation, especially when decisions get harder or more contentious.

We have found that our learning, particularly the learning of farmworkers in building cooperatives, is actions based. We learn best by doing. This means that we also have to learn generosity and flexibility, with the determination that we get to decide what are failures. To us, the only real failures are when we fail to act in solidarity with each other—doing something counter to what has been agreed upon together. All other "failures" or mistakes are opportunities to go back and do a closer analysis that can help us move in a different direction—maybe a direction we were supposed to take to begin with. We believe that a solidarity economy is a living possibility because we see the solidarity economy as a transitional system on our way to dismantling capitalism. While we do not know yet what the future economy will be, the more we practice and refine, the better our economic practices will be.

## Women-Led Cooperative Practice

Las Margaritas was a group of farmworker women forming a cooperative to sustain their families economically and improve their economic autonomy as women. They organized in response to family separations they experienced together as a result of a workplace immigration raid at Northwest Health Care Linen in Bellingham. They lost their jobs after the raid because they were undocumented. They were not allowed to find a job in another company, because they would be breaking the law. They wanted to do something so their children would not experience hunger and so they could meet the needs of their families. When they were asked what they had always wanted to do but never had the chance to do, the members of Las Margaritas shared stories about cooking healthy traditional food for their families and friends. They decided to focus on cooking and catering healthy food and educating the community about the importance of using organic and locally grown produce. Together, they worked to provide the basic resources their families needed while also creating a respectful, dignified, and peaceful community. They worked in cooperation with Cooperativa Jacal, which supplied them with fresh produce for the cooked-food market stand at the Bellingham Farmers Market on Saturdays. It was a practice to develop new models for how people, particularly women, could survive in our current food system.

Today their food workshops are recalled with enthusiasm, but Las Margaritas dissolved as a cooperative after one of their leaders was detained by US Immigration and Customs Enforcement in 2010. We continue to receive calls requesting catering from Las Margaritas for events as far away as Olympia, Washington.

One of the most significant legacies of Las Margaritas cooperative is the experience that members and volunteers had in witnessing the cooperative development processes. This experience inspired another group of Latina farmworkers to come together and establish their own organic tortilla-making cooperative, Hita Quiuci. Founded in 2013,

Figure 12.2. "Food Yes. Jails No." C2C antideportation campaign poster.
Courtesy of Community to Community.

Hita Quiuci was organized and is operated by Latina farmworker
women who identify as survivors of domestic violence. Another last-
ing legacy of Las Margaritas is the *pasaporte* curriculum, a unique
campesina-produced curriculum used to develop farming co-ops.
Although we are free to refine it, it is the closest we have come to a
curriculum that is culturally appropriate for farmworkers seeking to
build worker-run cooperatives in the United States.

It takes practice many times over, learning as much as we can each
time around and adjusting our way toward a solidarity economy. One
of the most important lessons we derived from Las Margaritas and
Hita Quiuci is the importance of including opportunities for women
to unlearn and heal from our internalized oppression. Although cooperation and cooperativism evoke an image of easy relationships based
on sharing, the reality is that we all internalize capitalism, sexism, and
racism. When we actively challenge the messages we internalize about
our own worth, the value of our work and of the work of others is
recast within an ethics of accountability to our relationships, and this
is not as easy as it might seem.

A major component of cooperative building is putting our best thinking into how competition works and gets in the way of cooperation, and why it exists in the first place in our relationships with each other. And we found again that the best way to undo our ideas about the value of our and others' work was to practice working with others and practice working with ourselves. This curriculum teaches proven invaluable practices for participatory finances and equitable distribution of resources that are essential to building cooperatives in the future. We would not know what we know today about undoing internalized sexism and racism if C2C had not been witness and partner with the cooperatives that have grown with us.

Most importantly, we learned that everyone learns in different ways. For farmworkers and people who work with the land, learning has to be manifested in action. It cannot be gained solely from books, classroom lectures, or workshop sessions. We saw that many of us learn most effectively by doing. We acknowledge that we have generational histories of learning by doing: if something did not work, our ancestors tried something else until they knew that it was right. Cooperative development, like learning by doing, is flexible. We learned that cooperatives are all about the people forming the cooperative and creating their own processes based on their abilities and their own pace.

## The Future of the Food System: Family Well-Being and Ecofeminism

In 2014, ten years after Community to Community was formed, we began the year with strategic planning, assessing the current political moment and sharing with each other what our goals, beliefs, and commitments were toward transitioning to a just economy. We made commitments to build C2C as a self-sustaining center for the local solidarity economy in Whatcom County, and to steer the local food system toward justice, beginning with the struggles of farmworkers.

As women of color and people of the global majority, we have always understood liberation as the work to move our families, particularly our children, to be well. This means that our relationships to ourselves as powerful women must be rebuilt through ecofeminist practice. Our revolution is in process as we all continue to work to have everyone—our families and the planet—be well, joyful, and complete in an alterNative solidarity economy emerging from our labor, creativity, and unique forms of self-organization that transcend borders and challenge dominant concepts of citizenship and human rights. The autonomy and dignity that Indigenous farmworkers attain are in the end for us the best indicator of a society's respect for food sovereignty, and in the United States we clearly have a long way left to build the world we want.

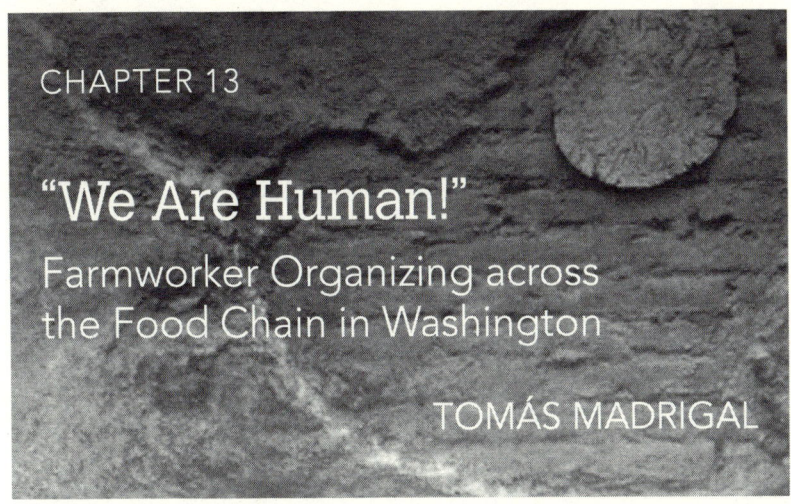

# CHAPTER 13

# "We Are Human!"
## Farmworker Organizing across the Food Chain in Washington

### TOMÁS MADRIGAL

*Our work, the work we do is life . . . because we plant the plants*
*that bear the fruit so that people can eat, that they can be strong*
*and healthy, to have the energy to survive. With our own hands*
*we plant, we sort, and we harvest the fruit. And so the work that*
*we do is life, the life of the entire country.*

—FELIMÓN PINEDA, vice president,
Familias Unidas por la Justicia

Present-day migrant streams and their re-emplacing communities force us to shift our politico-organizing perspectives beyond the limited parameters of regions, nations, and their borders. Triqui- and Mixteco-speaking migrants from Oaxaca and Guerrero, for example, have built a social and political network that spans the North American Pacific Rim from their roots in southern México to a trunk that extends from the border through California, Oregon, and Washington all the way to Alaska. This migrant network branches out to the Midwest, Eastern Seaboard, and the Deep South, cross-pollinating with all of the cultures both stationary and transient along the way.

My story starts with the formation of Frente Indigena de Organizaciones Binacionales (FIOB) in 1992 and successful activation of that network in 2006. That mobilization—in order to hold simultaneous demonstrations in many anchor cities up the Pacific Rim coast in solidarity with the uprising in the capital city of Oaxaca—sparked the imagination of transborder communities and changed the game in organizing across sectors. It sparked the seemingly spontaneous immigrant rights marches in multiple US cities that same year in response to the draconian H.R. 4437 passed by the US House of Representatives a year earlier.[1] More recent examples of this continuing mobilization include the port shutdowns along the West Coast in solidarity with Palestinians after the Israeli siege on Gaza in 2014. In forty-three US cities, students and youth have led the campaign in solidarity with the families of the forty-three disappeared students from the Ayotzinapa normal school in Guerrero, who were fighting against neoliberal cuts and educational "reforms" in México in 2014 when they were murdered. The cycle of current struggles includes the Not1More Deportation campaign, which continues to spark demonstrations across the Northern Hemisphere to end the mass incarceration, detention, and deportation of displaced Indigenous and migrant-stream food-chain workers.

All of this has happened despite (or perhaps because) of everyday conditions of "hyper-criminalization,"[2] as well as "super-exploitation, super-control, and hyper-surveillance."[3] The targets are mostly non-union immigrant farmworkers, food processing workers, restaurant workers, and grocery store workers. But in bold actions of ongoing open resistance, these food-chain workers have led successful multi-pronged campaigns, all of which have redeployed the civil rights–era slogan "I am a Man!" and translated it into the all-inclusive "We are Human!" This has been a slogan in popular movements since the massive immigrant rights marches of 2006, which have been met with an unprecedented level of repression for immigrants, including mass detentions and more than two million deportations under the Obama administration. The age of mass incarceration is ascendant with an increase in police murders, repression, and the persistence of racial profiling. The numbers and statistics are alarming to any movement facing so paramount an obstacle. It is easy to forget that we are winning despite these numbers.

This common refusal to be relegated to statistics and to give personal narratives to the numbers has arrested, if for a moment, the passage of immigration reform legislation (e.g., S. 744) that would all but give away many of the jobs held by immigrants to indentured labor across the food chain.[4] Immigrant workers, students, and detainees have effectively pressured the Obama administration to provide administrative relief in the form of "Deferred Action for Childhood Arrivals" (DACA) and the currently blocked "Deferred Action for Parent Accountability" (DAPA) through a diversity of tactics. Washington State farmworkers and other food-chain workers, social justice activists, and labor organizers have been involved in all of these campaigns.

The strategy behind food-chain organizing on the hemispheric scale, as demonstrated by the struggles along the Pacific Rim by Indigenous people, immigrants, low-waged workers, and their allies, has been based upon the developing concepts of food justice and food sovereignty. It is fitting that the vast majority of low-waged food-chain workers, both domestic and immigrant, can trace their own roots back to subsistence farming, sharecropping, and agricultural labor. Food justice, according to the Community Alliance for Global Justice (CAGJ), can be defined in the following way:

> Food Justice is the right of communities everywhere to produce, distribute, access, and eat good food regardless of race, class, gender, ethnicity, citizenship, ability, religion, or community. Good food is healthful, local, sustainable, culturally appropriate, humane, and produced for the sustenance of people and the planet.[5]

The definition of food justice is determined differently by each stakeholder, and so the above represents only the broadest gist of the idea. The opening epigraph for this essay offers an Indigenous Mixteco understanding of the concept by a recently formed independent farmworker union, Familias Unidas por la Justicia, based in Bellingham, Washington.

While food justice functions as the glue holding the movements together, food sovereignty is the goal that these movements are working toward. The Community Alliance for Global Justice offers the beginning definition: "Food sovereignty is the right of people to determine their own food and agricultural policies; essentially, the democratization of food and agriculture."[6] Organizations such as La Via Campesina and the Domestic Fair Trade Association (DFTA) have worked to define food sovereignty via principles and standards that they wish to uphold, while grassroots organizations such as Community to Community Development and Malcolm X Grassroots Movement are following the lead of Movimento dos Trabalhadores Rurais Sem Terra (MST). The Landless Workers Movement of Brazil has sought to build "solidarity economies" based on worker cooperatives as part of their practice of food sovereignty. The ability to embrace diversity and loath conformity has been a saving grace of the food justice movement, in which all of the people and organizations whose actions are described in this essay are involved.

The role of the hunger strike in the context of food justice will perhaps be the most difficult connection for most readers to make. In the simplest terms, the food-chain cycle ends the moment you place your eating utensil in your mouth. A hunger strike by an organized group of people disrupts the final stage of the capitalist food system supply chain. Similar to the strikes and boycotts of Familias Unidas por la Justicia, which disrupt the supply chain on the production and consumer ends, the hunger strike seeks to change production relations within the closed-circuit economy of the capitalist food system.

There were hunger strikers at the California prisons and in the detention centers across the country. Hunger strikes were staged in 2013 and 2014 as a tactic seeking to focus attention on the need to change the quality of and access to food. Outrageously predatory commissary prices were not the only target. The hunger strikers were also protesting wages: immigration detainees reported as low as a dollar a day, and state and federal prisoners make cents on the dollar for their indentured labor, which buys them access to commissary. There is a third connection: It is through the threat and experience of incarceration and detention that food-chain workers are often forced to accept lower wages because of their legal status or penalties for time away from work. Food justice can be theorized while remaining aware of the basic slogan "We Are Human."

This essay weaves narratives across the food chain emerging from the class struggles in Washington State to mark the unique contributions of diaspora workers to decolonizing dominant discourses of social

Figure 13.1. "I am human." Cornelio Ramirez during a 2014 march to Sakuma Bros. Farms. Courtesy of Edgar Franks.

movements by centering their political interventions around the production, preparation, and consumption of food. In this essay I examine the food justice work of Community to Community Development, which is fighting against unsafe working conditions in agriculture and against the arbitrary detention and "hypercriminalization" of farmworkers along the northern border. I review the rise of Familias Unidas por la Justicia in Skagit County and their defense via strikes, lawsuits, and consumer boycotts of farmworkers' rights to adequate health care and housing and against a regime of superexploitation, supercontrol, and hypersurveillance. Finally, I chart the recent campaigns of Northwest Detention Center Resistance (NDCR) to blockade deportation buses and stage three hunger strikes inside the Tacoma Northwest Detention Center to fight against deportations, indefinite detention, indentured labor, and the high cost of commodities inside the detention center. I hope to demonstrate the centrality of food and foodways in this newest of social movements.

# Beyond Labor Markets and Statistical Manipulations, "We Are Human"

*We are the names you will never know.*
*The faces erased from history.*
*We are the statistics on your fact sheets.*
*The fulfillment of your quotas*
*We are one link in a long chain of struggle.*
*We are the lion in a history written by hunters.*
*But we do not need to be known to create change.*
—Not1More Deportation poem, 2014

In mainstream Marxist terms, the systematic displacement of Indigenous people is often thought to result in forced migration or "re-peasantization." The persistence of non–market value economies is presumed to constitute an exception that will recede as an active location of class struggle. The concept of class struggle, even through the many peasant revolutions of the twentieth century, was reserved to what was termed the "productive sphere," comprised of those who sold labor for wages (or the proletariat) and the capitalists who imposed and profited from this exchange relation. The rigidity of the Western gaze made it difficult for most structural Marxists to see the reality of the revolutionary capacity of the reproductive sphere consisting of the activity of the unemployed and casually employed in non–market valued immigrant economies.

In this sense, none of the scientific economic interpretations of what was to be done were in conflict with the internal logic of imperialism or capitalism. The questions asked were instead focused on whether the limits or contradictions could be overcome so that an imagined "evolutionary process" of capitalist economic development could take place and we could emerge on the other side. This capitalist logic is evident in discourses of contemporary class struggles. In defense of his company's misuse of the H-2A guest worker program during a labor dispute in 2013, the CEO of Sakuma Bros. Farms, Steve Sakuma, said with confidence,

[The H-2A program is] a threat to advocacy groups, it's a threat to immigration reform and it's a threat to the path to citizenship. I get that. But the outcomes some of these advocacy groups are looking for are the same we're looking for: A stable, legal, cost-effective workforce.[7]

The traditional approach of labor market manipulation has consistently been to pit different classes of workers against each other. Even though this was the dominant perspective, there were minorities that believed that another path was possible, and even Karl Marx himself acknowledged the possibility of a different path for Russia.[8]

The production of distinction between classes of workers was observed as a historical process that was crucial to the development of industrial agriculture in the United States for hundreds of years.[9] Systems developed to produce difference among the classes have included the latifundia (large landed estate) surrounded by peasant farmers in Europe;[10] haciendas and plantations articulated to Indigenous and peasant communities in the Americas;[11] and the slave-based plantation economy that emerged in the US South.[12] The articulation of peasant workers into the capitalist economy is also seen as involving strategies using transnational seasonal migrant farmworkers,[13] or guest workers,[14] and the stabilization of a marginalized and vulnerable working poor via barrios, *colonias*, and company towns.[15] Eric Wolf argues that cultural distinctions are continuously produced and reproduced between classes of workers in industrial agriculture.[16] David Griffith and Ed Kissam argue that "the farm labor market has always been heterogeneous. Internal differentiation has rested, at various times, on ethnic background, residence (migrant vs. local), gender, attachment to farm labor, or legal status."[17]

David Montejano believes that "racism and racial exploitation, rather than disappearing with the march of capitalist development, appear instead as its intimate companions."[18] Regarding the making of contemporary racial orders, Montejano finds that the

commercialization of the rural order does not result in the formation of an unrestricted proletariat, that is, a body of agricultural workers who are free to sell their labor. Rather we find a working class that is tied to the land through the use of non-market criteria and sanctions—through violence, coercion, and law.[19]

One of the ways that US policy helped to develop capitalist large-scale agriculture for the world market in California was by keeping wages artificially low by facilitating through policy a perpetual shift in the ethnic makeup of the entire labor force.[20] This allowed capitalist agricultural firms in the United States to externalize the cost of the reproduction of labor power in two ways: first, by keeping local workers in marginalized communities via unstructured labor markets, such as *colonias,* that supply a pool of available labor for surrounding agricultural enterprises;[21] and second, by depending on foreign labor power that is imported directly via contract and indirectly via informal social networking. This labor power was traditionally reproduced via noncapitalist modes of production in foreign domestic spheres.[22] These shifts were managed in the United States by public policy beginning in the 1920s, when harvest labor markets were formalized during moments of crisis and deregulated during periods of economic prosperity, producing a pattern of reliance on an informal harvest labor market,[23] which was also maintained through various racial projects.[24]

Racism and racial exploitation were omnipresent from the very beginning of these policy changes, as those industries relying heavily on recently emancipated former slaves were summarily excluded from the gains of the American working class at the turn of the twentieth century. Even as late as 1935, when industrial workers in the United States made unprecedented labor rights gains, agricultural workers and domestic workers were excluded from the Wagner Act, the Taft-Hartley Act of 1947, and again from the Landrum-Griffin Act of 1959. Citing a presidential commission, Justin Akers Chacón described the Bracero Program as

> a throwback to a bonded labor system that, according to the Presidential Commission on Migratory Labor, supplied growers with, "a labor supply which, on one hand, is ready and willing to meet the short term work requirements and which, on the other hand, will not impose social and economic problems on them or on their community when the work is finished. . . . the demand for migratory workers is thus essentially two-fold: to be ready to go to work when needed; to be gone when not needed."[25]

This expectation that human workers in agriculture can be treated as a disposable and bonded labor market continues to be deployed to this day. In Washington State, growers like Steve Sakuma, quoted above, unquestionably shared this perspective, an antiunion and inherently racist point of view that is amplified by grower lobbying organizations including the Farm Bureau and the Washington Farm Labor Association. All are special-interest stakeholders who have made an economy out of the transportation, contracting, and deployment of bonded labor, no less and no more akin to the slave traders and plantations of the eighteenth-century United States. Commerce in seasonal farm labor is a lucrative business, and those involved all have names and addresses.

In response to a question by Representative Chris Reykdal of Washington's House of Representatives' Labor and Workforce Development Committee regarding the exorbitant tax subsidies and incentives provided for the agricultural industry, Scott Dilley, spokesman for the Washington Farm Bureau, said,

> It would be nice if we could extend some of those tax preferences to other businesses as well . . . those preferences have grown up over the course of generations. Comes out of family farms, most farms in Washington state are still family owned, they may be corporations, because of tax liability purposes, but they're still family owned, and they're multi-generational. Those preferences are needed in order to keep ag competitive with other farmers across the country.[26]

The Bracero Program, similar to these tax subsidies and the land tenure system that preceded it, provided a labor incentive for growers that was directly linked to this mentality of making US farms competitive in the world market. Just as in the slavery system of the US South's plantation economy, the farms that benefited most from the use of bonded labor were not the small-scale family farms that operated domestically within the borders of the United States, but instead those large-scale firms that produce export crops for a world market.

The wartime foreign labor contracting program that came to be known as the Bracero Program expanded in the years after World War II, spurred by the mass deportation of Mexican nationals during a campaign known as Operation Wetback in 1954. This immigration enforcement policy decision, coupled with the bonded labor program, had the effect of driving wages down, and it also kept Mexicans from gaining a stronger foothold in the manufacturing industry in the urban centers of the Southwest, in particular Los Angeles.[27] I argue that the agricultural industry, and Mexican labor in particular, came to function as the whipping boy for all of the advancements made by organized labor in the manufacturing industries in the early twentieth century.

It is well documented that the Bracero Program lowered labor costs for growers, which allowed large-scale farmers to take home more of their earnings. It allowed growers to increase yields and produce more acreage, and this temporarily reduced the cost of food. It did so by depressing the wages of farmworkers, preventing the success of farm labor strikes, and encouraging growers to invest in lobbying for the continuance of special treatment.

The Bracero Program was terminated in 1964 due to popular resistance. Media campaigns, such as Edward R. Murrow's *Harvest of Shame*, which aired on national television the day after Thanksgiving in 1960, also drew attention to the living and working conditions of the domestic migrant workers who were being displaced by the indentured laborers. On the racial front, the rise of the civil rights movement in both the US South and Southwest also helped insure that the policies were rescinded (see figure 13.2).

Fifty years later, in 2014, we find ourselves in a similar historical moment. Over the course of half a century, many of the gains of the civil rights movement and the early wins of organized labor have been completely or partially rescinded. The rise of the prison industrial complex began almost in direct response to the gains of the civil rights movement, but even more directly as a response to the black power movement of the 1970s, when federal agencies, including the CIA, were directly involved in the assassination of emerging leaders and in the attack upon the ethnic enclave communities through the introduction of crack cocaine and firearms, as documented by the release of COINTELPRO materials.

The militarization of the US-México border in the 1980s and the increase in immigration enforcement raids on workplaces that followed terrorized Latino immigrant communities until 1986, when President Ronald Reagan announced the Immigration Reform and

## Figure 13.2. *Bracero contracts vs. removals, 1943–64.*

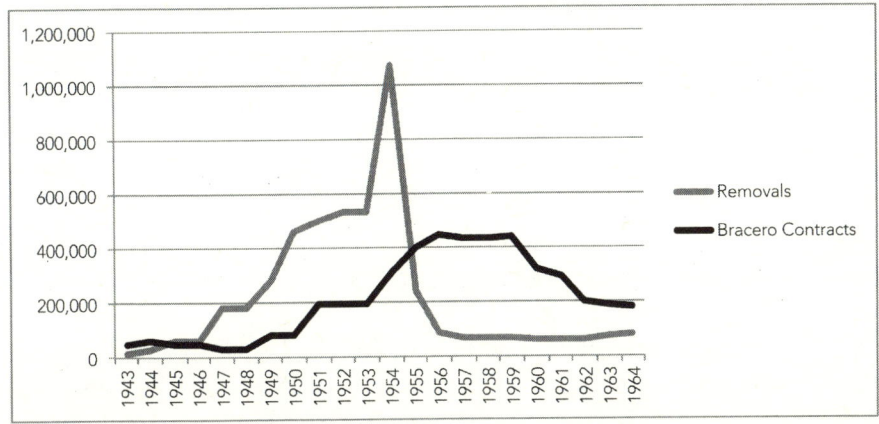

Sources: Elac 1961; *Wikipedia* 2016; Hernández 2010, 122–23.

Control Act (IRCA). The bill's Seasonal Agricultural Workers (SAW) program provided a pathway to citizenship for undocumented farmworkers, which had the effect of sabotaging major unionization campaigns underway by the United Farm Workers (UFW) at the time. Once again, immigration policy was used to undermine the union organization of farmworkers. The UFW had been able to pass the California Agricultural Labor Relations Act in 1975, on the heels of the civil rights movement, which allowed the farmworker union to have protection under the law and extended the rights of farmworkers in California. IRCA had the effect of flooding the labor market with eligible nonunion farmworkers, many of whom felt a sense of debt to the growers who had sponsored their pathway to citizenship under the SAW program.

Between 1983 and 1990 thousands of farmworkers in California lost thousands of union contracts due to policies of Governor George Deukmejian that limited the capacity of state labor enforcement agencies and due to increased competition for jobs after IRCA, as workplace immigration raids had less of an impact on the size of the agricultural labor force. A flooded labor market also meant an increase in domestic or internal migration. Washington's Latino population, for example, grew exponentially in this period, as did Latina/o populations in the Midwest and later in the Deep South.

The passage of the North American Free Trade Agreement (NAFTA) on January 1, 1994, also increased international immigration to the United States over a course of five years. By 2000 many Indigenous peasant farmers who had been able to sustain themselves on their own agricultural cultivation were being undersold by subsidized US crops, corn in particular, making it necessary for these farmers to sell their labor power to survive. Many of these Indigenous people came to the United States in debt, and this facilitated conditions of bonded labor that were no less than indentured servitude. This debt peonage system emboldened US growers to rescind many of the gains that had been made for farmworkers since the civil rights movement, including protections against abuse, exploitation, child labor, and wage theft. The fact that many of the victims could barely speak Spanish and could not speak English made it difficult for this population to defend itself.

The suspension of civil liberties as part of the Patriot Act of 2001 paralleled this development in agriculture and also led to the completion of the militarization of the US-México border, a process that had begun roughly around the end of the first Bracero Program in 1964.

The militarization of the border and US foreign policy led to the detention of immigrants beginning in 1981, when economic refugees from Haiti and, later, Cuba were held in camps in Florida. Mandatory detention did not evolve until 1996, with the passage of antiterrorism and immigration reform acts. This trend continued through 2003 with the establishment of the Department of Homeland Security (DHS), when Immigration and Customs Enforcement (ICE) became responsible for Detention and Removal Operations (DRO). ICE has since opened and operates more than fifteen immigration detention centers, some of which hold entire families for indefinite periods of time as they wait for processing, court rulings, and deportation. This economy is an extension of the prison industrial complex and is the pinnacle of the use of immigration policy as a means of labor control.

When we look at the proliferation of and dependency on the use of a bonded labor program that depends on highly regulated H-2A visas and compare the number of these visas to deportations in the United States, it appears that we are at that moment just before a shift into a formal labor market economy, which would undo many of the gains made by immigrant workers toward achieving a unified class

**Figure 13.3.** US H-2A contracts versus US removals, 1998–2015.

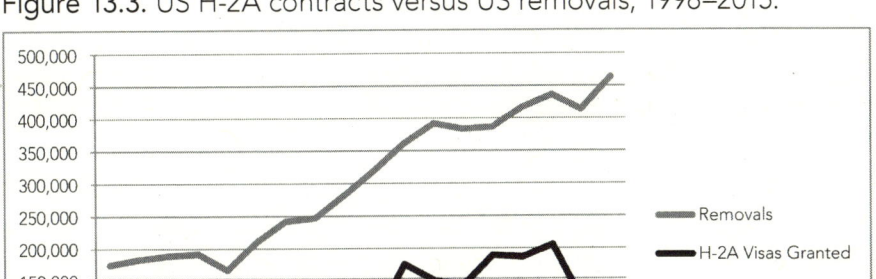

Sources: US Department of Homeland Security, *Yearbook of Immigration Statistics*, Table 39, https://goo.gl/dccMrL; US Department of Labor, FY Reports, 1988–2015.

composition and collective working-class power across the food chain (see figure 13.3).

It is fitting that we begin our narrative with the struggles of immigrant workers in Washington State that emerged in 2013 as part of a wave of mobilizations across the food chain that included farm laborers and nonunion workers in the restaurant, fast-food, and grocery industries, and also included hunger strikes in prisons, jails, penitentiaries, and, later, detention centers over access to food, mistreatment, and low pay. All of these mobilizations had organizing histories that extended back ten years and beyond, coming from labor rights, civil rights, immigrant rights, farmworker rights, gender justice, and resistance culture organizing work on multiple fronts. Though the history of these struggles does not need to be known to create change, as the Not1More Deportation poem in the epigraph states, knowledge of this history provides a reference and catalyst for other communities engaged in decolonial processes to raise class consciousness, and to build on the knowledge that our anonymity does not mean we are alone.

The explosion of farmworker and immigrant worker mobilizations in Whatcom County and Skagit Valley did not emerge in a vacuum, and we can learn about the context of international encounter, strategic storytelling, and successful campaigns through the history of Community to Community Development (C2C), an ecofeminist women-led grassroots political organization based in Bellingham, Washington, which was founded in 2004 by Rosalinda Guillen and community stakeholders.

## Community to Community Development

I open this section with an extended excerpt from field notes developed while I worked in Burlington and Bellingham and with C2C:

> It was a Saturday; Nayra was eight months pregnant with Harold, her fourth child. Her entire family worked together, pruning raspberry bushes in Lynden, Washington. Nayra and her husband would do the heavy work while Eddy her nine-year-old son and Eva her seven-year-old daughter would trim only around the edges and tie the anchor lines for their parents to complete afterwards. Enrique, her youngest was a year and a half old, he had stayed with a babysitter so that Nayra could work with the family. This was their routine on weekends because there was no school. Both Nayra and Pablo wanted to teach their children the value of work, and as a reward they promised to take them to McDonalds for lunch. Around noon, they left the raspberry fields and were on the way to McDonalds for a family meal.
>
> As they drove down Guide Meridian, the main rural road in Lynden, Nayra noticed that a police car, . . . almost out of sight behind them, turned around and began pursuit. She noted that they had followed all the street laws and were not driving erratically, so both she and her husband Pablo were surprised when they saw the lights behind them. They quickly pulled over and the Lynden police officer, noting they were Mexican, asked if they spoke English. Pablo communicated with the officer in broken English, his accent was more Mixteco than Spanish, but the officer insisted that an interpreter was necessary. The Lynden Police officer called the Border Patrol for language assistance and they responded quickly. Upon arrival, the Border Patrol officer began grilling Pablo in Spanish. By then, the children were frightened, they began to cry uncontrollably, at which point the Border Patrol officer ordered them all out of the car and escorted them into his patrol car.
>
> Nayra did all that she could think of to intervene, in anticipation she had called her niece Fernanda, a US citizen, while the

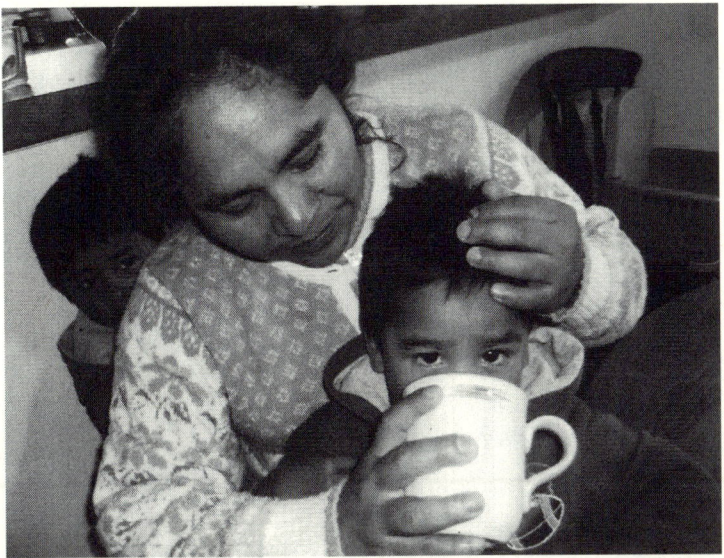

Figure 13.4. Nayra and her children Harold and Enrique, 2012. Courtesy of Rosalinda Guillen.

officer interrogated her husband. Fernanda arrived by the time they were in the patrol car and did all that she could to explain, but the entire family was detained and taken into custody by the Border Patrol.

While in detention Nayra experienced harassment and shame from a Chicano Border Patrol agent, she and her children were jailed together in a temporary holding cell, while her husband was separately jailed at a Sumas Border Patrol Field Station. Hours passed and Nayra ultimately signed a voluntary deportation order under duress, though she had held out longer than her husband Pablo, well into the evening. Shifts changed at the station, and only the two senior field officers remained. One of the field officers began to make conversation with Eddy and Eva, who were in their work clothes, overalls and muddy work boots. He asked Nayra if the children worked with her and her husband in the fields, and to him, she explained their labor, and that they only wanted their children to learn the value of work, and how they had nowhere to go on the Mexican border. Her story touched him, knowing that the children had not eaten lunch and had not been offered water or food while they had been in custody. He asked her if she had somewhere to go, and Nayra gave him Fernanda's phone number, which she had memorized. The agent then proceeded to tear up her voluntary deportation order and fill out alternate paperwork. Nayra, Eddy and Eva were released close to 1am into the custody of Fernanda with an order to appear in immigration court. Pablo was soon after, deported.

Nayra is one of the strongest leaders in Community to Community Development, and the organization helped her enunciate a compelling personal story that drove a campaign to change immigration enforcement practices. She was the lead plaintiff in a class action civil rights lawsuit against the Whatcom County cities of Lynden, Sumas, Blaine, and Everson for their local law enforcement practices of using the Sumas Border Patrol station as a dispatching hub and using Border Patrol officers as interpreters during routine traffic stops, making it difficult for immigrants to access emergency services without risking detention or deportation along the northern border.

Nayra's experience did not happen out of the blue. Immigration and Customs Enforcement (ICE) had carried out two major workplace raids in the city of Bellingham prior to her detention. The first was at Northwest Health Care Linen on August 30, 2006, and the second at Yakato Engine Specialists on February 4, 2009. The first ICE workplace raid resulted in the arrest and detention of twenty-six workers, and the second raid resulted in the arrest of twenty-eight workers by a multiagency SWAT team of seventy-six heavily armed agents in full gear. In the second case, oddly, it was the employer who reached out to Rosalinda Guillen of C2C, who, with the support of the Immigrant Solidarity Committee she had founded to work toward professional law enforcement action against the vigilante Minutemen Project stations along the northern border, was able to have three of the women released on humanitarian grounds to care for their young children.

With these three women, whose families had become members of C2C, the Las Margaritas cooperative and the Raíces Culturales Youth Empowerment Program were established as part of grassroots community organizing. The families led the organization in holding regular vigils at the Tacoma Northwest Detention Center, where the fathers of the families had been taken after the ICE raid. Member-led cooperatives at one point included a farmworker-run organic farm on three acres near Blaine, Washington, which together with Las Margaritas advanced the vision of Guillen and colleagues at C2C, who note on their website:

> We don't just fight for justice, we come with a solution. . . . Cooperatives are a basis for empowerment of immigrants and farmworkers. . . . This is a way out, by following the seven principles. . . . Social problems create the need and cooperatives can be the solution.

Guillen first encountered the cooperative model at the World Social Forum in Porto Allegre, Brazil, where she presented on the plight of farmworkers in the United States in her capacity as a high-ranking officer of the United Farm Workers, which she served as after securing the sole farmworker union contract in the State of Washington in agreements with Chateau Ste. Michelle and Columbia Crest wineries in 1995. Even the name "Las Margaritas" honors the local solidarity economy that was established in Porto Allegre as an umbrella of women's cooperatives and microenterprises.

Later on, Nayra's nieces would pilot a tortilla-making cooperative, given the Mixteco name of Hita Quiuci, and an educators' cooperative with the name of the Canopy Collective. These cooperatives would set the groundwork for developing a solidarity economy center curriculum designed specifically for farmworkers, which was developed over years by Guillen with the support of community stakeholders.

But a forward-looking stance by the grassroots organization was by no means its only avenue of effecting social change. Nayra's story would birth a member-driven community campaign to end racial profiling in Whatcom County that included Nayra's civil rights lawsuit and a media campaign that joined efforts to stop the expansion of secure communities, a law enforcement policy allowing different agencies to share information, which led to an increase in the detention and deportation of low-priority undocumented immigrants, as Nayra's story illustrates.

## Racial Profiling and Harassment along the Northern Border

Between 2006 and 2012 Community to Community Development collected intake forms from victims of racial profiling that were referred to the organization by other members and community organizations. One America (formerly Hate Free Zone) also conducted a more limited study, which produced a report in 2012. Figure 13.5 compiles and compares the number of reported traffic stops leading to DHS detention and/or deportation by the initiating agency using data from both organizations and data that I collected between 2011 and 2013. Most alarming to me is how the majority of cases of routine traffic stops leading to detention occurred in Lynden, the city where Nayra's family was stopped.

Figure 13.6 further outlines the locations where this harassment occurred based on the reports. The majority of the racially motivated harassment targeted at Latino immigrants from 2004 to 2012 took place during quotidian activities such as driving to and from work, at home, at the workplace, while shopping, and even randomly on the street. Based upon these compiled data, Community to Community Development filed a civil rights complaint holding that the cities of

**Figure 13.5.** Traffic stops leading to detention by initiating agency.

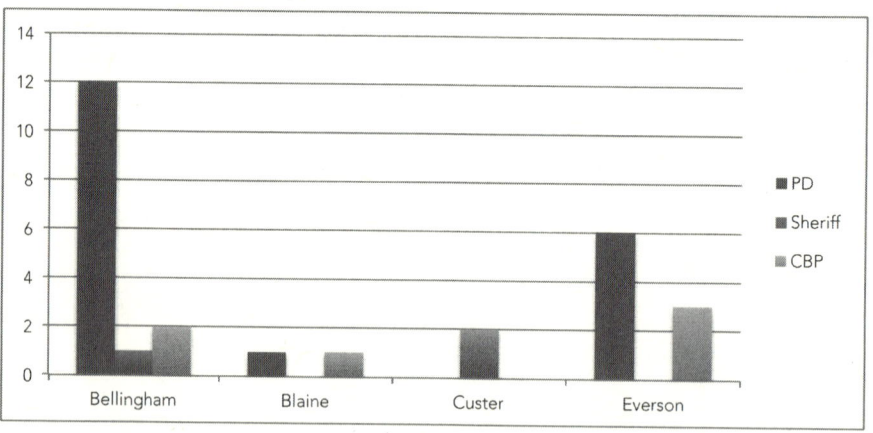

Source: Author's field notes; Community to Community intake forms; One America 2012 report.

Everson, Lynden, Blaine, and Sumas systematically violated the civil rights of Mexican farmworkers, including Nayra's family.

The lawsuit sought to extend the member-driven campaign for law enforcement agencies to be governed by federal and statewide civil rights laws that had banned the use of racial profiling in policing since 2000.[28] The lawsuit held that Mexican farmworkers and undocumented immigrants were entitled to access professional law enforce-

**Figure 13.6.** Reported locations of harassment in Whatcom County, 2004–12.

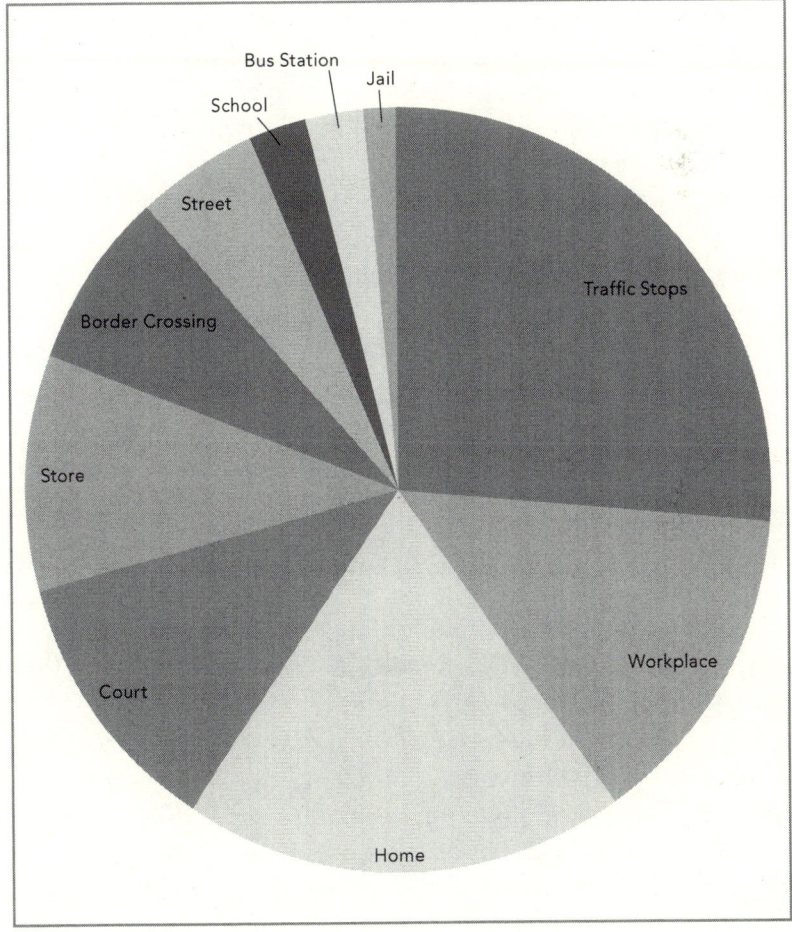

Source: Author's field notes; Community to Community intake forms; One America 2012 report.

ment and emergency services without fearing detention or deportation, which was severely limited through the practice of dispatching
through the Sumas Border Patrol Station and the practice of calling
upon CBP agents to interpret during routine traffic stops.

The successful civil rights lawsuit was one of many that were filed
along the US border regions. In Washington State, this included similar lawsuits in Spokane to the east and in Forks on the peninsula.
These lawsuits sought to strengthen the legislative project of limiting the use of the program Secure Communities by law enforcement
agencies in Washington State because of the proliferation of harassment based on racial profiling policing tactics. However, the legislative
campaign to end Secure Communities met failure in the Washington
State Legislature, dying in committee in the spring of 2014. The failure
of state legislators to act upon a popular bill sparked a series of direct
actions that encouraged detainee-led hunger strikes at the Tacoma
Northwest Detention Center, the topic of the final section of this essay.

After winning the civil rights lawsuit, the US Department of
Justice conducted a site visit to Bellingham on February 7, 2013, where
they announced a new executive memorandum by President Barack
Obama that took effect on November 21, 2012, and declared that

> if a federal, state or local law enforcement organization other
> than another Department of Homeland Security (DHS) com-
> ponent requests CBP assistance based solely on a need for lan-
> guage translation, absent any other circumstances, those requests
> should be referred to a list of available local and national transla-
> tion services such as those provided by the Interagency Working
> Group on Limited English Proficiency.[29]

This memo corresponded directly to the complaint filed by Community
to Community Development and its allies across the state and along
the borders of the United States.

A second memo had been issued on January 18, 2013, announcing
that schools, places of worship, community centers, and hospitals are
locations that require "careful consideration and planning" in regards
to enforcement actions, thus discouraging Border Patrol agents from
harassing people in their everyday activities. These early civil rights
wins led to the complete elimination of the Secure Communities program by memorandum on November 20, 2014, but only after wide-

spread mobilizations by low-waged workers, both immigrant and citizen alike, across the food chain.

Prior to the mass mobilizations by immigrant detainees, including those within the detention centers, was the emergence of Familias Unidas por la Justicia, a community of Mixteco and Triqui migrant farmworkers who forced the reluctant agricultural industry to recognize their transformation into an independent farmworker union by utilizing existing state labor law and by seeking the support of grassroots community organizations, including Community to Community Development, and by securing the support of traditional labor unions, such as the AFL-CIO.

## Familias Unidas por la Justicia

On July 11, 2013, more than two hundred farmworkers observed the first of seven work stoppages during the 2013 harvest season against their employer Sakuma Bros. Farms in Burlington, Washington, just two hours north of Seattle. Contrary to reports that the stoppages were spontaneous, this community of Indigenous migrant farmworkers reported having stopped work at least once a season during the decade prior to this historic date.

Most of the previous work stoppages had not gained more than the firing of the instigator. Medical anthropologist Seth Holmes wrote of one strike on the farm in 2004 where he had intervened together with local clergy and immigrant rights advocates and gained a temporary piece-rate increase along with a company memo against harassment in the workplace.[30]

The 2004 strike was significant because it established Marcelino Raymundo, one of Holmes's central informants, as the spokesperson for all of the farmworkers. In the years following, Raymundo would be called upon by Steve and Ryan Sakuma to "placate" the farmworkers if a work stoppage lasted longer than a day. Cornelio Ramírez recalled that in 2008 he had stood up to an abusive foreman named José, who had tried to fire him. Ramírez, a Mixteco-speaker, recalled that the mestizo foreman José mistreated all of the Indigenous farmworkers, in particular the Triqui-speaking farmworkers. Ramírez argued in 2013 that "La huelga de Marcelino no saco nada" (Marcelino's strike did

Figure 13.7. Two hundred farmworkers confront Ryan Sakuma (*center*), president of Sakuma Bros. Farms, during the first July 11, 2013 strike. Courtesy of Tomás Madrigal.

not make a difference), referring to the negotiations documented by Holmes in 2004.[31] In contrast to the top-down negotiations of 2004, Ramírez joined nine other coworkers on a negotiation committee formed on July 11, 2013, in an open-air assembly that occurred just after the confrontation pictured in figure 13.7. There the two hundred farmworkers and their families came up with a fifteen-point list of demands that the ten-person negotiation committee was to address with their employer.

The demands required the reinstatement of Federico Lopez, a raise in the piece rate, the removal of an electronic scanner, the return to paper punch cards instead, that intimidation in the workplace stop, the removal of the foreman, better treatment of workers (including respect, sanitary living conditions, better cabins, and not to be yelled at), the ability to take sick leave, that employees be treated with dignity, that foremen stop intimidating workers, that the company pay transportation costs for any migrants who are fired, that guest workers not be hired since they did not hire them to harvest strawberries, that their

ORGANIZING

employer pay overtime, and that the employer make it easier to secure child care by reporting hours worked on their pay stubs.

Ramírez, though one of the shorter farmworkers, is not one to back down from advocating for himself and his family. He has worked across the food chain, having picked mandarins and pruned vines in California, and having picked berries and processed seafood in Washington. He is one of the fastest skilled pickers of mandarins and blueberries and takes pride in his work. In 2011, Ramírez was injured on the job in Delano, California. He was paid a settlement for his injury because he advocated for himself under California labor law. He observed, "En California es diferente" (Things are different in California). It is fitting that the primary work stoppage that occurred at Sakuma Bros. Farms during the 2014 berry harvest season revolved around the constructive firing of Ramírez on August 11, 2014.

The embattled berry farm refused to hire unionized farmworkers until ordered to do so by a Skagit County Superior Court judge on June 30, 2014. To claim a loss, the Sakuma executives decided to give away thirty acres of strawberries to the public in a free self-pick that same weekend they were forced to rehire and house Ramírez and his family.

In the weeks following, Ramírez reported a high frequency of closed audience meetings between antilabor consultants Mario Vargas and the farmworkers in reaction to the daily presence of a union access team led by President Ramon Torres near the workplace. On August 3, 2014, Ramírez had replayed a recording of himself standing up to antiunion consultants and advocating for his union, Familias Unidas por la Justicia, as a member of the negotiation committee. In the recording, Mario Vargas, an HR consultant, argued that the UFCW Local 5 access to the Sakuma-owned NorCal Nurseries, Inc., was insignificant and that the farmworkers should not join Familias Unidas por la Justicia, because they were lying about its significance. In the recording, Ramírez asserted that the consultant had no right to tell anyone not to join a union, that it was their choice. In the background another female farmworker argued that instead of talking about her union, he should talk about raising the piece rate. The antiunion campaign did not stop at closed meetings. Ramírez's foreman took him aside one-on-one to suggest that he resign from the union, promising future benefits of siding with the corporation. Needless to say, Ramírez was cited with warnings for speaking up for himself and his union. He was also cited, he claims falsely, for not washing his hands before picking. In this way, Ramírez was constructively fired based upon the arbitrary enforcement of a warning system.

The farmworkers' strike and threat of further litigation led the company to rescind the warning structure as the farmworkers demanded, and as Cornelio Ramírez negotiated on behalf of the union during his exit interview with Ryan Sakuma. During this strike, Rhett Searcy and his younger brother Ryan Searcy deployed the use of heavy machinery to keep the striking farmworkers from the union access team, who at the time was relegated to public property.

On October 28, 2014, after union counsel presented as evidence a dispatch call by Rhett Searcy, Sakuma Bros Farms security consultant, to local law enforcement, threatening the use of violence against the farmworkers to defend the company's private property, Skagit County Superior Court judge Susan Cook granted Familias Unidas por la Justicia an injunction against the farm's policy of denying full access of union representatives to company property.[32] During the trial, Steven Sakuma testified that he did not authorize Searcy to do so on behalf of his company and that he did not know what he hired Searcy to do, only that he hired him because of his experience.

As one can see, the unionization campaign by formally non-unionized migrant farmworkers has relied heavily upon the deployment of worker strikes and litigation. The union has made history by settling two wage theft claims with the company, one for $6,000 in back wages to youth between fourteen and twenty years old over the three-week strawberry harvest of 2013, and a second one, an out-of-court settlement for $850,000 for the denial of paid rest breaks and lunch breaks over the previous three years. The farmworkers' union has also successfully protected their concerted activities from retaliation and interference in the form of injunctions based on Washington State's Little Norris-La Guardia Act that provides protections to any employee from their employer in Washington State. But even those accomplishments are only the tip of the iceberg.

Having launched a berry boycott on August 19, 2013, by a unanimous vote by 246 Sakuma berry pickers, Familias Unidas por la

Justicia was able to encourage multiple grocers to pull Sakuma-labeled fruit off the shelves. The following year, Sakuma Bros. Farms increased their contract to pack fruit in containers with Driscoll's labels and also packed in unlabeled containers to attempt to avoid the boycott, which caused the farmworkers to encourage grocers to stop selling Driscoll's-label berries, a campaign that was met with success at several cooperative grocers across the United States.

Another avenue that was opened because of the farm labor mobilizations involved legislative oversight. Farmworkers were invited to present at a special legislative work session held in Everett, Washington, on November 14, 2013, which led to the demolition of the worst of Sakuma Bros. Farms' cabins under the supervision of Senator John McCoy, whom the farmworkers honored during a legislative reception held in Olympia in 2014.

On February 11, 2014, members of Familias Unidas por la Justicia held a demonstration at the rotunda at the Washington State Capitol on Latino Legislative Day. The demonstration evolved into a direct action inside the rotunda in support of the Trust Act against Secure Communities. The Washington Trust Act died in committee shortly after and sparked a second direct action at the Tacoma Northwest Detention Center led by Maru Mora Villalpando and members of the Not1More Deportation coalition on February 24, 2014. This action sparked hunger strikes inside the Tacoma Northwest Detention Center, as we will describe in the next section.

Familias Unidas por la Justicia, under the leadership of Ramon Torres, sent caravans in support of the hunger strikers inside the Tacoma Northwest Detention Center on March 15 and 20 and on April 6, and later joined the Not1More Deportation contingent in the Seattle May Day workers' rights march in 2014.

On June 5, 2014, Governor Jay Inslee appointed a farm work group to advise him on possible administrative solutions to the issues brought forth by the uprising at Sakuma Bros. Farms, as proposed by Scott Dilley on November 14, 2013. Governor Inslee had authorized $50,000 in hotel vouchers from state tax payers administered by José Ortiz of the Tri-Parish Food Bank to subsidize Sakuma Bros. Farms executives' refusal to house union farmworkers until the matter was settled on June 30, 2014, via injunction. On October 16, 2014, Governor

Inslee appointed Felimón Piñeda, vice president of Familias Unidas por la Justicia, to the farm work group.

The independent farmworker union became recognized by the state through litigation as well as by joining the ranks of the Washington State Labor Council, AFL-CIO. The employer, Sakuma Bros. Farms, also had recognized the union by settling the $6,000 wage theft over strawberries, agreeing in writing to a new piece-rate process on July 25, 2013, and by reaching a written agreement promising no reprisals on August 14, 2013. Early on, it appeared that the company was negotiating in good faith with the emerging union, but it was short-lived. Within weeks of operation, both agreements were broken by the company, and the first to be broken was the piece-rate agreement on the day of the no-reprisals agreement. After the arrival of 175 H-2A guest workers at Sakuma Bros. Farms during the last week of July 2013, it became clear that the farm had used the agreement in order to secure the fulfillment of the contract with no intention of honoring the negotiations, evidenced by the company's immediate denegation of the agreement once the H-2A guest workers were secured.

The US Department of Labor investigated the violation, but the agency's primary role as an enforcement agency is to bring the company to compliance, rather than to discipline the corporation for illegal practices. The silver lining for the farmworker union was the closer scrutiny of the corporation's application the following year for over 438 H-2A guest workers in 2014. The application was defeated through an organizing campaign of Familias Unidas por la Justicia leadership, which communicated transparently with regulating agencies and their employer about their commitment to work for the company where they had worked for over a decade. Rank-and-file members of Familias Unidas por la Justicia signed letters of intent to work and attached copies of their Sakuma Bros. Farms identification cards and employee numbers from the previous year.

After the company lost the ability to deny the rehiring of union members for missing work during strikes through litigation and was ordered by the court to continue to house migrant farmworker families, the corporation withdrew its application for guest workers.

The independent farmworker union continues to make history and to build solidarity across the food and commodity chain.

The union participated in picket lines in solidarity with health care workers in their struggle for unionization, with Walmart workers in Bellingham, Burlington, and Walla Walla on Black Friday. The union boycott had banners present for the historic climate march in New York, and leadership also took a public stance against the state murder of young people in Guerrero, México, where many of the rank and file are from. The Frente Indigena de Organizaciónes Binacionales (FIOB) recognized the union as a sister organization during their organizational congress in 2014. On December 14, 2014, they made history once again by electing negotiation committee representatives that represented each Indigenous group, most hometowns, women, and young people, from Washington and California.

As amazing as this community of farmworkers is, they are not the only group of immigrant workers who have organized themselves and changed the world as their own advocates. As alluded to in the previous sections, a major site of struggle for immigrant food-chain workers in Washington State has been the Tacoma Northwest Detention Center.

## Northwest Detention Center Resistance

I begin this part of the story again by drawing from field notes taken during direct actions:

> José Moreno spent six months at the Tacoma Northwest Detention Center, he was held on detainer for a DUI because of Secure Communities.
>
> He lives and works in Renton, making a living as a restaurant worker. He currently busses tables, but had worked his way up from a dishwasher, moving on to prep cook and then to his current position. Though he was absent six months, he was rehired because of his work ethic, he currently earns $11.00 per hour + tips as a part-time restaurant worker and spends the rest of his time as a spokesperson for the NWDC Resistance, an organization that emerged from the struggle at the Tacoma Northwest Detention Center.
>
> José is from México. He originally came to the U.S. to work with his brothers in Florida and moved with them along the eastern migrant corridor to Pennsylvania where they worked

construction jobs. He made his way to Washington because he decided he wanted to live apart from his brothers who weren't completely aware that he was gay.

By the time that José arrived at the Tacoma Northwest Detention Center, he did not have anyone to deposit money on his behalf for commissary, let alone to advocate for him from the outside. As luck would have it, the day he was transported to the Tacoma Northwest Detention Center, the GEO transport he was on was one that was being blockaded by a group of #Not1More Deportation activists. He took the experience with him as he was processed and would spend the next six months at the NWDC.

Once inside, José had to rely upon his ability to make light of situations, observing and code-switching to fit in with the different groups at the detention center. He observed that his pod was segregated into four distinct groupings. Of these, the Guatemalan people were treated the worst and as personal slaves to the Salvadoran group. José, like most of the detainees had no special allegiance to any of the four groups and was able to move around between them. He navigated carefully to ease tensions and to trade in the informal economy that was developed by the detainees inside the detention center.

José observed that everything had value inside the detention center. He, for example would save butter, jellies, bread and milk from breakfast to make himself a snack overnight and trade the milk to the "Cafeteros" who liked to drink coffee at night for a cup of ramen noodles from commissary. The economy also involved different types of services between detainees that they brought with them from their experience as service, food and commodity chain workers on the outside.

Based upon his observations, José categorized new detainees in the following way and adjusted his approach toward befriending them accordingly: 1) those who slept all day, 2) those who kept to themselves, and 3) those who did anything they could to make the time go by.

The detainees occupied themselves primarily by playing soccer and dominos. Domino matches would start at the crack of dawn and also contributed to the economy. After a conflict over the noise the dominos made on the metal tables, they figured out that the noise the bricks made was muffled if they used a blanket as a table cloth.

After a while, José built a support network around him through the friendships he made with other detainees. The main person who he remembers was nicknamed "el tio" (uncle). He had earned his nickname because he had the habit of calling everyone he met "sobrino" (nephew). El tio provided the muscle to back up his thinking and words when addressing other detainees.

José's inside organizing model was to build relationships with the detainees. José would make sure to be the first to greet new arrivals with a gift of ramen noodles, knowing that his act of compassion would contradict the isolation he knew from his own experience and because he was aware that many had not eaten up to 16 hours while they were being transported and processed by ICE and later GEO.

His outreach to new detainees became the basis of a detention center solidarity economy based on friendship, because the new detainees would reciprocate José's gesture and some would also do the same for new arrivals. When José would come across the type of people who would try to sleep all day, he would playfully go to their beds and say, "¡No seas puto, salte a jugar!" (Don't be a fairy, come out and play!) Often the juxtaposition of an openly gay man calling you a homophobic slur was enough to contradict their depression. José was also one of the few detainees who would take the time to sit and talk one on one with those who kept to themselves to make sure they did not feel completely isolated, he often used his sense of humor to connect with these men.

Despite his best efforts to organize, fights did often break out inside his pod. The stress of being detained indefinitely, often for minor offenses, and being placed in the same holding facilities as folks actually involved in organized crime created a sort of pressure cooker as mistreatment from guards and other inmates based on segregation would create conflict. These fights would blow up over quotidian, seemingly small things such as someone cutting in line or something even more petty. When fights broke out, José would run to his cell, though he had the protection of people like el Tio, pod fights were a free for all that everyone used to let out steam and the safest place in that environment was inside your cell.

Being that there was plenty of time to talk, José used that opportunity to persuade detainees to select more appropriate targets than each other. One of the first issues that he organized for the benefit of the pod involved sanitation. He organized the members of the pod to orient detainees not to relieve themselves all over the bathroom floor and on the toilet seat. He argued that it was a hygiene issue because detainees could not afford to get sick and unsanitary communal bathrooms contributed to the spread of hepatitis and e-coli amongst the population. His experience as a restaurant worker no doubt was useful in explaining the consequences of poor sanitation.

Although el Tio was the official "leader" that the rest of the detainees followed, it was José who would do all the talking. He explained that he would speak to the detainees on their level, reminding them that they were not at their home and shaming them for being filthy, explaining the consequences of poor hygiene.

This is how detainees organized themselves as a pod before launching a series of hunger strikes, the first of which lasted 57 days and which José was able to lead and observe on the inside as a detainee and upon his release as a public spokesperson on the outside, joining the very people who he had seen upon arrival at the Tacoma Northwest Detention Center.

NWDC Resistance is a confluence of immigrant stakeholders including detainees, their families, immigrant rights activists, immigration lawyers, faith communities, and advocacy organizations. Their unique inside/outside organizing approach demonstrated the complexity and the coordination that it took to place a tremendous amount of pressure upon the federal government, the Department of Homeland Security, and the private prison contracting firm GEO Group. In their own words,

> NWDC Resistance is a volunteer community group that came together to stop deportations earlier this year at the now infamous Northwest Detention Center in Tacoma, WA under the Notimore campaign umbrella, and which then supported three hunger strikes organized by immigrants detained there calling a stop to deportations and better treatment and conditions.[33]

The main organizations involved in the leadership of NWDC Resistance include Maru Mora Villalpando and her consulting firm

Latino Advocacy, LLC; Sandy Restrepo with Colectiva Legal del Pueblo (People's Legal Collective); and the "Colectiva de Detenidos" (Detainee Collective), which was made up of several volunteers including Angelica Charazo, José Moreno, and Verónica Noriega. Beyond these three core organizations, there was also a broad base of supporters from different faith, labor, and community organizations as well as private individuals and families who decided to become involved in the movement for immigrant rights after seeing or hearing or reading about the mobilizations on television, radio, and newspapers.

The work of the leadership on the outside came from a long line of organizing for immigrant rights that spanned back over a decade.[34] Both Maru Mora Villalpando and Sandy Restrepo, for example, were former members of "El Comité Pro-Reforma Migratoria" (Committee for Immigration Reform), which was formed in Seattle after the WTO demonstrations in 1999 and later changed its name in 2008.

El Comité was plagued by a series of internal splits, which led to Mora Villalpando's departure in 2003. After serving as lead organizer for the "Tenth Annual March for Immigrant Rights" in Seattle, with Washington Community Action Network (WCAN), she founded Latino Advocacy, LLC, in 2010 to create a new independent and sustainable grassroots-organizing model. Mora Villalpando cited sexism as the major reason for her leaving El Comité to find new pathways of community organizing. Due to the unchanging dynamics inside the organization, Sandy Restrepo also left El Comité in 2013 to cofound Colectiva Legal del Pueblo.

Both Mora Villalpando and Restrepo played pivotal roles in catalyzing the mobilizations at the Tacoma Northwest Detention Center. In her capacity as a consultant, Mora Villalpando had traveled the United States, meeting with different communities of immigrants through dignity dialogues and community-organizing "101" workshops. Mora Villalpando covered the underserved areas of Oregon and Washington most extensively.

Mora Villalpando participated in the creation of what came to be known as the "Dignity Campaign," a coalition of organizations that took up the difficult task of challenging the well-funded Comprehensive Immigration Reform (CIR) campaign for S. 744, which if passed would have transformed and increased the guest worker visa program in the agriculture and technology industries and extended the guest worker program to include the service and construction industries through the establishment of the new "W" visa. Under this bill, most immigrants would have been excluded from pathways to citizenship.

Another of these key relationships was with the National Day Laborers Organizing Network (NDLON), which helped launch the Not1More Deportation campaign. Marisa Franco, the lead organizer for Not1More Deportation, led local shutdown actions against ICE, including the one at Tacoma Northwest Detention Center. Mora Villalpando in her capacity as a lobbyist had become well known at the state capitol. Anti-immigrant and anti–working class special interest groups formed what has come to be known as the "Washington Compact," and this group had Mora Villalpando and other, mostly female-bodied advocates, including Rosalinda Guillen, blacklisted from the business of immigration reform.[35] The groups comprising this so-called Washington Compact included several conservative and neoliberal policy NGOs and lobbyists for big agri-corporate growers, chambers of commerce, and law enforcement associations, and their influence was left unchecked by potential allies in Congress.

The Washington Compact modeled other special interest group–funded initiatives across the nation spearheaded by the law enforcement, chambers of commerce, and grower's lobbies, who were set to gain a considerable amount of capital with the passage of S. 744 in the form of federal government contracts and formalized labor markets for their stakeholders. S. 744 was defeated by the Not1More Deportation campaign, by the broad coalition's willingness to return to the media-savvy tactics similar to those used by Act Up in the 1980s, and by contributing its "be your own advocate" organizing model to the fray, which was much more attractive to immigrant workers than settling for the scraps that the compacts promised through CIR.

Beginning in 2012 when the Trust Act bill was introduced to the Washington State Legislature, Mora Villalpando lobbied heavily for its passage. Since many congresspeople were already aligned with the special interests outlined in the Washington Compact, H.B. 1874 was killed in committee. The Trust Act was perceived as negatively affecting the interests of the law enforcement lobby and any businesses that depend upon the mass detention and processing of immigrants.

This sparked a shift in the strategy that Mora Villalpando and her supporters used from individual lobbying to attempting to hold insiders accountable to the communities they represented. This included a public action against Congressman Frank Chopp for blocking the passage of the bill in 2013. Inspired by the uprising of immigrant farmworkers at Sakuma Bros. Farms, which Mora Villalpando had actively supported, along with the numerous Noti More Deportation direct actions she had witnessed in Arizona and other places, Mora Villalpando decided it was time to bring direct actions led by undocumented immigrants home to Washington State.

By the time that this strategy shift emerged, Restrepo was already supporting Mora Villalpando in the campaigns against Tacoma Northwest Detention Center. As an immigration attorney, Restrepo worked with the detainees inside. She would later play a pivotal role in linking communication between organizers inside and outside the guarded walls.

On February 24, 2014, Mora Villalpando and nine immigrant rights activists formed a blockade against GEO buses transporting detainees to their deportation. Restrepo and Angelica Charazo volunteered as part of the legal team for the action. Mora Villalpando and another activist had risked their own freedom as undocumented immigrants by engaging in the direct action, and together with seven other activists, they demanded an end to the over two million deportations that have occurred during the Obama administration. According to an article by *Yes! Magazine* that covered the blockade, news of the mobilizations reached all the way to the White House because of the hunger strikes that followed. On March 13, 2014, "the president announced a review of current immigration enforcement policies."[36]

José Moreno happened to be on the bus that was blockaded by the immigration activists. According to one account, Moreno "returned and told the other detainees about the support they had received from outside. Inspired, the detainees decided to organize a hunger strike."[37] Moreno helped organize the March 7, 2014, action and was released shortly after. He joined Mora Villalpando, Restrepo, and many others on the outside. Once the Colectiva de Detenidos was established, they gave the clear and justifiable demands. Led by Ramon Mendoza-Pascual, Paulino Ruiz, J. Cipriano Rios Alegria, and others who remained inside, the Colectiva de Detenidos demanded that the United States: (1) stop deportations, (2) allow bond so that detainees could fight their cases in their homes with their children, (3) end deportations for parents of children or spouses of citizens, (4) resolve cases more quickly, whether for asylum or other reasons, with a reasonable bond amount. The detainees themselves noted,

> [M]any people are in the detention center without bond and without any resolution of their asylum or other type of case, many times for months without the right to a bond because they are subject to mandatory detention. Without a bond we spend months, even 1 to 2 years locked up without knowing what's going to happen to us and our families and without being able to economically support our families, causing them to fall deeper into poverty or uncertainty, without knowing what will happen with our future or that of our U.S. citizen children and spouses.[38]

They further argued: "We believe that we deserve the opportunity to demonstrate that we want to be in this country legally and to contribute to this country" and emphasized that they were on hunger strike for their families' sake.

Hunger-strike leaders actions inside the detention center led to retaliation by GEO guards. Paulino Ruiz was transferred to a faraway detention center in a move to quell the uprising, and Ramón Mendoza Pascual and J. Cipriano Ríos Alegria were both confined to solitary as hunger strikes were staged twice more inside the detention center and again outside of the detention center by Mendoza Pascual's wife, Veronica Noriega, and by a new encampment in the fall of 2014.

On May 8, 2014, José Moreno and Mora Villalpando met with Suzan DelBene, Zoe Lofgren, Ted Deutch, and Joe Garcia, members of the House Judiciary Committee. They shared the plight of the hunger strikers and other immigrants, pushing the legislators to pressure the president on detention and deportation policies. They also met with DHS executives, including Esther Olavarria, who was the special counsel to DHS director Jeh Johnson.

This inside/outside organizing, together with the collective strength of multiple stakeholders across the United States working in unison, led to yet another executive order by President Obama on immigration enforcement, announced on prime-time television on

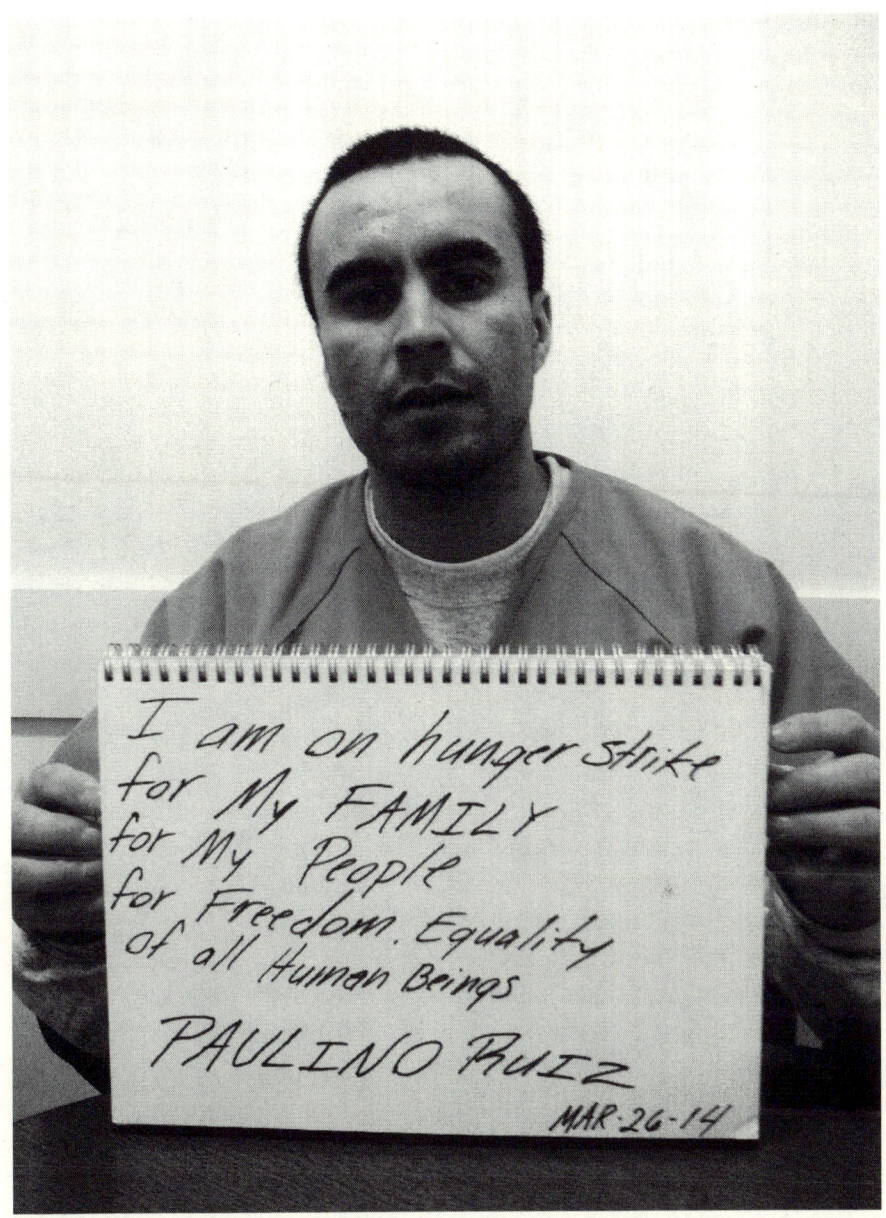

Figure 13.8. Northwest Detention Center hunger strike leader, Paulino Ruiz.
Courtesy of NWDC Resistance.

November 21, 2014. This action established Deferred Action for Parental Accountability (DAPA), expanded Deferred Action for Childhood Arrivals (DACA), and eliminated the Secure Communities program. It also funded further militarization of the border. Meanwhile, on December 10, 2014, the Colectivo de Detenidos received the Human Rights Organization award from the City of Seattle's Commission on Human Rights.

## Conclusion: "In order to live, we had to die."

Since the emergence of the second Zapatista movement in the jungles of southern México in 1994, the struggle for human dignity for Indigenous people in the greater Americas has gained a face. In veiling themselves, the Zapatistas brought down the veil of neocolonialism, exposing the continued imperialist expansion of the United States into México through the North American Free Trade Agreement (NAFTA) by saying, "¡Ya Basta!" (Enough!), in the only language they believed military operations would understand: armed insurrection on January 1, 1994.

The designated speaker on behalf of the Zapatista Command, Insurgent Subcommander Marcos, would say about Indigenous people, "A nosotros más nos olvidaron, y ya no nos alcanzaba la historia para morirnos así nomás, olvidados y humilados" (They forgot about us, and we had no history that could go beyond our deaths, forgotten and humiliated). Marcos continued: "Y miren lo que son las cosas porque, para que nos vieran, nos tapamos el rostro; para que nos nombraran, nos negamos el nombre; apostamos el presente para tener futuro; y para vivir . . . morimos." (How interesting that, in order to be seen, we had to cover our faces; in order to be named, we had to forsake our names; we gave up our present in order to have a future; that we may live . . . we had to die.).[39]

Many of the immigrants, the detained, and migrant farmworkers engaged in the actions and events described in this essay came from the ranks of these forgotten peoples, and though not all of them are Zapatistas, what we see in the historical record that I have presented here is that they bring something much more radical than an ideology, which is the practice of engaging in dynamic solidarity and non–

market value network economies in a diverse variety of settings, from the most supportive, as in the case of Community to Community Development, to the most repressive, such as the case of Familias Unidas por la Justicia and of course the Colectiva de Detenidos (Collective of Detainees).[40]

Solidarity economies are revolutionary in the face of globalized capitalism because these depend upon the formation of thick social relationships that rebuild the fabric of community and humanize the forgotten, the invisible. Solidarity economies are entrenched in the reproductive spheres of immigrant and farmworker communities; they are cross-pollinated by a wide variety of variants as cultures encounter other cultures and find commonalities in their difference and in their humanity. This process is politically recomposing the externally imposed class divisions and is challenging the nexus of power of the global capitalist project, which depends on unfettered control over labor markets and competition. This is, as Devon G. Peña has stated, "Solidarity outside and against predatory economies."[41]

I have to admit on reflection that at the beginning my studies with farmworkers, the scope was too narrow. Extending my analysis to the food chain opened new dialogues, but the new class that has been politically recomposed by immigrant workers is much bigger; it is beyond transnational, it is indeed global. We have been privileged to live long enough to experience and witness what global working-class composition looks like, and we have the benefit of having the ability to compare it to those moments in history where decolonial and peasant revolutions changed our world for the better, precisely because they were global in scope.

In my conversations with Familias Unidas por la Justicia's vice president Felimón Pineda, a Mixteco farmworker, he repeatedly emphasized that "farmworkers are not farm tools," we are human beings.[42] His words were the same rallying cry uttered from the mouths of other low-waged workers across the food chain the same year of the Familias Unidas por la Justicia uprising, and "We Are Human" was used years earlier by massive immigration, labor, and antiwar marches, which have continued, at least in Washington State, since November 30, 1999. It was a slogan repeated by both prisoners in 2013 and detainees in 2014 as they refused food and sacrificed their health in order to reclaim their humanity. As Marcos once said, "In order to live, we had to die."[43]

Food, from seed to table to action, is a project of life, as Pineda emphasized in an interview with Luke McKinley in early 2014.[44] The aims of global capitalism and capitalist development in contrast are "proyectos de muerte" (projects of death), as the late Juan Chavez Alonso of México's National Indigenous Congress (CNI) argued at the first Forum in Defense of Water in Vicam, Sonora, in 2010.[45]

In México, a popular understanding of the ongoing mass murder of Indigenous people by state forces—which has only increased over the course of the US drug wars involving counterinsurgency tactics deployed against Indigenous sovereignty and autonomy movements—refers to the deaths as having *autores intelectuales* (intellectual authors or designers). For the United States, this intellectual design has been part of the settler colonial logics of empire builders since the initial Yankee expansion under the force of Manifest Destiny and the Monroe Doctrine. In México, there has been a pattern of project after project promoting rapid mass industrialization combined with de-indigenization, efforts aided by the United States and other nations through a process long ago developed by the *científicos* (scientists) advising the Mexican dictator Porfirio Díaz. More recent iterations of this colonial project have included United States and Mexican, but US-trained, legal scholars, anthropologists, economists, and other social scientists advancing the green revolution agenda of Norman Borlaug clothed in the Western development ideology of Walt W. Rostow. This process accelerated during the Cárdenas *sexenio* (six-year presidential term). It certainly continued and expanded under the Salinas and Zedillo administrations. During the Zedillo sexenio, the anthropologist Arturo Warman was assigned the task of designing a plan to implement the dismantling of México's *ejido* system. The Zedillo administration's chief prosecutor, Jorge Madrazo, designed the plan to cover up the Acteal massacre just seventeen years ago. Such intellectual authors of capitalist projects of premature death have always had names and addresses. In the United States in the wake of the most recent report on the use of torture by the Central Intelligence Agency (CIA), we find the intellectual authors are also active because they have been allowed to act with impunity in the post-9/11 institutionalization of a state of exception that normalizes the persistence of crimes against humanity and especially the most vulnerable.

We are still faced with the enormous task of undoing the legacies of modernity and postmodernity as equally pernicious ideological projects of capitalism that stripped us of our humanity. Despite the cold calculations of capitalist megaprojects of death, the administration and surveillance of populations, and the racialized crafting of labor markets, the statistics and spin narratives veil the human and ecological costs of market-driven ideologies. These are the same antagonisms that over a century ago set the ground for eugenics and advanced genocide on an industrial scale. Ayotzinapa, Palestine, and Ferguson are but extensions of these legacies of settler colonialism.

Roxanne Dunbar-Ortíz dates this historical shift to the nineteenth century as the settler colonial capitalist mode emerged and was exported from the United States to other so-called developing countries as part of the rise of the American empire.[46] These intellectuals have designed a method and procedure to "streamline" the process of systematically apprehending, detaining, removing, and otherwise controlling and managing a target population; the current form of the sovereign ban denies the names, erases the human being, and reduces their traces to numbers fed into algorithmic equations to chart the strategic turns of this state of exception. The thing about fighting nations, bands, and tribes of people, as Dunbar-Ortiz pointed out during a book reading in Bellingham in 2011, is that Indigenous culture has a way of not confining itself to just one "tribal" population. Our ability to forge relationships and to transmit culture across differences has always been our strength in an instance of humanity as Indigenous peoples. Community solidarity economies, which abate the rule of the commodity form and surplus value, have sustained us in ways that capitalism will never be able to produce on its own, no matter how hard the effort is. In the face of our struggles, capitalists will always have to seek out new "labor markets" to continue the regime of exploitation.

Rosalinda Guillen recently described this transmission of our strength and resiliency as the spiritual component of labor organizing—a cultural flame that kindles the goodness of humanity to remember those lost to the class wars and to spark new rebellions in the hearts and imaginations of those who face the same experiences, who have survived the same violence, who are driven to make the world anew.[47] It is a fire that cannot be governed, managed, planned, or quantified, as this is the immanence of life itself. A closer examination of the attacks against our communities, by legal and extralegal means, by vigilantes, private interests, and the state, makes it clear that the target of these attacks is those relationships maintained through our solidarity economies. Our communities are under constant pressure from neoliberal regulation, and the struggle is far from over, but we are gaining ground in the practice of workers' self-organization, a process that is informing and shaping all future prospects in the allied food sovereignty, food justice, and food autonomy movements.

CHAPTER 14

# Organic Intellectuals and Direct Action
# Fifty Years Past Chicago's "War on Poverty"

PANCHO McFARLAND

I teach a "Class and Stratification" course for the Sociology Program at Chicago State University.[1] In class we focus on inequality and the global capitalist economic system, and critique writings on alternative economic systems. In class, we examine the problems of inequality caused by the capitalist economy and then focus on our own city, Chicago, as a means to understand our places in the economy as working-class people of color. To learn about ourselves in Chicago, I use a text written by the Chicago Grassroots Curriculum Taskforce, *Urban Renewal or Urban Removal?* This is the first volume of eight, and the authors include voices from activists, teachers, parents, long-time residents, and professors. This is an experienced grassroots group of dedicated organic intellectuals. I am using the term *organic intellectuals* in exactly the sense that Antonio Gramsci intended when he observed in his *Prison Notebooks*:

> Every social group, coming into existence on the original terrain of an essential function in the world of economic production, creates together with itself, organically, one or more strata of intellectuals which give it homogeneity and an awareness of its own function not only in the economic but also in the social and political fields.[2]

Like the authors of the curriculum report, I seek to enunciate grassroots community-focused discourses on inequality, poverty, violence, illness, and the myriad other maladies associated with neoliberal capitalism.

While my first writings on Chicanas/os focused on hip-hop, these share continuity with the current work because both proceed from the perspectives of the working classes and people of color.[3] My theoretical approach proceeds from the insight that many hip-hop emcees are the organic intellectuals of this urban youth music "subculture." They follow in the footsteps of the griots, poets, *tlamatimines* (Mexica orators), *corridistas* (corrido composers or performers), and novelists of previous generations. Chicago-based musician-intellectuals present critical perspectives of society through a language and logos familiar to our youth in Black and Mexican-origin communities.

My analysis of Chicago fifty years after the beginning of the War on Poverty uses the narratives of Chicago musicians to enunciate the ever-shifting ground of revolutionary subjectivities. Many Chicago musicians/organic intellectuals provide vivid descriptions and criticisms of the global capitalist system and the failure of US economic and social policies to rid our city and our world of poverty. Since we are still living in a segregated and degraded environment, many of the same problems of inequality and poverty continue to plague Black and Chicano/Mexicano communities in Chicago. Today, to an extent inspired by the poets and musicians of previous decades, the local urban food justice and sustainability movements are using Indigenous placed-based traditional ecological knowledge (TEK) and African heritage knowledge (AHK) as the means to struggle against poverty and its effects and to build the alternatives people want and need to maintain their health and community well-being.

## 1964

So I'm thinking the War on Poverty, 1964, Chicago, inequality, organic intellectuals—Curtis Mayfield! The Chicago-bred genius, musician, composer, bandleader, and lyricist seemed to have his finger on the pulse of urban Black Chicago and the civil rights movement when he released "Keep on Pushing" in 1964, the year that Lyndon B. Johnson declared his war on poverty. Filled with hope, Mayfield with his band, The Impressions, sings:

Keep on pushing
I've got to keep on pushing (mmm-hmm)
I can't stop now
Move up a little higher
Some way, somehow
'Cause I've got my strength
And it don't make sense
Not to keep on pushin'

The Impressions describe the emotional and spiritual strength that many Black Americans had at the beginning of 1964. There was a sense of hopefulness and resolve that made much of the Black community "keep on pushing." There was a drive to keep the struggle going until the movement's goals were achieved. The song is performed in a gospel style, connecting the Black liberation struggle to an African/African American humanist Christianity and spiritual struggle.

## Black Revolt and the War *with* Drugs

By the close of the decade and into the 1970s, poverty in the cities seemed to be getting no better. Resistance was at a high, and so was repression. The murder of Fred Hampton by the Chicago Police Department in 1969, and the subsequent dissolution of the Black Panther Party by the FBI, and other repressive paramilitary operations were conducted by the same government claiming to be serving the poor through the so-called War on Poverty. While urban poverty abatement projects developed from War on Poverty legislation, slum conditions persisted in much of Black Chicago. Many Blacks in Chicago used literature and music to examine the problems. In 1969 Chicago organic intellectual Sam Greenlee published *The Spook Who Sat by the Door*, a novel about armed resistance to government repression and extreme urban inequality.[4]

High levels of unemployment led many to occupations in the illicit markets. Further, the drug economy played an important role in inequality during the 1970s and in controlling rebellious Black communities. Ward Churchill and Jim Vander Wall argue that the great heroin epidemic of 1971–72 was a means of undermining the Black Liberation struggle:

The flood of heroin into U.S. ghettos, meanwhile, appears to have been calculated to narcotize the country's then-burgeoning black liberation movement in much the same way that LSD and other hallucinogens were employed to undermine the white "new left" movement a few years earlier.[5]

Black poverty combined with drugs and few fair-wage employment opportunities produced a volatile mix. The hopefulness and resolve faded for many in Chicago's Black communities.

The music, the images, and the message changed. I think about the War on Poverty, the 1970s, Chicago, resistance and repression, the drug epidemic, and organic intellectuals. Curtis Mayfield once again comes to mind. In the smash movie *Shaft*, he performs the hit song "Pusherman," and approaches the drug problem cautiously from different perspectives. Mayfield sings:

Two bags, please.
For a generous fee
Make your world what you want it to be.
Got a woman I love desperately.
Wanna give her somethin' better than me.
Been told I can't be nuthin' else
Just a hustler in spite of myself.
I know I can break it.
This life just don't make it.
Lord, Lord, yeah!
Got to get mellow, now
Gotta be mellow, y'all
Got to get mellow, now
I'm your mama, I'm your daddy
I'm that nigga in the alley.
I'm your doctor when in need.
Want some coke? Have some weed.
You know me, I'm your friend,
Your main boy, thick and thin.
I'm your pusherman.

The song discusses the thoughts and life of a reluctant hustler. The protagonist is trapped in a cycle of wanting better for himself and his female partner, racist stereotypes of him as worth little, and the drug epidemic that sends many Black, white, and other addicts to him to fulfill their "need." We can easily read an indictment of capitalism and its weapons of racism and the liberal War on Poverty. The psychological effects of racist colonialism like that suffered by working-class Black Americans are highlighted in the song. The protagonist seems to have a self-esteem damaged by racist stereotypes of poor Black men. Poverty and racism conspire so that the hustler believes his woman would be better off with someone else, someone more than a hustler like himself. In another passage in the song he describes himself as a "victim of ghetto demands." He sees himself, the drug addicts he serves, and the communities affected by poverty as the losers under capitalism. Yet, to survive and have the semblance of a livelihood, he has to do some of the dirty work of capital. Mayfield continues:

> Ain't I clean, bad machine
> Super cool, super mean?
> Dealin' good for The Man.
> Superfly, here I stand.
> Secret stash, heavy bread,     ·
> Baddest bitches in the bed.
> I'm your pusherman.

To look stylish ("clean") and present himself with an air of superiority ("bad machine," "super mean"), he must be good at dealing drugs for white bosses. The reference to the machine alludes perhaps to the regime of signification that produces the stigmatized and hated role. Since he is superior and can thrive in the violent world of drug dealing, he has high-quality drugs ("secret stash"), lots of money ("heavy bread"), and numerous beautiful women ("baddest bitches"). He is cast as a misogynist model of Black working-class masculinity.[6] In the racist organization of the capitalist United States, very few legitimate opportunities for success are available to him. Instead, like many Black men in Chicago during the 1970s, Mayfield's Pusherman turns to the illegitimate opportunities available to him. In his position as a great drug dealer, he is able to find some dignity as a human being and man.

## The Turn of the Century and the Rise of Hip-Hop

The city and its citizens experienced crack epidemics, a rapid rise in the prison population, the War on Kids of Color, a.k.a. the War on

Drugs, and welfare reform, imposed by the extant neoliberal regime. So "gangsta rap" emerged as the music that chronicled and critiqued it all in the 1980s and 1990s. In the new millennium we continue to see high rates of Black poverty, illness, and unemployment. There are few and diminishing educational opportunities, food insecurity, violence, and repression. We have had environmental devastation and climate chaos, the housing crisis, and continued "urban renewal," as the demolition of Chicago public housing and dispersion of its residents to far south and suburban communities illustrate. This has been accompanied by gentrification in neighborhoods such as the West Loop, Pilsen, and Englewood and communities fed up with violence, drugs, and poverty. The community peace movement, food justice and sustainability movements, grassroots education projects, the revolutionary Left, and the like have continuously fought the aftermath of the so-called War on Poverty and the capitalist economic system.

The situation that many of us struggle against is chronicled and explained by contemporary Chicago-based musicians. Today, a musical voice of young Chicago includes Chance the Rapper. This organic intellectual talks about pushin' too. His song from 2013 is "Pusha Man/ Paranoia." Late in the song he raps:

> They merking [murdering] kids. They murder kids here.
> Why you think they don't talk about it? They deserted us here.
> Where the fuck is Matt Lauer at? Somebody get Katie Couric
>     in here.
> Probably scared of all the refugees. Look like we had a fucking
>     hurricane here.
> They be shooting whether it's dark or not. I mean the days is
>     pretty dark a lot.
> Down here it's easier to find a gun than it is to find a fucking
>     parking spot.
> No love for the opposition specifically a cop position
> Cause they've never been in our position.
> Getting violations for the nation, correlating, you dry snitching.
> I've been riding around with my blunt on my lips
> With the sun in my eyes, and my gun on my hip.
> Paranoia on my mind, got my mind on the fritz
> But a lotta niggas dying, so my 9 with the shits.
> I know you scared, you should ask us if we scared, too.

I know you scared. Me too.
I know you scared, you should ask us if we scared, too.
If you was there, then we just knew you'd care,' too.
[Verse 5:]
It just got warm out. This is the shit I've been warned about.
I hope that it storm in the morning. I hope that it's pouring out.
I hate crowded beaches. I hate the sound of fireworks.
And I ponder what's worse between knowing it's over and
    dying first.
Cause everybody dies in the summer.
Wanna say ya goodbyes? Tell them while it's spring.
I heard everybody's dying in the summer so pray to God for a
    little more spring.

In this narrative we can see how poverty conditions continue fifty years after the War on Poverty began. In many cases, continued violence, multigenerational hopelessness, and the availability of firearms have left entire communities in more precarious conditions than in 1964. In some Black West Side and South Side Chicago communities there are more boarded-up buildings and abandoned lots than occupied housing and businesses. Chance observes that it looks like the aftermath of a hurricane. He provides us with a thick description of the conditions that many of his peers suffer today and describes fear and sorrow resulting from neoliberal urban-planning policies. Chance offers a vivid description of much of the Black South and West Sides of Chicago. Interestingly, he argues that Black youth have been deserted by the rest of our society. In the wake of this neglect, much of Black Chicago saw the development of food junkyards. A lack of access to healthy food (one indicator that a community is a food junkyard) follows the capitalist and governmental desert of many sections of the city.[7] It seems little has gotten better since 1964.

One thing is dramatically different in Chicago compared to 1964. The demographic composition of the most exploited sector of the labor force in Chicago has changed. Today, the system of illegalized immigration creates a class of worker that is superexploited with few rights or resources with which to resist their circumstances, not unlike the Black population under the US apartheid system, from sharecropping to Jim Crow. This illegalized immigration regime creates a superexploited sector of the working class. This superexploitation has the effect of disciplining the rest of the workforce to accept poor conditions and wages as well as of muting worker dissent. Additionally, the superexploited labor of illegalized immigrants lowers the price of necessary goods, easing consumer anger. Young organic intellectuals such as Juan Zarate and the 1.5 generation in Chicago are documenting the circumstances of illegalized, mostly Mexican immigrants.

Zarate's most insightful examination of living in a poor, Mexican/Mexican American community in Chicago is the song "El Santuario" (The Sanctuary) from his 2008 compact disc *El Sacrificio* (The Sacrifice). He describes a multigenerational economic problem in much of Mexican Chicago: a lack of "good" jobs with living wages, job security, and benefits, and the ubiquitous presence of gangs, drugs, crumbling infrastructure, few educational or recreational opportunities, and the like. Despite the conditions of political and social ecological chaos and disturbance, he finds ways to recognize the barrio and the people living there with genuine fondness. He welcomes us to his barrio saying:

> Bienvenido, compa
> al único lugar que nos entiende
> Al veces es el peor, cabrón
> pero aquí nos acepta
> (Welcome, relation,
> to the only place that understands us.
> Sometimes it's the worst, cabrón[8]
> but here we are accepted)

The barrio as a vibrant, living community is often a difficult place to live, but at least there he feels a part of a community. He is accepted. The people of the barrio are essentially good. The circumstance of living in poverty in a racially segregated postindustrial city is the principal force leading to violent experiences, like the too common acts of gun violence and murder in Chicago. He raps:

> Nacímos con el santo de la espalda,
> morímos con balazo en la espalda
> (We were born with a saint on our back
> we died with a gunshot in the back)

"Born with a saint on our back" signifies the tattoo that adorns the backs of many Chicano street youth and organizations as well as the notion that they were born blessed by saints. The death by gunshot signifies Zarate's critique of this violence. The violence of capitalism forces Mexicans/Chicanos to the margins of society and into dangerous barrios. For many in Chicago's barrios, gun violence leads to death. Zarate's song also serves as a call to remember that—like the many poor and marginalized social groups throughout Chicago's history—people in his community fight back in a struggle to live with dignity. Zarate claims that his peers in the barrio are born with a resistant spirit: "alma de Zapata" (soul of Emiliano Zapata).

## Lessons from the Revolution in the Urban Garden

Economic statistics show that poverty is as common today as in 1964 and that the wealth gap has grown, not decreased. Almost all socioeconomic indicators show what organic intellectuals in Chicago already know: Little, if anything, is better.

This economic system cannot be reformed but rather needs to be replaced with something much more humane, equitable, and sustainable. The capitalist economic system requires inequality and poverty.[9] In addition, the political system of "representative democracy" serves the extremely wealthy at the expense of the working poor. These are the true lessons of the War on Poverty and the War on Drugs. To end poverty, violence, and other urban social problems, capitalism must be eliminated. In addition, the corporate-controlled top-down political system should be replaced with a truly democratic, locally focused, and nonhierarchical coupling of ecological and social systems. This revolutionary change is being modeled within the spaces of the urban food justice struggle.

The future is being organized in worker and consumer cooperatives and member-run food-producing cooperatives like community gardens and mini-farms. Food justice revolutionaries challenge and reject the capitalist values of accumulation, excessive consumption, competition, material worship, consumerism, and greed. Capitalist values are being replaced by cooperation, shared labor, conviviality, and hospitality. The ontological precepts of Indigenous peoples across the planet illustrate this: *In Lak'ech,* which is Mayan for "you are my other self"; *Ubuntu,* which is Nguni B'antu for "I am because we all are"; and *Mitakuye oyasin,* which is Lakota for "all my relations"—all enunciate respect for the "unity of diversity within difference," as my colleague Devon G. Peña states. Additionally, food justice movement members engage in everyday acts of resistance and autonomy and the formation of alternative networks of working-class self-management and self-valuing, free of the prison tethers of the commodity form.

I argue that we can circulate and grow this type of social organization by fusing two approaches to autonomous struggles. First is the wisdom of our Indigenous, place-based ancestors, who organized themselves communally with an understanding of themselves as one people among many interdependent peoples and other beings, the spirit embodied in *Mitakuye oyasin.* Those Indigenous traditions and relationships to nature that provide ecological democratic means of living should be passed on to our children and future generations as active forms of "knowing and being in the world." Placed-based ecological knowledge provides the wisdom necessary to be successful stewards of the land and to create self-sufficient communities. The ecological ethics of our Indigenous ancestors are a means to a sustainable lifestyle that provides ecological services to a diverse array of species.

In addition, the insights of contemporary communists and anarchists can be helpful in developing a more just world. A great deal has been learned from centuries of struggle against capitalism and its efforts to rule through the imposition of hierarchy and domination. The numerous communes, revolutions, temporary autonomous zones, and experiments in libertarian communist living have taught us much about human potential and the difficulties of organizing freely associated self-reliant and resilient small-scale societies in the midst of capitalist domination.[10]

These anarcho-communist social organizations espouse democratic governance in the direct participatory forms. In this organizational form, the "economic base" revolves around community needs and follows Marx's tenet, "From each according to his abilities, to each according to his needs." In such a direct democratic polity, food—and all the productive, reproductive, and creative (artistic) needs of society—can be created through the daily lived collaboration of communities within worker self-managed cooperative facilities, delinked from the production and circulation of commodities, through a process long-proven capable of unleashing the collective energy of freely associated producers and their communities of practice and residence.[11]

Lessons from our ancestors and today's anarchists are influencing our organizing activity against poverty in Chicago. Autonomous zones pop up, dissipate, and reemerge all over the city. Our work with economic and racial justice–focused people in the local food movement illustrates this well. The food justice movement involves a number of separate but interconnected organizations, individuals, and groups. The movement is organized in a rhizomatic fashion where there is no centralized locus of power/knowledge. Chicago's autonomous zones show that the power of collective action is dispersed, decentralized, and ever shifting in its spatial and cultural locations. Each node has its own decision-making process over activities, including how to relate to the rest of the food justice movement. Community gardens, farmers' markets, backyard gardener networks, and the like sprout up from the energy of many individual community members, families, and entire neighborhoods and are nourished by formal and informal civic and mutual-aid organizing activities.

In part all this is driven by the growing hunger for freedom from the dictatorship of the so-called free market and for emerging forms of direct democracy and anticapitalist livelihoods that resonate with the many pleasures of working in urban community or backyard gardens. The neighborhood garden I work with always involves democratic decision making. We decide through a long and often messy process of consensus how to use our plots and share the resources needed to produce food—the seeds, rootstocks, water, good soil, cultivation practices, and so on. We tend the garden fully aware of the communalist values of our ancestors. We remember and reinvent their worldviews in our garden and mini-farm work. Together we decide what is best for our community—what to grow, how to grow it, how to prepare it, and how to share the bounty. Through shared cooperative labor,[12] we increase our community's wealth of good food, healthy soil, and diverse seed and rootstocks. We use only organic ecological methods on our land because we know that all our relations must be present and thriving in the microecosystems we create. We reject private property and the private ownership of the means of production, namely, the land, tools, and seeds required to feed us. We reject hierarchies and the violence represented by the attack on all forms of diversity. Our gardens are acts of direct action against the capitalist system and for an autonomous, free, and well-ordered network of communities of practice and cohabitation.

While this is the ideal and most common organizational practice in the gardens and mini-farms of the Chicago food justice movement, the actions of individuals often fall short. Greed or self-interest can lead some leaders in the movement—perhaps in the midst of the emotional stress of struggling against capitalist deprivation—to attempt to control or dominate others. However, the most effective groups, communities, and networks insist on constantly examining and correcting bourgeois, racist, sexist, heteronormative, or other hierarchical behaviors.

The crops and ethos sowed and harvested in the garden illustrate the possibility of a more just and equitable future for us. They represent viable alternatives to the violence, greed, and ecological catastrophe that are by-products of capitalism and the state that exacerbate the second contradiction of capitalism by dismantling regulatory protections and disinvesting in the so-called social sectors. The community garden serves to show us the wrongheadedness of the so-called War on Poverty, which is really a War on the Poor, and similar cynical reforms that appear increasingly bankrupt despite the logos of the neoliberal dismantling of the social sector. Only through direct action, emerging from our own alternative community institutions, can we generate the biopolitical methods and strategies of self-determination required to end poverty and start to address the many maladies of the myriad forms of capitalist structural violence facing the working classes and poor in Chicago and the rest of the world.

Still, there are negative stigmas associated with Africa, Africanness, slave labor, indigeneity, and migrant farm work. This degrading racial project has driven Mexican@s, Chican@s, and Black Americans to assimilate into the US capitalist ethic and value system while rejecting anything that is "too African," "too Indian," or "too old-fashioned."

Figure 14.1. Abandoned factory and train across from the Sacred Greens community garden. Chicago, like Detroit, has seen community and "guerrilla" gardens occupy abandoned and vacant lots across the inner city. Courtesy of Pancho McFarland.

Exposing young people and adults to the pleasures of working with the soil and the great histories of agricultural innovation and genius of Africans and Native Americans can begin to "re-place" their consumer identities with empowered, rooted identities.

One summer at the Roseland Community Forest Garden, one of the young Black gardeners declared, "I feel like a slave." I responded, "Amina. It's hot, right?"

The high humidity and ninety-degree temperature at 10:00 a.m. made work difficult. Amina and a few other students affirmed that the weeding and seeding of a large raised bed made them feel "how slaves must have felt on Southern plantations." After a brief discussion, the group developed a consensus that the hot, humid air and hard work were unbearable even as they smiled and joked. I took this opportunity to suggest that this work and slavery were, in fact, the exact opposites. "This work is the exact opposite of slavery. The community garden can provide true freedom. Know what I mean?"

When no one responded, I continued: "What made slavery horrible was not the agricultural work but the relations of work. In fact, the work that slaves did and migrant farmworkers do today is some of the most dignified, noble, and important work that can be done. We are fed from their labor."

The small crowd of students in the conversation grew as their attention was piqued by this line of argument regarding slavery. Their understanding of the slave and sharecropping eras was that manual agricultural labor itself was one of the central horrors of slavery. Repositioning agricultural work as a highly dignified craft was something that most in the discussion were eager to explore.

I continued with the impromptu discussion on dignified versus alienated labor, slavery, racism, and liberation: "The problem with slavery was twofold: First, a human owning another human is obviously wrong. Second, is the superexploitation of a body. Their work and bodies, which we know are dignified, become degraded as the products of their labor are claimed by the owner for self-enrichment. The owner now owns the amazing, wonderful product made with the blood, sweat, and tears of another's labor . . . the owner steals the product of slave labor. If the slaves owned the products of their labor, it, first, would not be slavery nor would the labor seem undignified and horrendous."

Some took the long, deep pause for reflection as an opportunity to get back to the, in some ways, easier labor of weeding and harvesting a bed of snap beans. Others wanted to continue the conversation in the shade.

"As we have been learning in class, we have very little control over the products we consume, especially food. What we eat and how it is produced are determined by others. Having no self-determination is the very definition of slavery. Ya see? Then, without the ability to determine what and how you eat (remember food sovereignty?), your health and, by extension, your life and that of your family and community is controlled by someone else. Their concerns for profit will always determine what they provide. Your health goes unconsidered."

Connecting the dots, Amina responds, "That's like when we did our community maps. We found that there were no good stores in Roseland. But in the place where the better-off live, they got rich white people there, they have that store with the health food and organic."

Deuce adds, "That's because they don't care about Black people. All they care about is making money."

"Right!" I interjected, and then asked them a series of questions: "So, how can community gardens and a local food economy solve the crises of food access with all the junk food and lack of access to real food in the hood? What about food sovereignty, when the community determines the production, distribution, and consumption of our food and the ecological impacts of our decisions? How can the work we do in the gardens help us achieve greater freedom? Remember when we all stood around and discussed what we wanted to plant in the four new beds we had just weeded?"

The students nodded, and so I added, "How did we decide?"

Brian said, "We talked about the food we liked and about what would grow this late since its late in the summer."

"*That* was a collective, community process! We decided. We are growing it. Not wage slaves on a corporate plantation. We had control over our labor and the decisions about what we will eat. Freedom."

The conversation continued in this vein for a few more minutes, until it was time to resume our weeding and harvesting mustard and turnip greens. Those who took some vegetables home also took a new way of seeing the much maligned greens that generations ago helped Black people survive the atrocities of slavery, sharecropping, and lives of limited opportunity and structural violence under Jim Crow segregation.

Every season members of the Green Lots Project and the Roseland community plant okra, tend it, admire its beauty, and eventually feast on it. We also learn from the practice of working with okra. Our time with okra teaches us about its habits, the soil quality of our little plot, and the necessary conditions under which the crop flourishes and generously supplies us with sustenance. It can also teach us about humanity.

Okra was domesticated in the West African savanna more than four thousand years ago. During breaks from the difficult work of harvesting and cultivating okra, I replay for fellow gardeners what I have

Figure 14.2. Jackie Smith (*right*) of the Green Lots Project. Smith is among a cadre of Black and Latina/o activists who lead the urban agriculture movement in Chicago. Courtesy of Pancho McFarland.

just learned from the work of Judith A. Carney and Richard Nicholas Rosomoff in their book *In the Shadow of Slavery: Africa's Botanical Legacy in the Atlantic World.*[13] Doing my best to represent Carney and Rosomoff's arguments, I explain that Africans were the first to domesticate livestock and domesticated thousands of plants, many of which have become important for the global food supply. It took Africans millennia to carefully experiment with wild plants and develop those that they found to be beneficial for nutrition, medicine, livestock fodder, and crafts, such as instruments, ladles, and containers.

During the Atlantic slave trade and plantation slavery, Africans developed their foodstuffs as a means of survival in conditions of extreme deprivation. Armed with their vast botanical and agricultural knowledge, slaves transplanted domestic African plants and adopted Amerindian crops. African slaves and their progeny planted impor-

Figure 14.3. Sign dedicating Roseland Community Forest Garden to neighborhood children. Courtesy of Pancho McFarland.

tant staple crops that continue to constitute the basic diets of the African population in the Americas. Yams (*nyam*), plantain, bananas, okra, black-eyed peas, rice, and coffee all have strong connections to African farmers, seed savers, and plant breeders.

We also plant a lot of beans. The *frijol* offers its lessons of Mesoamerican horticultural and life arts and sciences. The *frijol*, *maíz*, and *calabacita* (Three Sisters) share space in the garden alongside collard greens, kale, and fingerling potatoes. We see the complementary nature of species. The Three Sisters serve as a metaphor for the possibilities of human groups to solve our problems through cooperative and complementary relationships. The psychic and physical violence

Figure 14.4. Roseland neighborhood children showing off some of their potato harvest. Courtesy of Pancho McFarland.

experienced in Roseland requires such a response. The violence to our bodies that many of us experience due to lack of access to quality food in urban "food deserts" also requires community mobilization to feed and heal ourselves.

These ancient foods, like the traditional African staples, form the foundation of a nutritious diet that can prove once again useful in this new era of colonially imposed deprivation.[14] Knowledge of the horticultural technologies and practices of our ancestors is empowering decolonial knowledge. Through conscious use of our indigeneity, we can decolonize and ground our identities in the cultures of our grandparents. We can challenge the disempowering racialized identities imposed on us through the psychic violence of the institutions of colonial domination, including schools, entertainment, mass media, and many religious institutions.

## Some Further Lessons for the Liberation of the City

What are the overall lessons from our progenitors—including Mayfield, Zarate, and Chance the Rapper—for today's organic intellectuals working the revolution in the garden? These organic intellectuals of Chicago's barrios sharpen our analysis of urban social problems using place-based, situated knowledge and speak with and for our communities, providing a consistent stream of subjectivity of resistance enunciating a desire for a more peaceful and fuller life in relation to place. They help us understand how a particular "ethic of place" developed in Chicago fifty years after the War on Poverty,[15] when state repression in the form of the police and the criminal justice system, and the assertion of biopower strategies included the spread of food inequality, food apartheid, and limits to human services. These governmentality strategies illustrated the failure of late liberal political and economic strategies and a transition to the neoliberal attack on struggles for freedom of the colonized, the "barrioized," and people of color more generally. Mass incarceration and the privatized prison industry continue as a technology of elite biopower seeking to discipline Black and Brown bodies so that whatever is left of our civil rights protections, alongside the hollowed-out remnants of "War on Poverty" programs, affirms the need for new forms of struggle to achieve autonomy and justice for people of color.

The latest iteration of this elite containment strategy is the War on Drugs, which once again returns us to the violence that Mayfield discussed in "Keep on Pushin'" so many decades ago. Drugs and the supposed war on them have wreaked havoc on public health in Chicago and other urban centers. The violence experienced and examined by Chance and Zarate in their songs results directly as a subaltern response to this containment strategy. Whole swaths of Chicago's South Side, including Chatham and South Chicago where Chance and Zarate were raised, are minefields of violence lacking health care facilities, healthy food outlets, and opportunities for creative development of the people of color who inhabit these neighborhoods. Entire blocks in the South Side are filled with empty lots overgrown with weeds and abandoned buildings turned into drug houses and hangouts for the gangs.

Mayfield's declaration "make your world what you want it to be" and his exhortation to "keep on pushin'" and Zarate's invocation of "el alma de Zapata" both provide inspiration and instruction in the face of continued repression and oppression of people of color and the poor. Zarate's analysis of his barrio includes both a lament concerning the consequences of colonialism and racism, and a celebration of community resiliency through heritage knowledge passed on and adapted for urban conditions.

Here is a pivotal source of hope: Zarate's neighborhood is one of many that pulses with a cultural ecology developed in México and transferred to Chicago over the past four decades and brims with a youthful energy exemplified by the widespread embrace of activist hip-hop performance, which we often hear wafting through the air in our urban gardens. The hope provided by Chicago's musical organic intellectuals is that with native ingenuity and inspiration, neighbors will continue to push back against capitalist enclosures and transform abandoned lots into green oases filled with healthy, collectively grown produce. The Indigenous "soul of Zapata" can thrive only if we are able to decolonize urban space in order to end both police- and self-inflicted violence while addressing capitalist domination as the root cause of hunger and malnutrition, psychic depression, and other public health problems. We are already reterritorializing the city by reclaiming abandoned and neglected spaces for the growth of community gardens and the thriving diffused networks of collective urban agriculture.

CHAPTER 15

# Sin maíz, no hay país

Mesoamericans and Civil Society
in the Defeat of Monsanto

ADELITA SANVICENTE TELLO
AND ARACELI CARREÓN

(TRANSLATED BY DEVON G. PEÑA)

We start by observing: Corn is the most precious fruit of the relationship established between people and nature and springs from
the bowels of the Earth Mother.[1] Next to *maíz* (maize, corn), our
culture flourished, creating a vision of the world that made possible the emergence of a civilization. Upon domesticating corn, our
Mesoamerican ancestors invented agriculture as art, science, technology, and knowledge, and with it they settled down and, thanks to the
generosity and alimentary value of the cereal, liberated social time

Figure 15.1.
Mexican
landrace maize
varieties,
CIMMYT collec-
tion. Photograph
from CIMMYT
Digital Archive.

for art, astronomy, poetry, and religion. In short, maize enabled the creation of a civilization.

In this way, women and men made corn, but at the same time corn also made human beings in the broadest sense because Mesoamerican achievements built a civilization with its own system of thought, culture, worldview, and pantheon. Maize is the center of creation, and its life cycle makes human life possible and forms part of Mesoamerican community life in all its aspects. As Guillermo Bonfíl Batalla states, "Corn is our invention. And corn, in turn, invented us."[2]

Today this system of thought that was capable of giving rise to a splendid civilization confronts the greed of a few corporations that view the concentration of the capabilities to generate seeds and exercise control over them as a gigantic business. This possibility offers corporations a degree of political power unprecedented in the history of humanity with profound implications and serious risks for the economy, environment, and life. Through biotechnology the territorial disputes of a given quality of capitalist expropriation and exploitation have moved to the molecular level, and through the insertion of genes and the use of patented techniques, the corporations have established private rights over genetic resources, with special attention to corn, given its civilizational character and its essential role today in helping to feed humankind.

However, the true inventors and custodians of maize seed diversity clearly sense this threat and are acting to preserve maize under their vision of communality. Over the past ten years, smallholder farmers have been organizing seed exchange fairs and developing the means for the safeguarding and safekeeping of heirloom seeds: Across the country, these efforts confirm that the best defense of corn is the protection of this creative Native diversity.

In this way, we consider that over the past few years in México we have experienced more than the plunder of our genetic resources and in addition are facing a serious political and legal dispute over the integrity and future of maize. On the one hand are the rural populations and Indigenous people of Mesoamerica who claim their legitimate right to sow, eat, and freely reproduce maize. Alongside them, scientists, environmental organizations, and human rights groups have joined in raising the alarm over the implications for food safety and environmental risks of the production of transgenic maize and the concentration of this technology in the hands of a few companies, which signifies theft for Indigenous and peasant communities. The dilemma arises between keeping our sacred plant as an important part of world heritage or permitting the appropriation of maize by means of genetic transformation and capitalist exploitation.

The adversaries are the biotechnology corporations and agribusiness monopolies in cahoots with the Mexican government, who are stripping corn of all the meanings, rights, and knowledge that link it to the peasant and Indigenous communities, and seek to transform it into a commodity, the pivot around which they seek to guarantee their profits in global food markets. Our struggle has managed to stop the indiscriminate planting of transgenic maize in México, which was predicted multiple times by the companies.

## Maize: Artifice of Civilization

Since its creation as a domesticated plant, maize has been more than a scared plant; it is intimately attached to human life. For Mesoamericans, corn is a demiurge, a humanized plant, a god incarnate, a member of the family-community that one speaks, sings, and prays to for blessings; corn is clothed with love and care. Maize in Mesoamerica is an animated object, whose life and existence cannot be explained without accounting for the place it has in the life of the Indigenous and smallholder farmers who sow it. Corn absolutely depends on them to reproduce, and vice versa: the Indigenous and peasant life is impossible without corn. Their symbiotic relationship is evident to such a degree that the two can be said to be only one, that one is not without the other. As Alfredo López Austin says, "Thanks to this link, both actors took on a very different nature. The union permeates all the way down to the molecular level of intimacy as corn was remade to be more useful to man and less useful unto itself; this is a quality of its domestication. This also domesticated man, modifying his social character."[3]

Maize is a subject, territory, cornfield, corn plant, seed molecule. The significance of this plant as the central organizer of space, community activity, the cosmos, beliefs, and ceremony imbues it with a protagonist's role in the construction of this civilization. The creators of corn erected around this plant a vision of the world that made possible our civilization. The coevolution of a civilization linked to a plant carried a worldview with it, a particular vision and construction of knowledge that is largely reflected through the legends described by López Austin and Enrique Florescano. For example, for the Maya, "the myth of the cosmogonic creation was essentially an agricultural myth, a celebration of the germinal powers of earth and water symbolized by the blossoming of the corn plant."[4] We are made from maize, says the *Popol Vuh*, the Mayan sacred book. According to the legend of Quetzalcoatl, maize and all the plants that are companions to the cornfield—chia, beans, and cotton—all sprouted from the gods. Even the gods were made of maize itself, which upon death was buried and revived as "Homshuck" among the Popoluca.

Myths, the precursors of history, speak of the importance of this sacred plant, and its wild predecessor has kept its Indigenous name, *teocintle*. Brenda Solares in her book *Madre Terrible* (Terrible Mother) writes:

> The Maya linked the cycle of agrarian life with the symbols of the succession of power. Just as the Corn God dies and is reborn with each planting, so too the royal blood links the corn seed and the dead king to his successors. The grain or seed that makes possible the cyclical rebirth of the corn plant is the equivalent of human blood that is transmitted by heredity and in the same way ensures the continuity of the lineage.[5]

Getting into the anthropological description of maize means to unravel the intimate relationship between maize and women and men who allowed the development of this wonderful plant. We will find teocintle in the Indigenous histories narrated in the myths, in the time when this transformation happened thanks to domestication, work done by the ancient inhabitants of Mesoamerica that involved a deep understanding of nature.

This then begins the process of domestication as part of a body of knowledge that allowed for the development of agriculture and civilization. Indigenous observation, experimentation, and labor achieved the transformation of this wild grass relative, whose cobs originally had only a few grains, to cobs now covered with hundreds of individual kernels.

The domestication of maize is what grounds the culture and opens the possibility of creating all the artifacts needed for it to be established. Mexican corn can almost be considered an "artifact" itself since it originated from and is dependent on human hands to reproduce, diversify, and adapt to changing conditions.[6] Careful sustained observation and deep knowledge of the environment, especially of the plants that germinate edible grains, allowed Mesoamerican farmers to transform this plant. Agriculture—the culture of the land, work, and peasant technology—becomes the artisan craft knowledge capable of producing what feeds us every day.

It was also in this region of the planet that agriculture was invented with the domestication of maize, which is one of the three greatest inventions of humankind, alongside the use of fire and the development of machines.[7] Agriculture develops the basis of this civilization and allows for the appearance of a wonderful plant with unique features that will allow it to be dispersed throughout the world and become over a few centuries the "grain of humanity":[8]

> Our wise and skilled ancestors achieved that important transformation even if today few acknowledge or understand the activities and knowledge of the culture of agriculture. In is only in the past century, that this achievement has gained recognition: These First Mexicans . . . carried out the great achievement of the reproduction of a plant . . . they transformed a native wild grass called teocintle, the "grain of the gods."[9]

*Cinteotl* is "the God of the Cob" and is the most important of all the deities. At this point it becomes important to recognize that teocintle is the ancestor of maize.

In fact, it has only been a few years since Mexican scientists, officials, and the civil society at large began to recognize, with some difficulty, something that has been acknowledged abroad for some time: "These [Indigenous farmers] can be given credit for having produced the maximum morphological change of any cultivated plant and have adapted maize to the broadest geographic range among the important crop plants."[10] This establishes another indissoluble bond: the link between the farmers' knowledge and work and the diversity of corn seeds.

## New Colonialisms of Capital

The conquest of the Americas meant the separation of subject and object as corn was torn from its civilizational context and traveled to Europe. Corn was brought to Europe and then to the rest of the world without the knowledge necessary for its use as human food. When the grain was consumed without using the process of nixtamalization,[11] corn began to cause pellagra in people who consumed it in countries where it was planted. From that time, corn was considered a grain suitable for animal feed but not for human food. Outside America, far from those with which it coevolved, the civilization that created it, corn was embraced as a generous plant, producing grain, but was not seen as a fundamental pillar of human life.

When our corn was taken out of the civilization that gave rise to it, the subject and object were split; corn lost its "humanity" and became an object, a traveling plant that produces grain and seed. But in Mesoamerica, "the cyclical view of the grain that is buried as a seed to give life remained alive, like the original thread that weaves Mesoamerican civilization together for thousands of years. In the Indigenous and peasant communities in México, corn never lost its humanity."[12]

Corn as object, however, went away with its own history. Whenever a culture recognizes seeds are the reservoir of life and of history, we appreciate their ability to store nutrients, such as genetic information that enables the reproduction of most vegetables, and their ability to protect the history of our collective knowledge. The development of desirable characteristics in corn through observation, crossbreeding, protection, and the exchange and search for seeds by Indigenous farmers over the course of ten thousand years of agriculture is summarized in the characteristics of the seeds. They are the storehouses of our history: "This is an accumulation of tradition, of an accumulation of knowledge about how to work those seeds."[13] The knowledge contained in them, ultimately, is the product of an ancient collective effort of humanity, which has sought through the same process to flourish as a species.

If we follow the evolution of the seeds throughout the history of humankind, we must necessarily also recount the knowledge and technological development linked to them, as well as the social conditions of production that determined these. In this manner, within the seed we find not just the genetic material that allows for the reproduction of the plants, but also the knowledge that humanity has accumulated over the centuries and through which it has attained a level of scientific and technological development.

It is for this ability to encapsulate life, history, and knowledge that the seed has become the object around which the dispute over corn has focused. While the whole corn plant is codified for its botanical characteristics, adaptive advantages, and potential use and productivity, the seed is the link that is considered strategic to appropriate all the maize in the food chain and ultimately the global food system, given the importance of our cereal in global food chains of our time.

The seeds contain a deep understanding of nature that fits into a system of thought and ultimately a worldview that involves a different relationship with nature. The appropriation of the seeds themselves leads to the expropriation and the usufruct of this ancient knowledge of humanity that has been created, re-created, and enriched over ten thousand years by a variety of ancient cultures, which keep it alive, to this day, in an effort that serves both human and plant life.

The current dispute over corn involves the attempts to transform the plant into a commodity, a process that would subject its most intimate structure of genes, with the intention of patenting, so that our sacred plant, farmers' knowledge, and the rural way of life linked to it are subordinated to the discretion of the laws of the market and the vagaries of capital. Technologies enable the appropriation of seeds and the ancient knowledge linked to enabling their productivity, resulting in the transformation of a common resource into a commodity, whose ultimate function is to generate profits. In this respect, the imposition of technologies acts to promote a privileged way to appropriate the value of nature.

In the twentieth century the so-called green revolution, involving the use of Mendelian genetics to develop improved varieties from crosses of selected parents, allowed for an unusual increase in food production across many parts of the world. The seed, subject to crossbreeding to produce hybrids, requiring another set of inputs (fertilizers, pesticides), in this manner came to play a central role.

Maize breeders and México were key because of the role they have played in the development of modern, highly productive varieties of corn in the Americas and especially in the so-called corn belt of the United States. Therefore, "the ecological aspect, biodiversity, and classification of Mexican maize varieties are of interest not only for crop improvement, but also for geneticists, and now for genetic engineering and the agricultural biotechnology industry."[14] Today, corn is the most studied plant on the planet.

In 1953, after Watson and Crick unraveled the structure of deoxyribonucleic acid (ADN in Spanish, DNA in English), in which hereditary information is stored, the door to the deep knowledge of the genetic structure of living things was opened. Toward the end of the century, the techniques of genetic engineering made possible both the modification of this structure in living organisms and the introduction of genes from another organism, which was a breakthrough opening up new possibilities for human beings.

The development of knowledge of genetics and its application led to a special interest in the sources of gene diversity. Countries like México, with particularly high levels of variability detected in both the environment and species, became a fundamental target of this interest. Environmental economists hold that in general countries that have a wealth of biodiversity also tend to lack technology, so the process of appropriation resides in valuing nature through a process that created so-called natural capital. Capital now has to look for new colonies to invade and exploit to continue the process of accumulation, a process known as the "new colonialism of capital."[15]

The very name "Plant Genetic Resources for Food and Agriculture," used as official terms by international organizations like the UN Food and Agriculture Organization (FAO), testifies to the new conceptualization of nature. The FAO clearly states that "while genetic diversity represents a 'treasure chest' of potentially valuable traits . . . [this diversity] is threatened and special efforts are needed to conserve this type of resource *in situ* and *ex situ*, so as to develop a solid capacity for its use, particularly in the developing world."[16]

In parallel fashion with the advances of Western science in the field of genetics, interest has increased in the Western legal and ideological view of seeds and biodiversity as forms of "natural capital." The genetic material itself is not the only object of Western scientific research to anchor capitalist interests. Scientific research as a capitalist productive force posits knowledge itself as property and, with utter disdain, views all the ancestral Indigenous knowledge as "traditional," "old," and "marginal." Thus ancestral heritage is currently besieged by Western science because of the possibilities for profits that this exploitation could deliver.

The territories that have not yet been sacked and looted and were considered outside the scope of commerce are the new territories in dispute: genetic resources, ancestral knowledge, the spaces where there is still a nature to "explore and sell"—not by nation-states but by transnational corporations. Having exhausted the possibilities of the industrial era, the market has "discovered" a new form of wealth: genes, which now are considered the "green gold" of the so-called Biotech Century.[17]

The new technological paradigm creates novel forms of scientific research governed by the logic of the market and consolidated by specific forms of property called intellectual property rights (IPRs). This regime makes possible the appropriation of the seeds and their associated knowledge and value-added products by transnational biotechnology companies, which constitutes what Armando Bartra calls "a rent on life."[18]

However, capitalist greed faces a central contradiction: On the one hand, by seeking to control agriculture, presumably to feed the world through the appropriation of seed, capital imposes limits on Indigenous, peasant, and local food self-sufficiency. On the other hand, it is in these marginal interstitial systems, which refuse to disappear, where seed knowledge flourishes, and so capital must still find a way to access these systems to extract future surplus value from seed varieties. Today, in full acceptance of climate change, modern biotechnology has shown its limitations: for example, biotechnology could not change, much less control, conditions as complex as those of the response of plants to water stress. In contrast, varieties adapted to drought conditions have been achieved by communities with their own systems of research, repeating what they did with teocintle a thousand years ago, and presenting a number of resistant varieties with promises of real (Indigenous) alternatives for adaptation of plants to adverse environmental conditions.

## Transnational Seed Corporations in México

In the 1970s, the government in México had almost a virtual monopoly on research and reproduction of original basic seed, registered and certified by the National Institute of Forestry and Agricultural Research (INIFAP); the monopoly was extended to the distribution and sale of improved varieties through the Productora Nacional de Semilla (National Seed Producer) (a.k.a. PRONASE). The participation of the private sector was limited, until trade negotiations with the US and Canada led to changes in the regulation in 1991 and opened the doors to private companies on equal footing with PRONASE and INIFAP. Today, "the public sector participates with only 6 percent of seed production . . . [while] foreign companies dominate the seed market by managing more than 90 percent of the capital managed in this sector in the country each year."[19]

Although there are still 350 certified producers in México, the national seed market, mainly involving commercial hybrid crops, is actually split among two transnational seed companies, Monsanto and Syngenta, with a small part, as mentioned above, controlled by the Mexican government. Forty percent of the seed certified for use in México by SAGRAPA (México's Department of Agriculture) is Monsanto seed.[20]

According to the National Seed Association, A.C. (AMSAC), the value of the seed market in México represents about one billion dollars, of which 80 percent is absorbed by twenty companies, which include Monsanto, Dow AgroSciences, Syngenta, Pioneer, and other transnationals, with the participation of the domestic enterprises México Royal International Ceres Group, Aspros, and Seeds Conlee Mexicana. Between 60 and 70 percent of the sale of hybrid corn seed is carried out by transnational corporations.[21] Alejandro Espinoza and Antonio Turrent Fernández further note that

> the participation of the private sector in the Mexican seed industry has changed dramatically in the past decade; in 1970 the

participation of the private sector in corn seed sales was about 13 percent, while in 1993 it was 90 percent and is expected to increase to . . . 96 percent in 2002.[22]

In the area of seeds, Monsanto controls 60 percent of the total market for hybrid corn seed and has sales worth $110 million. The earnings of the corporation with their agrochemical sales revenues amounted to $250 million dollars annually.[23]

In 2003, with the disappearance of PRONASE, Monsanto began to dominate the business of commercial seed sales in México. And by 2006, México imported 18,842 tons of maize seed, an amount 157 percent higher than in 2000, according to statistics from the Secretariats of Agriculture and Economics.[24] Cruz López Aguilar, leader of the Confederación Nacional Campesina (National Peasant Confederation) between 2007 and 2010, confirmed that 90 percent of the national seed market is dominated by ten transnationals.[25]

An example of the growing predominance of the private companies over public institutions is the fact that 74 percent of the titles for varieties of maize seed granted by SAGARPA between 2004 and 2008 belong to private companies (see table 15.1). This is confirmed by two of the most important investigators of INIFAP, who found that of the total "improved seed varieties" available in México, 92 percent belong to transnational private companies, 5 percent belong to other small businesses, and only 3 percent consist of free varieties from the public stocks.[26] The expansion of corporations within commercial agriculture, covering the activities that the state has abandoned, is evident. However, the adoption of improved and certified seeds is not only low but has decreased in recent years, as shown in the figure 15.2.

In contrast, estimates on the importance of native maize seeds to smallholder production range from 60 to 85 percent. Among the various sources there is variation in the data. According to Monsanto, only 40 percent of the surface area planted with maize involves commercial seed, and the rest is for self-consumption.[27] However, others report that 75 percent of the national surface dedicated to the cultivation of maize is based on native landrace seeds.[28] In this sense, 60 percent of the production of grains in México is not a product of industrial agriculture but rather is from peasant agriculture and

**Table 15.1.** Title holders of GMO maize seed patents, México, 2004–8

| Institution | 2004 | 2005 | 2006 | 2007 | 2008 | Total |
|---|---|---|---|---|---|---|
| INIFAP | 13 | 8 | - | 6 | 1 | 28 |
| Pioneer | 25 | 22 | 4 | 5 | 2 | 58 |
| Monsanto | 6 | 8 | - | 5 | - | 19 |
| Dow Agrosciences | - | - | - | - | 2 | 2 |
| Subtotal | 44 | 38 | 4 | 16 | 5 | 107 |
| % public | 30 | 21 | 0 | 38 | 20 | 26 |
| % private (corporations) | 70 | 79 | 100 | 62 | 80 | 74 |

Source: Based on data files from SAGARPA 2008.

**Figure 15.2.** Production and sales of certified seed (in tons), 1998–2006.

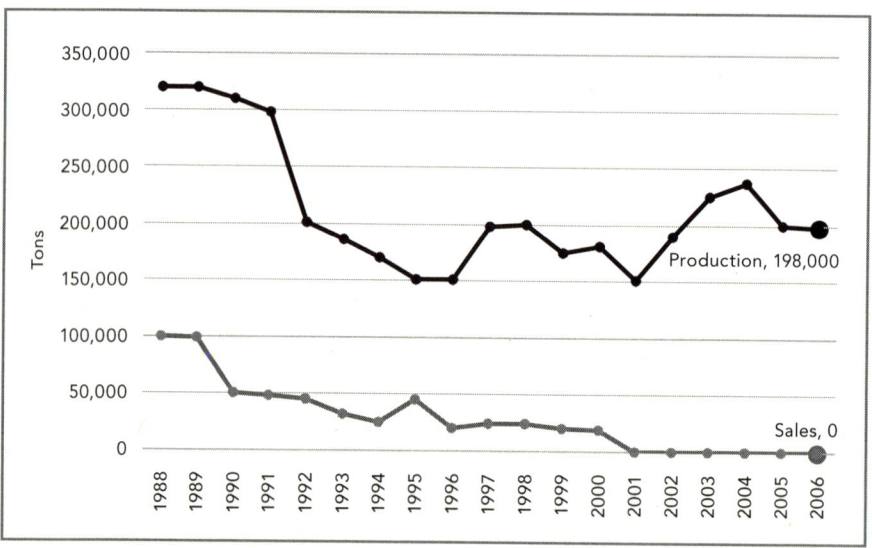

Sources: For sales of seeds, Consejo Nacional Agropecuario and PRONASE; for production of seeds, SNICS, 1983–2004, http//www.sagarpa.gob.mx/snics.

Indigenous subsistence farming, which produce 18 million tons of corn annually on 8.5 million hectares; 70 percent of the corn crop is from native seeds.[29]

With the domination of the hybrid seed market, these transnational companies play an important role in the destruction of the national seed market, but they also seek control, appropriation, and exploitation of genetic resources in Indigenous and peasant areas where lands have been abandoned, which in effect excludes producers from engaging in productive activity. As already happens in the United States, the domination by a few corporations of the hybrid seed market in México would allow such companies to perform absolute and relative monopolistic practices to unilaterally exercise substantial power in this relevant market. The impact that such monopolistic practices in the seed industry could have on the entire food production chain could be very serious, especially for the more fragile links in the chain: peasant and Indigenous farmers, consumers, and the environment.

Given the relative marginal use of commercial hybrid seeds by Mexican corn farmers, the organization known as Agrobio México—which receives funding from the transnationals Monsanto, Syngenta, Pioneer, and Dow AgroSciences—works to promote the adoption of the transgenic seeds and counteract the activities of the anti-GMO movement growing within civil society. Agrobio México projects that "transgenic seed will eventually absorb 80 percent of the market value which today is dominated by conventional hybrid corn seed in México, which is valued at more than of 2.5 billion pesos."[30] According to Fabrice Salamanca, the director of Agrobio México, a transgenic seed costs between 15 and 20 percent more than a hybrid, which is equivalent to about 600 pesos per hectare, but "higher productivity is achieved since traditional agriculture employs about eleven people, while sowing seeds genetically modified only occupies one person."[31]

The analysis of the evolution of the permissions granted for planting GM corn in México reflects without doubt the advancement of these monopolistic practices by transnational companies.[32]

Authorizations granted for the planting of transgenic crops in México are divided into two periods, the first corresponding to the period when the permissions were evaluated based on the Federal Law on Plant Health (known by the acronym, LFSV) and the norm defined under FITO-056. The second period began when the Mexican government turned to using the law on Biosafety of Genetically Modified Organisms (known by the acronym, LBOGM).

Table 15.2 shows that the permissions granted in the first period,

## Table 15.2. Permits granted under different juridical authorities, Mexico 1988–2009

| Field experiments approved under LSFV and NOM FITO 056 | | Commercial plantings approved under LBOGM | |
|---|---|---|---|
| Organism | Total | Organism | Total |
| Alfalfa | 3 | Alfalfa | 2 |
| *Baccilus thuringiensis* | 1 | Cotton | 157 |
| Banana | 7 | Maize | 40 |
| Canola | 1 | Soy | 30 |
| Canola oil | 1 | Wheat | 2 |
| Cotton | 113 | Total | 231 |
| Lemon | 1 | | |
| Lino | 1 | | |
| Maize | 34 | | |
| Melon (cantaloupe) | 7 | | |
| Papaya | 5 | | |
| Pineapple | 1 | | |
| Potato | 6 | | |
| *Pseudomonia sp.* | 1 | | |
| Rockcress (Arabidopsis) | 1 | | |
| Rice | 1 | | |
| Squash | 47 | | |
| Safflower | 2 | | |
| *Rhizobium etli* | 1 | | |
| Soy | 53 | | |
| Tobacco | 6 | | |
| Tomato | 26 | | |
| Wheat | 6 | | |
| Total | 330 | | |

Source: CIBOGEM 2011.

between 1988 and 2005, were only for experimental purposes with a wide variety of crops (twenty-three) and organisms (three), which suggests an intensive research effort, while those awarded after 2005 involved experiments with just five crops.

On the other hand, looking at the permissions granted in the case of maize (see table 15.3), we observe that although initially these were granted to public institutions, over the years they increasingly involved requests by transnational corporations. Ninety percent of the permits granted to experiment with transgenic maize in México between 1993 and 2011 were awarded to transnational corporations, and of this a total of 30 percent pertained to Monsanto requests. We note also that in recent years, there are only four companies that arise in these permissions with different names associated with Dow AgroSciences, Pioneer Hi-Bred (PHI), Syngenta and Monsanto. Only one experiment, the first permission granted, was for a national public institution (see figure 15.3).

Continuing our analysis of the permissions granted for experimental planting of transgenic maize in México, we note that as of April 13, 2009, solicitations for the experimental release of transgenic maize were subject to public query and commentary, and we were able to conduct a timely follow-up to these requests. Numerous respected scientists and diverse organizations have made comments based on Article 33 of the LBOGM. However, these objections and concerns have not been considered, resulting in the approval of all the permits.

Of the 106 permissions granted from 2008 to 2011, we found that in addition to benefiting four transnational companies, these were for experiments involving just ten "events" (the name given to each genetic modification by the corporations), and each of these corresponds to what the law (LBOGM) calls a "case" because the statute states that the evaluation is to be done on a "case-by-case basis." These ten events are presented in table 15.4. All involved so-called stacked traits, which means that there are at least two modification events in a single request.

Our attention is drawn to the predominant feature of these events, which is herbicide tolerance and resistance to certain insects, even though it has been widely publicized by the Mexican government and the companies that this technology would save the Mexican

**Table 15.3.** Permits granted for GMO maize experiments by institution, Mexico

| Year solicitation | Institution | Type of institution & financing | Number of experimental permits granted |
|---|---|---|---|
| 1993 | CINVESTAV | National; public | 1 |
| 1994 | CIMMYT | International; public* | 2 |
| 1995 | CIMMYT | International; public* | 1 |
| 1996 | CIMMYT | International; public* | 5 |
| 1996 | Asgrow Mexicana | International; private | 2 |
| 1996 | Pioneer | International; private | 1 |
| 1997 | Mycogen Mexicana (Dow Agrosciences) | International; private | 1 |
| 1997 | Monsanto | International; private | 3 |
| 1997 | CIMMYT | International; public* | 1 |
| 1997 | Asgrow | International; private | 4 |
| 1997 | Monsanto | International; private | 3 |
| 1997 | Híbridos Pioneer | International; private | 3 |
| 1998 | Monsanto | International; private | 1 |
| 1998 | CIMMYT | International; public* | 2 |
| 1998 | Asgrow Mexicana | International; private | 3 |
| 1998 | Híbridos Pioneer | International; private | 1 |
| 1999 | CIMMYT | International; public*' | 2 |
| 2005 | Dow Agrosciences de México | International; private | 1 |
| 2005 | PHI México | International; private | 2 |
| 2005 | Semillas y Agroproductos Monsanto | International; private | 3 |
| 2005 | Monsanto Comercial | International; private | 1 |
| 2009 | Dow Agroesciences/ PHI México | International; private | 15 |
| 2009 | Monsanto Comercial | International; private | 19 |
| 2010 | Syngenta Agro | International; private | 10 |
| 2010 | Monsanto Comercial | International; private | 6 |
| 2010 | Semillas y Agroproductos Monsanto | International; private | 9 |

| Year solicitation | Institution | Type of institution & financing | Number of experimental permits granted |
|---|---|---|---|
| 2010 | PHI México | International; private | 15 |
| 2010 | PHI México/Dow Agrosciences | International; private | 26 |
| 2010 | Dow Agrosciences de México | International; private | 1 |
| 2011 | PHI México | International; private | 5 |
| | | Total | 149 |

* Centro Internacional de Mejoramiento de Maíz y Trigo (International Center for the Improvement of Maize and Wheat). CIMMYT is an international organization with public and private funding.

Source: Based on data files from SENASICA, SAGARPA, and CIBOGEM.

Figure 15.3. Permits granted by type of institution.

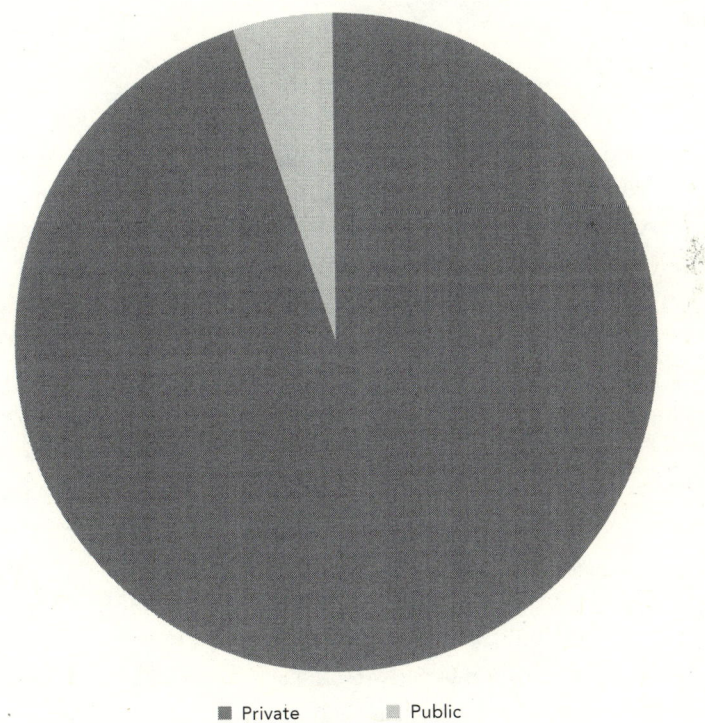

■ Private      ■ Public

Source: Based on data from SENASICA, SAGARPA, and CIBIOGEM.

Table 15.4. GMO events receiving permits, Mexico, 2005–11

| No. | Event | Characteristics |
|---|---|---|
| 1 | MON-00603-6 | Tolerance to glyphosate herbicide |
| 2 | MON-89034-3 | Resistance to Lepidoptera insects; tolerance to glyphosate herbicide |
| 3 | MON-88017-3 | Resistance to Coleoptera insects; tolerance to glyphosate herbicide |
| 4 | MON-00810-6 | Protection against certain Lepidoptera insects |
| 5 | MON-00021-9 | Tolerance to glyphosate herbicide |
| 6 | DAS-01507-1 | Resistance to Lepidoptera insects; tolerance to glyphosate herbicide (containing ammonia) |
| 7 | DAS-59122-7 | Resistance to Lepidoptera insects; tolerance to glyphosate herbicide (containing ammonia) |
| 8 | SYN-BT-011-1 | Resistance to Lepidoptera insects; tolerance to glyphosate herbicide |
| 9 | SYN-IR-162-4 | Tolerance of insect crop pest species |
| 10 | SYN-IR-604-5 | Tolerance of insect crop pest species |

Source: Based on CIBOGEM data files.

countryside by "improving yields," and, lately, by the possibility of conferring "resistance to drought." It should be noted that these features can hardly be accomplished through a genetic transformation, because they involve a more complex set of genes and environmental conditions including the response of the plant itself. In the United States, it has been shown that transgenic crops do not increase yields,[33] and even the companies have recognized this, pointing out "that transgenic crops that inherently increase yields do not exist in the market."[34]

In addition, it has been repeatedly shown that herbicide tolerance means the annihilation of the polyculture cultivation system of the milpa, a farming system in which different crops are harmoniously planted, for example, beans, chile, and pumpkin. We assume that when herbicide-resistant maize is planted, crops that accompany the corn are "weeds" that will be fought with the applied herbicide. Also, clearly the use of herbicides is increasing, as the experience of the United States and Argentina has shown. Charles Benbrook provides information showing that under transgenic technology in corn the use of herbicides increased by 30 percent, compared to conventional technology with non-GMO corn.[35] In addition, new studies are revealing that the herbicide glyphosate (Roundup Ready® is the version marketed by Monsanto) is highly toxic, as corroborated by a resolution of the International Court of Justice in The Hague.[36]

It is important to note that half of these genetic modifications of corn—five out of ten—involve a single company, Monsanto, which opens the possibility to monopolistic practices that this company could exercise for control of maize seeds through the establishment of patents. Another expression of extreme monopolization that is allowed and even encouraged in México is observed by reviewing the features of the 106 permissions granted during 2009, 2010, and 2011: 86 of the 106 permits involved at least one MON event, which is to say that 81 percent of the permits benefit Monsanto. Many permits for MON events are granted to companies other than Monsanto, suggesting that there are agreements between companies.

Researchers and members of the society have pointed out that there are a large number of scientific and technically supported arguments to deny permissions. Since 2009, when the first applications were filed, the movement has flooded the courts and regulatory bodies with detailed comments and analysis of impacts. We discuss some of these arguments here.

In statements responding to the call for public commentary, our scientists iterated that there are inadequacies in measures for ensuring biosecurity and biomonitoring. Our attention is drawn to México's weak system of biosafety, which should be strengthened prior to the approval of any stage involving experimentation with or commercialization of transgenic corn. In our analysis of public comments, we observe that critics noted that applications lacked essential information that would ensure the biosafety of native, improved, and hybrid varieties of maize (*Zea mays* ssp. *mays*), as well as their wild relatives (other species and subspecies of the genus *Zea* existing in México).

Meanwhile, Turrent, Garza, and Espinosa, investigators of INIFAP, explained the implications of this technology from the perspective of agricultural production. A very important point they emphasize refers to the need to measure with scientific precision, and not only empirical data, the effectiveness of events against pests. To do this, the scientists from INIFAP indicated that it is necessary to collect—systematically and comprehensively—populations of the *plagas-blanco* (target pests) in areas where the release of corn has occurred and study, in the laboratory and greenhouses, the insecticidal activity of the event experiments. This observation shows that it is not necessary to carry out experiments in the field involving releases to test these events. Indeed, biosafety norms require that such studies should be done *before* approval of environmental release of the events.

It has been repeatedly noted that it is necessary to carry out these experiments outdoors; in the list of trials authorized during the period from 1995 to 1999 we see that this was done in at least three of the events for which the government authorized permissions. If these events were already tested in México before the establishment of the current moratorium on the planting of corn, there is no need to experiment again.

In fact, the INIFAP investigators detected flaws in the experimental designs, such as lack of information, which is fundamental to evaluate the supposed utility of these crops, as well as information that is both useful to increase the national expertise on the biosafety of these genetically modified crops and to determine their agronomic utility. In their commentary, they pointed out that "the implied assumptions of the series of experiments in requests for release of transgenic maize submitted for consultation by the Servicio Nacional de Sanidad, Inocuidad y Calidad Agroalimentaria (SENASICA, National Service for Health, Sanitation and Food Safety and Quality) experimental programs are incomplete, irrelevant, and misleading to the universal (commercial release) application of the knowledge derived from experimental comparison."

Over the three years being considered here, it has been apparent that rather than conducting true experiments to test the utility of this technology, these corporations are only minimally complying with the procedures laid down in the Mexican law on the biosafety of genetically modified organisms for commercial planting of GMOs. Additionally, their goal seems to be to establish transgenic corn in these pseudo plots in México so that they can become a kind of foci of infection from which the transgenic technology can disperse, without the possibility of a return from escape events.

## The Popular Defense of Maize

The process of enclosure and appropriation of maize has arrived at a key point as a result of the eruption of GM crops in México. Since 1999, we note that there have been many different processes of awareness, knowledge, organization, and work in defense of the collective property in maize seeds, particularly in Indigenous and peasant communities. The dispute over corn has taken shape through a process of recognition, appreciation, and reaffirmation of the inalienable collective ownership of seeds, as well as the peasant and Indigenous knowledge linked to them.

The knowledge, action, and collective agreements regarding this issue very quickly evolved from a point in 1999, when then-undersecretary of SAGARPA Victor Villalobos expressed the view that civil society organizations knew nothing of biotechnology and therefore should not participate in discussions.[37]

At the end of the last millennium, when the debate on GMOs was growing across the world, México began an effort to gain knowledge and acquire information. The Colectivo Ecologista-Jalisco organized the first events to this end in March 1998; the first was called "Citizenship Before the Intellectual Property of Seeds and Medicinal Plants,"[38] and the forum "Corn, Patents and Citizenship" was held in the Morelos State Congress in August of the same year. At both events, there were already warnings about the serious implications of the use of this technology for México's Indigenous and peasant peoples, as well as about the risk of abuse by corporations of corn as a fundamental aspect of the nation's heritage.

In 1999, Greenpeace-México unveiled the start of the negotiations of the Biosafety Protocol, in Cartagena, Colombia, expressing concern over the introduction to México of thousands of tons of transgenic corn from the United States. This organization started a public campaign for the protection of maize in its center of origin.[39] That same year, the Permaculture Network and another 120 organizations asked President Zedillo for a moratorium on GMOs in all their applications, particularly for maize. The moratorium was rejected due to the international commitments of México, including NAFTA.[40]

However, in what is considered a great achievement, which influenced relevant studies of and discussions about maize, on September 3,

1999, the National Committee of Agricultural Biosecurity (CNBA) announced that it was suspending acceptance of applications for experimental planting of genetically modified corn. This led to the first moratorium on the cultivation of transgenic maize in México; it was called *de facto* since the committee did not have a legal (juridical) basis and was doing so under the agency's own rule-making authority.

Looking to have a greater presence in the political arena, in 2000, the biotechnology companies created the Mexican office of the industry organization Agrobio under the auspices of DuPont, Monsanto, Aventis, Novartis, and the Mexican financial group Pulsar/Savia. Agrobio is incorporated as a civil partnership with the possibility of participation in the public realm with the objective of promoting biotechnology in Mexican agriculture. This new organization urged the federal congress to complete the regulatory framework on biosafety. Discussion of the Law on Biosafety of Genetically Modified Organisms (LBOGM) was initiated in 1999 with a forum organized in the Senate of the Republic.

On the other hand, the Mexican debate became an international discussion, when in 2001 the presence of transgenic maize was detected in plantations of native corn in the mountains of Oaxaca. Encouraged to work in the area by the Zapotec-Chinanteca Union, the researchers Ignacio Chapela and David Quist published their findings on the presence of transgenic maize in this area in the journal *Nature*. This article unleashed a controversy not only at the national level but also at the University of California, Berkeley, where the researchers work. This investigation led some of the affected communities to request, in 2002, that the Commission on Environmental Cooperation (CEC) conduct and publish the study "Maize and Biodiversity: Effects of Transgenic Maize in México."[41]

The results of this study as well as the wider debate on the initiative of the LBOGM that took place in the Chamber of Deputies during 2004 marked a watershed in the fight against transgenic maize, since the process included a wide range of social actors. The debate concluded like a waft of steam with the approval of the LBOGM, under pressure from the biotechnology industry, without attending to most of the social concerns raised by the scientific and Indigenous communities.[42] Thereafter, the issues posed by transgenic corn ceased being the exclusive concern of scientists, experts, and researchers and became a matter of public interest with increasing visibility in the media, particularly the new electronic (or social) media.

In March 2005, the LBOGM, named "Monsanto Law" by the Oaxacan communities, was published in the official journal of the Federation. Despite the clear momentum by lobbyists from the biotechnology industry, and thanks to social participation in the Congress, this law included a series of biosafety mechanisms to safeguard the quality of México as a center of origin of several species, notably corn. Other mechanisms include the regime of special protection of maize under Article 2, Fraction XI of the LBOGM, and although it is not established clearly that all México is a center of origin of corn and this definition is left pending, article 88 of the law does ban transgenic animal and plant species.[43]

As indicative of the social outrage signified by approval of the LBOGM, the response led to the strengthening of civil society's organization and discourse against GMOs. A number of noted Mexican scientists, as well as intellectuals, joined this mobilization. It was also the starting point of a massive discussion in which Indigenous and peasant communities took the initiative to defend their maize varieties as a vital common good through their own autonomous processes.

Since then, the work in relation to GMOs has spread and consolidated in various lines of action: the type of action that began as a creative response to the process of appropriation of the seeds. Various processes of recognition, revalorization, and reaffirmation of the inalienable collective ownership of seeds, as well as the knowledge linked to it, have multiplied as well in peasant and Indigenous communities. This has resulted in many experiences at seed libraries (banks) and exchange fairs at all levels: local, regional, and even national, such as the "Fair of the Milpa" that took place at the UNAM in May 2010.

Another action that has strengthened the movement is the lifeline scientists have provided in this fight with the creation in 2006 of the Unión de Científicos Comprometidos con La Sociedad (Union of Scientists Committed with Society, a.k.a. UCCS), which has brought together recognized and respected scientists without conflicts of interest. They have pledged their effort to develop fundamental research to study and understand the risks involved in this technology. During 2006 UCCS managed to stop permits for transgenic maize, showing illegality incurred as a result of executive action. Since then various groups have asserted that the Mexican government has systematically violated the feeble LBOGM.

Various organizations have promoted a line of legal work to follow and monitor the planting of GMOs. This work has meant deploying different legal strategies and resources: demands of *amparo* (due process protections), proceedings and formal complaints against public officials, popular demands, protests and demonstrations, and even a constitutional controversy. In addition, the movement organizations participated in public forums determined and required by the law. The result is difficult to evaluate. While it is true that the juridical regime and legal personnel have not recursively refused to support the merits of the anti-GMO arguments in all cases, we have been able to stop the advance of transgenic crops in México. The question of legality presents an impediment to the movement since the government constantly refers to the need to give legal certainty to individual stakeholders like the transnational corporations.

Finally, we have worked in a very interesting way on the diffusion of information from different perspectives and in different settings: taking advantage of new communications technologies, we have used these to quickly and efficiently circulate and share information. In addition, these tools have been used to summon specific actions that are timely and can multiply and diversify.

That our decentralized network has carried out the national campaign *Sin maíz no hay país* ("Without maize there is no country") demonstrates the usefulness of coordinated action among different organizations from different sectors with different specialties, knowledge, and skills, creating synergies and joint projects with the Indigenous peoples most directly impacted by GMOs through local authorities with their own organizational forms.

We believe that the process has contributed to the re-creation of a sense of community that was latent in the cities' inhabitants and has allowed urban populations to reassess and give a new meaning to the countryside through activities such as the "Evening Against Transgenic Maize" held in the México City Zócalo (Central Plaza) in

2009 and the "National Day of Maize," which has been held since 2009 in different zones of the country and beyond in the United States and other places.

The experiences of the first years of public discussions on GMOs showed that when confronted with the economic power and media control of biotechnology corporations, it is useful to form alliances and confluences between peasant, Indigenous, environmental, human rights, consumer, and civic organizations and academic scientists without conflicts of interest. The one enduring quality of the struggles against transgenic corn in México is the diversity of proposals, speeches, and strategies; all are determined in a decentralized and nonhierarchical manner without central coordination of all efforts.

The defense of maize as the iconic heart of the common heritage of México has been achieved from various trenches, strengthening identity and collective organization in rural, Indigenous, and urban communities. This exercise of collective rights for the defense of maize fits within the conception of communality described by Floriberto Diaz in which the relationship with nature encompasses many different heartfelt senses of place.

## Communality: "We are seeds inhabiting the corncob"

Maize not only allowed the emergence of Mesoamerican biodiversity, but it has allowed Indigenous peoples to survive the ravages of capitalism, from the earliest years of conquest and colonization to the processes of "agro-suicide" in our neoliberal times.

With the vital support guaranteed by native corn, Indigenous peoples and the Mesoamerican civilization survived the assaults and plunder of capital. Maize allowed the Indigenous peoples of America to survive the imposition of an alien way of life, governed by capital, and they thus kept alive distinct modes of life, social organization, and production, autonomous from the logic of capital. Guillermo Bonfíl Batalla notes, "The testimonies of this long process of civilization surround us in all directions. We always have before us a material trace, a way of feeling, or of being, a name, a food, a face."[44] To this day, cultural and religious expressions and festivals held by many Mexican towns are governed by agricultural activity and are focused on corn.

While in the dominant Western scientific context the study of communal ownership, management, and use of common property resources (CPRs) is relatively new,[45] the notion of community and common property is a deep-rooted institution among Mesoamerican peoples. CPRs are defined as "aspects of life that, since ancient times are accepted as collective property and have remained marginal to the process of commodification and, more recently, of real submission to capital."[46] According to Elinor Ostrom, these institutions still offer us tools and intellectual models necessary to understand the problems associated with the regulation and administration of natural resource systems, as well as the reasons why some institutions work in some and not in other environments.[47] Commenting on private and common property, Vandana Shiva argues that

> there are major differences between ownership of resources shaped in Europe during the enclosures movement and during colonial takeover, and "ownership" as it has been practised by tribals and farmers throughout history across diverse societies. The former is based on ownership as private property, based on concepts of returns on investment for profits. The latter is based on entitlements through usufruct rights, based on concepts of return on labour to provide for ourselves, our children, our families, our communities. Usufruct rights can be privately held or held in common. When held in common, they define common property.[48]

In México, different forms of appropriation, exchange, organization, and creation of various resources persist. Property is not the central concept that governs the ownership and management of resources, and nature is viewed as comprised by all the elements that compose and govern the particular relationship between humans and nature.

The best reference to this way of life is the Indigenous community itself. Floriberto Díaz, a Mixe writer, proposes to seek an understanding of this relationship from the Indigenous perspective,

> because we can understand different things and even contradictions. Reviewing the meanings in dictionaries shows us that at most, the concept (relating to commonality) will predominantly give an idea related to property. . . . For others, it is a simple aggregate of persons. . . . It's a community of arithmetic.[49]

Díaz elaborates on this alternative Indigenous explanation of a world in motion:

> For the indigenous person, the indigenous community is geo-metric, the elements that make it up are: a territorial space, a common history, a language, an organization and system of justice. . . . That is, an indigenous person does not understand community as only a set of things or people, but people with history, past, present and future, that cannot be concretely, physi-cally, defined but must be also seen as spiritually in relation to all nature. . . . A community first establishes a series of relationships between people and space, and in the second instance, among the people together.[50]

In México, the dominion over this land—which for capitalism is the key asset for its social reproduction, as is the Earth, the so-called natu-ral resources, including seed—rests in Indigenous and peasant hands. About 54 percent of Mexican territory belongs to Indigenous peoples and peasants.[51] This social property of land, derived from a historical background that covers centuries, established a unique context to the extent that these so-called natural resource domains comprise these territories. Eckart Boege says that

> the indigenous territories are true biocultural laboratories where, with important historical-cultural gravitas, where exchange between wild plants, weeds or ruderals (introduced species), and purely domesticated plants is still practiced. This extraor-dinary wealth is located in the territories of indigenous peoples of México.[52]

These factors give the reality of México a very particular complexity because when we point out that we are, with the rest of Mesoamerica, a center of origin and genetic diversity of corn, this statement has broad implications. On the one hand, this denotes the classic concept of a center of origin in terms of the "geographical area where there is a maximum of diversity of cultivars that coexist or coexisted with wild relatives";[53] on the other hand, this also includes "the process of interaction between humanity and nature, which started six thousand years with our ancestors, who transformed the wild ancestors, cen-trally *teosinte,* to invent and develop maize."[54]

Inscribed in the conception of communalism, maize is "the fruit of the most precious relationship established between people and nature, through the collective work . . . the *tequio* and springs from the maternal heart of the Earth." "De-kernalizing" (*desgranando*) this deep relationship, Floriberto Díaz points out:

> As I get to reflect on this, and make it part of the conception of life and death, our ancestors took many years. Through attentive observation of the movement of plants and of the stars, and in parallel fashion through the ability to look at themselves in the same way, ancient people became capable of developing knowledge of the plants and animals that inhabit the Earth.[55]

The Indigenous peoples and peasants of México maintain agricultural activity as the axis of their reproduction scheme, and continue to develop their own model to preserve and continue generating knowledge around seeds, preserving them and promoting their exchange for free or barter at local markets they control. Peasant and Indigenous practices learned from generation to generation maintain constant experimentation and improvement processes with which to achieve innovations and adaptations of crops to current productive and environmental challenges; these peoples are exerting, moreover, their historical right to the seeds above the capitalist enclosure that seeks to privatize them.

The geography of México favors the rapid differentiation, since it possesses various kinds of insulating factors. In addition, corn production conditions are extremely complex, and they are the result of an array of social, economic, technological, and natural variables. In terms of natural features, it is clear that "probably the most important feature of the production of native corn in México is its high degree of heterogeneity. . . . Mexican corn varieties are well adapted to changes in climate, pests, winds, humidity, low nitrogen, and acid soils."[56]

For centuries, farmers have used these natural advantages, and given the virtue of the open-pollinated quality of the crop, they have crossed cultivated maize with wild relatives and weeds, and in this way guided the evolution of new varieties adapted to their needs, preferences, and local environments. Managed by the farmers themselves, maize populations continue to evolve, raising the crop's performance and sometimes its resistance to adverse factors. The resulting wide range of varieties provides specialization across the many different habitats of the agricultural sector. There are special applications born of each new cycle of maize planting, breeding, and seed saving. This dynamically advances selection of native qualities across millions of parcels, as many as there are producers cultivating and selecting their own seed. This does not happen with the specimens preserved in so-called gene banks, whose condition remains completely static.

Being a center of origin of maize—with a prodigious diversity of shapes, textures, colors, behaviors, and geographical adaptations, which very few cultivated species can match—means that there are thousands of native varieties as well as wild relatives in our territory. This means that we are the natural genetic reservoir of corn.[57] Joining these peasant plant breeders, the most important corn specialists in the United States recognize that the Mexican countryside steadily continues to generate innovations by peasant hands experimenting in an empirical manner with crops and carrying out what is known as farmer-directed crop improvement.

The diversity of populations of maize that peasant farmers grow in Mexican rural communities is staggering. This struck even the growers when many neighbors in a demonstration plot gathered samples and exposed them to other varieties.[58] Native maize populations present features of great value in agronomic terms, and perhaps the most outstanding is the high degree of adaptability to multiple environmental conditions; maize is cultivated in different regions and climates from sea level to three thousand meters above.

México is, therefore, a key country for the preservation of the richness of biodiversity and agrobiodiversity crucial to agriculture and food in the world, and therefore is also a blank target for the interests of the large transnational corporations seeking to profit from the commons.

The forms in which we defend our sacred plant, accessible to the rest of humanity, with its multiplicity of senses, values, uses, and community development, will be key to resolving the dispute over maize and food on the planet, as well as to preserving the diversity of ways of life, knowledge, and human communities that offer alternatives to the dominant market model.

Resistance to the planting of transgenic maize in México is not only a defense of the reproduction of small producers: Peasants and Indigenous people, in addition to questioning the appropriation and control over seeds, are developing their own proper schemas for the defense and promotion of seeds, under their own community forms and institutions of self-management that allow them to survive the absolutism of the capitalist market.

One wonders if the research, experimentation, and improvement carried out by farmers around the cornfields—along with the momentum of community-based seed funds, fairs for exchange of seeds, GMO-free zones, networks and unions of peoples in defense of corn, networks of organic *tianguis* (from Nahuatl *tiānquiz(tli)* or "mercado"), festivals of gastronomy of the corn, workshops, forums, and meetings on Indigenous science conducted throughout the country— are only a strategy of resistance, or if these also build an alternative model to capitalism, since they today maintain and renew the subjects responsible for the generation and maintenance of the biodiversity, on which depends the future of the feeding of the world.

A civilizing proposal in all the meanings: From a cosmovision or worldview that involves a different way of conceiving the world of relations with nature, with plants; through a proposal for a science, technology, to guide the development of agriculture and produce our food; determining the food we consume and its quality, and including health options through medicinal herbs.

Corn created community, and the community made corn in a symbiotic relationship in which one is not explained without the other. This foundational bond is what allows the defense of maize to occur on a daily basis in diverse contexts, in which cultural rather than economic determinants hold sway. As Díaz observes, ultimately,

> the sacredness of the land does not support division or possession; neither of the people over the land and natural resources or of a certain people over others. The integrity of all the natural elements and living things explains the primacy of community and family, as opposed to the individual, and respects the community's interests, and the corn is respected in itself, but also deserves the respect of the community and its family. Sacredness is joy; understanding; mutual aid; the sharing of what each family has;

it is a gift. This sense of the sacred is why we do festivals and these events are originally linked to the work cycle with corn: The feast of fallow; the feast of sowing; the new maize harvest festival.[59]

Against this, in their book *Nuevos colonialismos del capital*, Davíd Sánchez and colleagues observe that

> capital and its corporations end up considering that the path can and must be subject to privatization because scientific activity (of course, this is corporate science) is the only social force that can presumably build and improve agriculture. . . . With this unique form of knowledge (the science of capital) transnational capital intends to monopolize the bases of life, devaluing others' traditional knowledge as independent of capitalist science and thus considering these incapable of producing innovations and knowledge. It does not recognize the key role of traditional knowledge or the legitimate rights of farmers, indigenous peoples and local communities even though, paradoxically, these are the main producers of knowledge and innovation in relation to the sustainable use of biological resources.[60]

This is the central dispute between systems of thought, as Indigenous scholar Díaz argues:

> This is what has been pursued since 1492, in some with more and others with less time, according to how the invasion of Europeans affected them. Some resisted and could continue to follow what had been learned; others had to hide it in the secrets of the night and in places far afield; and others were touched by such brutal violence that they forcibly abandoned and forgot all their knowledge, changing their beliefs and festivals for those that were imposed by missionaries.[61]

# CHAPTER 16

# Sodbusters and the "Native Gaze"

## Soil Governmentality and Indigenous Knowledge

### DEVON G. PEÑA

*In memory of Adelmo Kaber, friend and mentor,*
*who once said: "El frijol le da fuerza a la tierra."*
*(The bean gives strength to the land.)*

Wes Jackson is notable among the sustainable agriculture advocates whom I have heard recount a farmer's joke about a legendary encounter between a young pioneer "sodbuster" and an older Native American, who some claim was from the "Sioux" Nation (Dakota-Lakota-Nakota).[1] The encounter memorialized presumably actually occurred somewhere in mid-nineteenth-century North Dakota, but some versions place the rendezvous in Michigan, Minnesota, or Iowa. The genealogy of the joke remains murky. I have suspicions it may have first appeared around the time Frederick Jackson Turner announced the "closing of the frontier" and white settler farmers were busy reworking the land across Indigenous territories immediately west and east of the 100th meridian.[2] In the versions I have heard or read, the Native observer remains an amorphous figure: Gazing from the margin, he is presumably seeing a plowed field for the first time. Unsaid is a deeper history enunciated by the Native observer, which

I am going to presume embodied the first memories of being human in that place. From the edge, the "old Sioux" eyes the sodbuster's transmogrified perennial prairie, a land soon to be converted into uniform rows of pioneer staple wheat, corn, and barley. I imagine this moment—as it really happened somewhere in time and space—was for the observer a bearing witness to the violent refashioning of mostly wild (self-willing) land into the legible grid of furrowed disturbance. This was imposed by a newcomer's desire for order in what was likely viewed by the settler colonist as an unimaginably threatening and unfamiliar place that required treatment as dispossessed and hostile territory. In the typical retelling of the encounter, the white man's act of busting up the soil appears to puzzle the Native observer. From the sodbuster's perspective, the observer is shaking his head in what is presented as a state of bewilderment. The sodbuster—sometimes he even has a name, a Mr. Christensen in one version—notices the Native standing at the edge of the freshly plowed field. He reins back the sweaty oxen team, and the "heroic" single-spade moldboard plow comes to a squeaky stop with a hollow thud on soon to be exhausted soil. He ambles over and asks, "What is it?"

"Wrong side up," is the immediate reply as the observer points at the disturbed earth with lifted chin and a half shake of his head. From a decolonial standpoint, this signifies not bewilderment but mindful disapproval.

The conventional interpretation highlights the idea that the joke conveys a stoic and prophetic attitude. Gee whiz, the Natives had it right all along! Those sodbusters should have paid attention and recognized that soil is connected to all life and culture. Such nostalgia for the imaginary disappearance of a Native way of life essentially erases the inherently violent nature of the bloody expropriation, which Marx shrewdly called "primitive accumulation," although he understood this as an ongoing process rather than just a crime delegated to some distant past.[3] In the context of the end of frontier times, when the joke was first uttered, it seems fair to surmise that it served as a racist trope. It conveyed a sense of white settler superiority gained by virtue of self-proclaimed technical prowess, a reassuring sense of manifest destiny, and a revealing necessity to draw a sharp contrast between settler colonial mental attributes and the "ignorance" of an apparently "savage" and receding way of life.

"Wrong side up." When my colleagues retell the joke, they seem eager to illustrate that it was the sodbuster who was puzzled by this display of insight. Native soil knowledge was (and is) rooted in multiple generations of Indigenous place-based knowledge gained directly through resilient coinhabitation of ancestral territories. The sodbuster could not have understood the profound epistemological differences underlying the utterance, and so he would have failed to see that the joke was actually on him. So, we have a popular joke that seeks to teach us that the Native observer was wisely alerting the sodbuster to the idea that plowing earth, without the least restraint, bears toward agroecological catastrophe. The Native observer is said to recognize that forcing the soil out of place is an indictment of the sodbuster's disrespect for the land. The Native observer understood that overplowing makes soil more vulnerable to the loss of fertility; it causes soil erosion and compaction, and it degrades wildlife habitat and the diversity of flora and fauna. From the settler colonist's vantage point, the new plow-based agriculture eclipsed antecedent cultural norms. These norms remain hidden to the plow masters even today because they fail to understand the sources of spiritual and mythic obligations followed by many Indigenous peoples to sustain and protect the environment as a shared coinhabited place.[4]

The advent of the moldboard plow is recognized as one of the most transformative technological innovations leading to a more "modern" form of agriculture. It is also increasingly recognized as an invention that over time led to a massive increase in the scale and intensity of ecological damage associated with settler colonial and capitalist farming practices. The principal effects of overtilling were well known and avoided by Native American farmers,[5] and many still regard unrestrained sodbusting as a foolhardy violation of "Original Instructions" for the tender human use of land.[6]

Many of our interlocutors are trying to "revalue" Indigenous thought. Good enough. What they often fail to consider is how a particular racial arrogance underlies the identity location most directly tied to actual sociopolitical projects to enclose, dispossess, and displace the native inhabitants. This ideological milieu would have prevented the sodbuster from recognizing the soil knowledge of the Native observer. The sodbuster was embedded in the spirit of ecological counterrevolution of settler colonists who violently refashioned and repurposed Indigenous ancestral lands through enclosures that enacted the policy of "Indian removal." For the Indigenous peoples of the bioregions immediately west and east of the 100th meridian, the first wave of soil governmentality involved plowshares over agroforestry mosaics. The sodbusting settlers made use of the Homestead Act of 1862, a law designed to impose a more "legible" form of settlement and "development." This also provided legal cover for the homesteaders receiving title to unceded tribal lands redefined by the square grids of the cadastral surveys completed in accordance with the Land Ordinance of 1785. The sodbuster's settler colonial imagination could not invite open counsel: The presumed superior status of the knowledge steeped in his own European-American heritage and spatial-legal order prevented openness. He likely heard the response as quaint and superstitious. I am thinking: We are not ghosts of the primitive accumulation, and centuries of aboriginal resistance are reaching a new crescendo and challenging enclosure and land degradation.

## "Digging up" Indigenous Soil Knowledge

Today the sodbuster joke is shared in restoration ecology circles more than among soil scientists or farmers. I have overheard the joke at several recent academic conferences. I sometimes feel the restoration ecologists and conservation biologists who have shared the joke are performing for my approval: They enunciate the joke as a discursive act of solidarity designed to acknowledge the legitimacy of Indigenous knowledge. Good enough again. This is exactly what a broad swath of the anthropological discourse tries to do as well.[7] In matters of soil knowledge, much of this research focuses on documenting local soil classification systems (ethnopedology) through Indigenous farmer input. Other studies seek to "use modern technology to fully understand and validate traditional knowledge" of ecological processes in the soil (ethnoedaphology).[8]

Soil knowledge is now widely recognized by ethnoecologists as a form of traditional environmental knowledge (TEK) and part of a much broader stock of *situated* knowledge related to the uses of the environment by cultures in place and, for us, *sometimes out of place.*

This includes domains of nonlawyerly customary law and practices that inform the autochthonous organization and management of common resource assets, like soil quality. This adds a dimension seldom addressed in soil sciences discourse involving the contest over *land as property* versus *land as relation*—a conflict that becomes ever more acute under the conditions of capitalist globalization with the spread of neoliberal governmental regimes. The same is true of water. The forces that have shaped what constitutes legitimate knowledge of soil biodynamics also determine the legitimacy of particular sets of soil conservation and management policies. The regimes of soil governmentality place Indigenous knowledges and practices in conflict with modernizing settler colonial state formations and capitalist power over the enactment and implementation of policies and regulations affecting what gets qualified as legitimate soil knowledge.

From the vantage of decolonial discourse, the sodbuster joke betrays a more complex history of contradictions in the production of soil knowledge and reveals contempt of the Other for possessing "dirt-poor" knowledge. In "frontier times," the sodbuster would have likely thought that Native Americans were ignorant of agriculture. After all, they ate bugs, tubers, other wild plants, and even fungus— let's for now forget the "de-peasantized" culinary arts and the wonders done with offal and truffles. It is only in the white colonial imaginary that one can forget how the slaughter of bison accompanied the genocidal violence of the invaders against Indigenous peoples. You would probably eat bugs too in the aftermath of a white settler zombie apocalypse like this one. This also resulted in the displacement and erasure of millions of acres of polyculture gardens and shifting agroforestry mosaics that stretched across entire bioregions of the Indigenous corn belt.[9] A vital corollary of this settler colonial logic: They had no plows. Ergo, they were uncivilized. It may not have occurred to Mr. Christensen that when planting corn, beans, or squash, he could have thanked Native Americans (which for me always includes Mesoamericans), who gifted these cultivars to the settler colonists. If planting lentil, wheat, or alfalfa, Mr. Christensen could thank the Syrians, Arabs, and Persians. If planting cotton, he could thank the Culhua Mexica (Aztecs) or Egyptians, and if potatoes, then the Aymara, and so on across the many settler colonial nations that received these "gifts" from the world's Indigenous farmers spread out across Vavilov's centers of origin, where domesticated landrace crops grow in close interaction with wild relatives.[10]

Over time, the sodbusters' misguided practices led to massive soil erosion, compaction, eutrophication, and nitrification. These problems resulted in the creation in April 27, 1935, of the Soil Conservation Service (SCS) (under Public Law 74-46), which was established to address the cumulative effects of settler colonial farming practices in the aftermath of the Dust Bowl. This "new" scientific approach ignored the destruction of perennial native plant communities by indiscriminate plowing and the spread of monoculture habits. No one paid much attention when Native Americans first warned about the consequences of turning the world upside down. Native knowledge was reduced to caricature in cartoonlike images plying at the edge of racist end-of-frontier discourses, and, perhaps worse, is still today in the romanticized projections of deep ecologists and restoration ecologists.

There are many well-intentioned scientists and sustainability advocates who have made successful careers since the 1970s urging farmers to turn the world right side up again by adopting permaculture principles. These are truly seldom recognized as having Indigenous origins and analogs: perennial and annual polycultures, crop rotations with long duration fallows, intercropping with biodynamic and allelopathic companion plants, the classification and care of soils, the preparation and application of biodynamic soil treatment concoctions—all these and many more agroecological practices are results of Indigenous knowledge created in the centers of origin and well before the arrival of the fashionable, modern, and profitable advocates of biodynamics and permaculture.[11]

The predominantly white male discourse retells the sodbuster joke as if it suffices to say, "the Natives had it right." *Had* (past tense). The problem with this epistemic closure is that it ignores Indigenous peoples' continued presence in struggles for food justice and autonomy and to restore and apply our soil knowledge to ancestral, working heritage landscapes. It is true that the polyculture garden meadows, agroforestry mosaics, and other larger-scaled cultural ecological landscapes of Native homelands have in many places melted back into the Earth. Most were abandoned under conditions of severe intergenerational historical trauma that knocked people out of joint in time and space. But recovery is evident across the Indigenous corn belt in the struggle for food autonomy that is part of a broad social movement to restore Indigenous territories and regenerate eroded lands, while revaluing our place-based agroecological knowledge. Our movement seeks to relink these with associated forms of conviviality, participatory governance, and ceremony. These are biopolitical acts that create social and cultural change supportive of environmental, community, and self-healing.[12]

From the outside looking in, the Native gaze in the sodbuster joke presents a parallax view of our capacity to bear critical witness on matters of soil that first brought Indigenous eyes to look upon these acts of environmental violence. Today's monoculture farmers are still dreaming and waxing nostalgic in celebration of the "superior productivity" of sodbusting. They even adopt the latest models of "sustainable minimum tillage," but these models are tied to intensive fertilizer, herbicide, and pesticide treatment protocols concurrent with the use of transgenic crops. The Native gaze enunciates an uncertain fate for the settler colonial GMO farmers, for they repeatedly fail at the hands of the Gene Giants and will fare no better adopting RNA interference (RNAi) or "gene-edited" technologies. Today's sodbusters eagerly embrace remote-sensing technologies sold by the market-steered technocrats and obedient purveyors of Monsanto, Syngenta, Bayer, and DuPont "precision-farming" contracts who reside in the USDA field offices and dealers. Farmers are reduced to contract growers. The new "bio-serfs" keep voting neoliberals into office hoping they can get a bigger cut of the next round of farm bill subsidies and a shot at the next best transgenic "miracle" crops, which will surely have routine stacked-trait "events" approved by officials with ties to the very industries they "regulate."

Patrick Henry once declared, "The greatest patriot is the one who stops the most gullies."[13] If I may take that a bit out of context, the tragic environmental history of Euro-American agriculture was driven by white settler colonial farmers like our protagonist Mr. Christensen. His principal failing was to be possessed of a sense of superior methods combined with an utter lack of knowledge of the unanticipated and unintended effects of "modern" agricultural practices: Soil degradation, arroyo cutting, the exhaustion of the natural fertility of topsoils, hardpan soils that break plowshares, the loss of native bio-diversity, and diminished ecosystem integrity. Today's sodbusters believe GMOs will save the day. Soon they will use genetically engineered crops that can grow without topsoil, or better, on land contaminated by heavy metals. Is this our epistemological end point for soil matters? Hardly. As an act of epistemic disobedience, digging up Indigenous soil knowledge becomes the source of an agroecological revolution for worlds residing beyond the furrowed disturbances of the sodbuster's settler colonial imaginary.

## Indigenous Soil Knowledge and Epistemic Violence

In the 1980s, anthropologists and soil scientists began to center discussions around the science of soil ecology which they call "edaphology," from the Greek ἔδαφος, *edaphos* (bottom, base, ground). Cato wrote a treatise on soils circa 160 BCE, *De Agri Cultura*,[14] but similar knowledge was developed as early as the pre-Classic Maya period (2000–1000 BCE) and by the Mexica at Tenochtitlan-Tlatelolco (1248–1521). The Mexica classified soil into more than sixty varieties, which they described in terms of variations in the ratio of organic to stony material, depth of topsoil with recognition of distinct strata, permeability, erosive properties, compaction sensitivity, and color.[15] Mexica knowledge is striking because the scholar-farmers in the *calmecacs* ("line of houses," referring to higher education institutions) classified soil in a manner that anticipated the birth of soil conservation science in the United States by at least four hundred years.[16] Studies of Mesoamerican "folk soil taxonomy" were encouraged by a México-US team led by Barbara J. Williams and Carlos A. Ortíz-Solorio in the mid- to late 1970s.[17] A more recent extended discussion in 2009 hails "the incredible detail and knowledge of the Aztec soil classification system."[18] *National Geographic News* also published an article about a surviving ancient codex from 1540–44 pertaining to the Mexica site of Tepetlaoztoc, northeast of Texcoco.[19] The codex records each household and the number of members, the amount of land cultivated, and soil types as stony, sandy, or yellow. Evidence of this agroecological landscape still remains faintly imprinted on the land.

Figure 16.2. Culhua Mexica soil glyphs. Humboldt Codex Fragment VIII. Source: http://www.wikiwand.com/es/Sistema_métrico_mexica; accessed May 2, 2016.

The Tepetlaoztoc Codex has been widely examined in studies of Mexica ethnoedaphology and ethnopedology.[20] Few of these studies seem willing to address the extent to which Mexica teachers may have paired instruction in ethics with the practice of observing and experimenting with the biophysical and geomorphological properties of soil. Surely, serious damage had been done by the 1560s to the peoples and lands of México's Central Plateau. Accounts of the development of soil knowledge must be weighed in view of the environmental destruction and epistemic violence unleashed by coloniality. We can reasonably assume very few graduates of the calmecacs were still alive forty years after the fall of the Mexica Triple Alliance (1519–21). The vast bioregional infrastructure of aqueducts, dikes, dams, terraces, and erosion-control permaculture features had by the 1560s already fallen into wide disrepair. Many features were rapidly obliterated and recycled into settler colonial architectural and settlement projects wherever empire took hold. The infamous ill-advised draining of Lake Texcoco was part of this ecological apocalypse.[21] The knowledge relevant to the maintenance of these systems was likely compromised and degraded four decades after smallpox, measles, and Cortez's army defeated Cuauhtemoc's diminished force of Eagle warriors. This makes the Tepetlaoztoc Codex ever more precious.

Beyond antecedent status as a precursor science, Maya and Mexica soil knowledge included extensive references to human actions since these were perceived as capable of leading to uncertainty and unintended consequences, that is, "stochastic" effects in modern parlance. There is a deeper sense of respect of soils when truth is conceived as *Neltiliztli* (a well-grounded stability and well-rootedness).[22] This seems evident among Indigenous farmers today in the methods for use of different soils and how these are still codified by reference to biophysical concepts like *tepetate* (hardpan).[23] These concepts are based on empirical observations of biophysical indicators and align with instructions that transmit awareness of the role of anthropogenic disturbances in modifying soil quality. I can easily imagine an elder Mexica soil scientist instructing calmecac students on how the piece of hardened clay in their hands was created by an abusive and greedy farmer who violated these instructions. This melded scientific study with moral instruction on matters of soil and land ethics. A recent study verifies that this level of knowledge and instruction endures in the P'urhépecha communities of Michoacán.[24] We need to revive agricultural calmecacs today across the Indigenous corn belt.

Well into the 1990s many anthropologists and other social and soil scientists were still peddling the myth that the Maya were victims of an ecological catastrophe provoked by an allegedly rampant and ignorant "slash-and-burn" type of agriculture, which was presumed to have led to mass deforestation and demographic collapse.[25] Evidence to the contrary was ignored, including compelling studies of the resilient "Maya managed mosaic."[26] Such myths persist in an even more insidious form today. In a recent study on the history of soil science in México, coauthored by an anthropologist, the authors observe that

> soil knowledge in the pre-Colombian era was a noticeable attribute of indigenous people in México. A Mayan soil classification for the Yucatán Peninsula has been used by local people. The Mexica and the Toltecs [before] in the Central Valleys classified soils by land use and textures. Some names still persist today.[27]

Despite awareness of Indigenous soil science as a living method and practice, the authors simply note how the "modern" era of soil science in México starts in 1926, when the National Commission of Irrigation (CNI) was convened. The event

> brought American soil scientists to train the first agronomists on soil surveys required for the implementation of irrigation of lands. In 1929, the first Mexican scientific meeting, known as "The First Agrological College," was held in Meoqui, Chihuahua. This meeting is considered as the first formal activity in the field of soil science in México. . . . One of the major problems in the

development of soil science in México has been the lack of communication between the farmers *and scientists*. To alleviate this problem, some researchers have suggested that the ethnopedological knowledge should be incorporated into soil maps, since, in many cases, a map generated from ethnopedological knowledge is more precise and accurate than similar technical maps for management purposes.[28]

The authors repeat an obvious slip of Eurocentric historical periodization by using terms like *pre-Columbian*. More severe is the notion that Maya soil knowledge was just a "noticeable attribute." This trivializes the extent to which our ancestral civilizations invested major institutional efforts, intellectual resources, and communal labor toward matters of soil conservation and regeneration and watershed protection.[29] The first formal application of soil science occurred in the Mexica calmecacs and their Mayan antecedents, a fact routinely dismissed or overlooked.[30] Soil matters in the Mexica twin-island metropolis were serious enough to involve mobilization of hundreds of farmers, mathematicians, "diviners," landscape architects, and civil engineers among other specialists in the *tequio* (collective work) required to design, construct, and repair and maintain structures like aqueducts and viaducts, terraces, check dams, dikes, canals, ponds, and mounds. In rural areas of the Yucatán Peninsula, the same labor force had to tend numerous *rejolladas* (circular, highly fertile depressions resulting from gradual sedimentary deposition after the collapse of rock walls at the top of shallow cenotes), *bajadas* (natural low-lying areas with fertile soil), *xinampas* or *chinampas* (floating gardens), and agroforestry mosaics on a rather large spatial scale.[31] It would seem difficult to communicate over soil matters when the belittling of local knowledge of soils dead-ends with a concern for a "lack of communication between farmers and scientists" who emphasize incorporating these into their own maps. Can these two subject locations readily understand and respect one another in an environment permeated with the presumed intellectual and managerial superiority of academics who gaze upon the lands of the Indigenous commons without relational solidarity? This is binarism at its worst, and the assumptions on display here obscure how the Indigenous farmers *are scientists*.

Even among those recognizing the depth and breadth of Indigenous soil knowledge, most accounts of folk soil taxonomies fail to consider the bedrock Mexica precept that soil is a living organism. Soil has, in modern parlance, "biodynamic" properties. While unaware of microbes, nematodes, mycorrhizal bacterial colonies, and other microorganisms, the Mexica clearly understood the importance of human respect for the health of soils. The Mexica, like other Mesoamerican peoples, practiced *regenerative* agriculture—their cultural practices regenerated the natural conditions sustaining diverse soil organisms that define the capacities and limits of agroecosystems. Recycling human, animal, and plant wastes and debris, and fiercely dedicated to protecting drinking and irrigation water quality, the Mexica produced an urban agroecological revolution by redeploying ancient Maya *xinampa* agricultural techniques and recasting these within the massive hydraulic system of the lake district of Texcoco-Chalco-Xochimilco. The productivity in corn, bean, amaranth, squash, fruits, roses, gladiolas, poinsettias, and thousands of herbs and medicinal plants accomplished by the "floating gardens" of Lakes Chalco and Xochimilco would not be matched until many decades after the 1910 Mexican Revolution.[32] The true tragedy in this is that México lost food and especially maize self-sufficiency by the early 1970s, a history charting the loss of food autonomy beyond the scope of this essay.

This brief ethnohistoriographical account indicates that much of the anthropology that has been practiced in or dealing with Mesoamerican civilizations regarding the issue of soil classification, management, and conservation needs to be decolonized and "grounded" in Indigenous voices and epistemologies. Despite the pivotal work of esteemed colleagues like Barbara Williams and her team and Narciso Barrera-Bassols and his mindful collaborators, the study of soil knowledge in México and the United States largely continues to perpetuate epistemic violence. Too many researchers are simply out and about collecting "cognitive maps" from local placemakers in acts of appropriation without solidarity. A decolonial methodology of relational accountability works from within Indigenous epistemologies to seek an understanding of how observable patterns from place-based experiences can situate truth claims based on proven knowledge of place(s). These constitute a very different epistemological trajectory from those expounded through distanced normalizing observation.[33]

Indigenous soil knowledge is relational and intersubjectively based on a sense of coevalness. These differences are fully revealed as extant in the contemporary politics of neoliberal regimes of soil governmentality. This regime includes not just the rules of soil knowledge, management, and conservation but the active production of the "subjects" authorized to act rationally on these matters. One of the only ways I can practice the method of an alterNative anthropology of soil is to address these issues on the lands of a historic acequia farm in Colorado's San Luis Valley in the headwaters of the "Río Bravo del Norte" (Rio Grande).

## Devolution or Revolution in Soil Conservation?

The Acequia Institute (TAI) is a nonprofit charitable foundation dedicated to education and research to support environmental and food justice movements among Native Americans and Chican@s. The institute's specific focus is "water democracy" and "regenerative agriculture." The home farm of TAI is a classic *extensión* (riparian long lot) traversing the Culebra River bottomlands within the boundaries of the historic 1844 Sangre de Cristo *merced* (Mexican land grant).[34] Our irrigation water comes from "La Acequia de la Gente de San Luis de la Culebra," a.k.a. San Luis Peoples' Ditch (SLPD), which is the oldest adjudicated water right in the state of Colorado, with an appropriation date of 1852.[35] The Sangre de Cristo Land Grant includes uplands that were springtime hunting grounds for the Capote bands of the Mountain Ute First Nation well into the mid-1800s. Chicana/o land grant activists filed a lawsuit in 1981 seeking to reverse Jack Taylor's 1960 private enclosure of the land grant common.[36]

The lands of the TAI farm were originally deeded by Don Carlos Beaubien, the grant recipient, to Dario Diego Gallegos, as the founder of the settlement of La Plaza de San Luis de la Culebra, established in 1851. The deed shows the long lot went through several generations of partible inheritance within the Gallegos family and then into new ownership by nonfamily members. My sister, Tania P. Hernández, and I acquired the 181-acre parcel in February 2006 for purposes of establishing TAI and fulfilling the philanthropic wishes of our late father, Alfonso Carlos Peña. Two previous owners were Anglo-American families. One involved two generations that established a successful wool and mutton operation and commercial cauliflower and beer hops operation over a period of nearly four decades (ca. 1946–80). The other involved a retired Air Force colonel, who was widely regarded as an angry curmudgeon. He maintained a sheep and alfalfa-hay operation from 1984 to 1998 and was widely disliked for a stubborn disrespect of acequia (community ditch) customary practice for the sharing of irrigation water.

I knew the retired colonel well and interacted with him almost daily for more than five years when we both lived in the area. During one of our many conversations, he told me something that reminded me of the sodbuster's attitude. One morning he came by the local coffee shop after a dispute with the mayordomo (ditch boss) over the rotation schedule for the allocation of water on the SLPD. Fuming, he said, "These damn farmers. They've gotta modernize . . . Pretty backwards 'round here. Just too plain lazy. Too set in their ways . . . These ditches? Hell they date back to medieval times!"[37] He went on: "They should convert to sprinklers and corrugated pipe. Get more efficient. Save water by, you know, paving the canals. It's all about becoming more modern . . . It's not about race like you were asking earlier."[38]

I will skip dwelling on the retired colonel's apparent racial prejudices in the construct of "lazy and backwards Hispanic" farmers and dispense with the fact that some social scientists have expressed similar views in peer-reviewed journal articles.[39] I had an even more practical and pressing problem: The sodbusters were gone, but they left a heavy imprint on this land. There was damage from a now-removed center-pivot circle sprinkler. We had just acquired the land in 2006 from a multigenerational Hispano farmer from the Española Valley who had himself acquired the farm for an alfalfa and hay operation and supplementary grazing range in 1999. TAI inherited the curmudgeon's seventy-five-yard-long center-pivot mechanical irrigation sprinkler installed in the 1980s. We irrigated the patchy alfalfa and bromegrass field remnants with the center-pivot that first season in 2006 and had significant diesel fuel, labor, and maintenance costs. The field was pockmarked with prairie dog burrows. As planned, we stopped using the sprinkler the following year (2007) during the annual April to October irrigation cycle. We have since continued doing the best we can to realign and reexcavate the abandoned and badly damaged acequia network. I spend a lot of time "changing water" on those large meadows, an activity I cherish. We finally dismantled the six-ton mechanical centipede-on-wheels and removed the last of it in 2010. In 2009, we restored full use of acequia gravity-driven flood irrigation to the north-end *vegas* (meadows), and in 2012–14 started reseeding with organic and conventional non-GMO alfalfa; this is our principal asset as far as money-making operations go. We also have started implementing permaculture features to slow the movement of water over the land and along the lateral and *espinazo* (spinal) ditches by planting native orchard trees and brambles to reduce the potential for erosion while creating habitat and an edible landscape. Finally, we have effectively restored the cottonwood-alder riparian forest along the quarter-mile stretch of river meanders that bisect the long lot. This restoration has led to the return of native medicinal and edible plants and bushes including rose hips, wild asparagus, and oshá.

The San Acacio Culebra acequia bottomlands are resilient and not too sensitive to erosion, but flood irrigation methods present many challenges. The land still has fairly deep soil horizons ($\geq 1.8$ m., or 6 ft.), but restoration work is affected by swales and the presence of Pleistocene streambed depositions of gravel and river stones that lie too close to the surface, especially around the former location of the old center-pivot circle. These conditions presented an opportunity for us to experiment with soil regeneration by working with the acequia gravity-driven deposition process. This process is evident across our bottomlands wherever the fields receive windswept dust from surrounding mesa-top volcanic soils or fine sediment transported from the mountain peaks and cirques through flood irrigation practice. There is little evidence of compacted clay lens (*tepetate*). We do not have any gullies. The main challenges on this northern upper-elevation half of the farm are the concentric grooves produced by the wheels of the old center-pivot sprinkler and the state of the acequia network for the alfalfa hay fields with potential for arroyo cutting, given the decades of inattentiveness and disrepair. It is a challenge to move water such a long distance to flood irrigate the furthest fields, depending solely on gravity, which of course is great renewable energy but can erode the surface of the ditches, creating gullies.[40]

The myriad issues with center-pivot agricultural sprinklers are legendary. The one that reminds me the most of the steadfast arrogance of the sodbusting plow master is that in our area the sprinklers encourage prairie dogs to burrow more profusely into the irrigated fields. Flood irrigation by acequia techniques keeps burrows to a minimum. The critters tend not to locate in flood-irrigated fields because flooding makes the burrows uninhabitable. Modern sprinkler irrigation is more like a long, steady rain. I have seen prairie dogs showering under evaporating mists. Studies show that mechanical sprinklers are less efficient than acequias at delivering water to the crops because of aerial evaporation, especially in our high-altitude alpine desert environment. The differences in soil erosion control and the effectiveness of getting water to crops are rather striking.[41]

The retired colonel kept insisting the mechanical sprinklers were superior to acequias, and he was solidly backed up by the local agricultural establishment controlling the water and soil conservation districts across the San Luis Valley. I thought he was just too worn-out and could no longer invest the long hours and skilled labor required to master the art of flood irrigation. He belittled the methods of the "Parciantes" (farmers with water rights on a ditch). Yet, as soon as he lauded the superiority of mechanical center-pivots, he followed with complaints about rising fuel costs for diesel. He lamented how he had to run the sprinkler a lot longer than it takes for acequias to irrigate a comparable field. This created scheduling conflicts with other irrigators. He complained mightily about the high cost of maintenance and the many hours spent driving long distances to acquire expensive parts for repair jobs. Sometimes, unable to do the job himself, he was delayed because the repair mechanic couldn't do the job in a timely manner. The sprinkler became his personal maintenance nightmare and caused all kinds of misery for the SLPD since these breakdowns disrupted the customary timing of allocations to the different irrigators. Once he claimed the sprinklers were better because they reduced soil erosion, a position echoed in much of the USDA rhetoric.[42]

He reluctantly acknowledged that the sprinkler, circling the field on large tractor-like wheels, was producing erosive features in concentric grooves. Wherever the wheels traced a path through the ground, after more than a decade of use, they cut deeply incised rings of compacted soil, creating channels that redirect water flow, thus causing uneven distribution.

Repairing this damage has been a major task of our work to restore healthy soil conditions by reintroducing acequia flood irrigation to these meadows. We do so without increasing the exposure of Pleistocene features wherever these are already at the top of the soil horizon. At one point, one of the local Natural Resources Conservation Service (NRCS) soil technicians recommended we "laser level" the swales. It was suggested that we should then replace the acequias and deploy corrugated pipe to irrigate on a flattened, more uniform landscape. TAI rejects uniformity as a basic permaculture principle and declined the proposal. Leveling of any sort would expose the underlying ancient riverbed gravels. This is a lesson I learned from *los animalitos* by observing the piles of Pleistocene rubble around the entrances to the prairie dog burrows. The San Luis Valley has its share of sodbusting monoculture farmers who continue to abuse the land. We don't need to add to the load. The results of past abuse are apparent in many places on the TAI farm: It can be seen in the *barrancos* (eroded banks) along the Culebra River. Over the decades, wherever sheep and cattle trampled the ground and cleared the edges of native vegetation, riverbanks collapsed in large chunks of topsoil with rootstocks from willows, alders, and rose hip bushes. These long ago washed out into the river, leaving bare walls exposed to the river. Further out into the middle *vegas*, the effects of poorly timed grazing produced a pattern of hummocks (*cespedes* or *mogotes*). This area marks the transition from the riparian zone to *las vegas de en medio*, the middle meadows hosting native grasses and flowers watered by subirrigated flows from the upstream acequias. We are engaged in a practical battle with the ghosts of the sodbusters by repairing damaged riverbanks and restoring soil health.

## Resisting Soil Governmentality at the *Almunyah*

Our first ten-year plan (2007–17) for the lands of TAI emphasizes the restoration of riparian areas, stabilization of acequia networks, and repair of the hummocky meadows. There are numerous "invasive" species in Colorado, and our watershed is no exception. Our lands are located far down the Acequia Madre in an area not yet overwhelmed by noxious plants. In Colorado, many plants originating in the Asian steppes are the "scourge" of farmers and ranchers. At TAI, we reject the concept of "weed" but share the concern of restoration ecologists for keeping the balance in favor of native plant associations. Among the domesticated plants, we are proponents of many "naturalized exotics" including heirloom potatoes (Mountain Rose, Sangre) and *habas* (fava beans).[43] Leafy spurge, Canada thistle, and Russian knapweed are some of the "noxious" plants spreading in the San Luis Valley agricultural districts and are not easily contained, let alone eradicated. These troublesome species are unwanted because they are toxic to most livestock and reduce the quality and output of alfalfa hay production. A chief concern is how these species displace native plants and thereby affect habitat for many living organisms. These noxious arrivants are prone to dominating the landscape like a settler colonial monoculture. This happens more readily in, or is in any case associated with, soils that have suffered considerable disturbance from human activities. These noxious plants are the biological baggage and ecological legacy of global sodbusting empires.[44] The restoration of Indigenous soil knowledge must be accompanied by active ecological restoration to "exorcize" the biological analogs of colonialism and the degradation of the land. I have approached our work by borrowing from Western scientific concepts in restoration ecology, conservation biology, and biodynamics and aligning these with antecedent Indigenous knowledge. Every day I work with irrigated land, I am acutely aware of the value of the soil knowledge of ancient Mayan, Mexica, Mixtec, P'urhépecha, Zapotec, and other ancestral Mesoamerican civilizations. This knowledge continues to shape our approach to working with the lands of our *almunyah*.[45]

An additional source of our epistemic disobedience rises from our objection to how these "noxious invaders" are usually treated, under the framework of a chemical "warfare"-against-weeds paradigm. For the past fifty years in our district, the USDA, through the local office of the NRCS, has announced that it was "launching an all-out war against these noxious invasive weeds."[46] The "war" generally involves rapid deployment of herbicide treatments, including Monsanto's Roundup Ready® brand formula for glyphosate. In the summer of 2007, I watched, like the Native at the edge of the sodbuster's field, while USDA teams targeting leafy spurge sprayed the herbicide on test plots by the high school athletic fields. Acequia farmers declined the offer of similar treatments to control willows and noxious weeds, principally Russian knapweed. We opted to use goats on a few patches and riparian strips within the perimeters of two SLPD long lots. The leafy spurge and Russian knapweed in the USDA test plots returned in 2009 as defiant fairy circles around the edges of the sprayed patches, but the entire area was recently razed to build a new public school complex. The goat treatments had the desired effect, and willow and the Russian knapweed retreated from the acequia farmers' test plots. Some NRCS technicians are warming to the use of goats, but most remain committed to the modern chemical treatment protocol. Our goats are for now gone, but the arrivants still pose the threat of reestablishing a toxic presence.

This story about weeds is suggestive of changes occurring in the relationship between the federal governmental regime (NRCS) and local acequia farmers. This entire episode reflected the top-down process of neoliberal devolution and changed discretionary planning at the local soil conservation district level. This was unimaginable two or three decades ago, when district board members and technicians were mostly white men from outside our community. Today, the local NRCS office includes a skilled multigenerational acequia farmer and has hosted a series of three progressive, sympathetic white women as technicians. All are limited by the shackles of federal policies, but the office now seems more open to acequia farmers as a unique cultural community deserving of respect for their Indigenous soil, weed, and watershed knowledge.

Permeating social interactions between acequia farmers and the NRCS is the fact that the process is subtly contested. One vision, the top-down one, allows the local NRCS some planning and design autonomy but within strict budgetary limits and subject to requiring individualized contracts in outreach with "underserved" and "under-represented" farmers. Local farmers enter into agreements with the NRCS for acequia infrastructure projects. Acequia farmers can apply for EQIP (Environmental Quality Incentives Program) grants. The associations of *parciantes* make effective use of these to improve acequia infrastructure, like *compuertas* (ditch head gates) and other water diversion, soil erosion, and sediment control installations.[47] However, the restoration ecology work on the *almunyah* cannot be supported under current federal rules for programs like EQIP. We face the continuing work of converting that meadow from sprinkler to flood irrigation. Repair work to restore compacted soil under the concentric grooves of the old sprinkler system is complete. We still must realign the network of *lindero* (perpendicular) and *sangria* (bleeding) acequias. To maintain our autonomy from governmental entities, we have so far relied on the privilege of being able to use the institute's endowment income to invest in these improvements; this is not an option for most of our neighbors.

There is another set of problems beyond the apparent current scope of these USDA soil conservation programs. We wish to rely on permaculture practices to slow down the movement of water through the more badly damaged north-side hay meadows. In our second ten-year plan, we seek to anchor and buffer the more erosive slopes and swales at the north end with a system of *ancones* (terraces), *alamosas* (cottonwood tree lines), and *bordos* (raised berms) planted with native fruit trees, like sand cherry, chokecherry, and gooseberry or oshá (*Ligusticum porteri*, a.k.a. Porter's lovage) and other medicinal and biodynamically active herbs. The plan will take time but should restore the soil horizon above the patches of ancestral riverbed gravels that have been exposed at the surface by decades of excessive plowing, inappropriate and poor irrigation practices, and overgrazing.[48]

There are always contested ambiguities presented by how the USDA works locally to implement programs from the top down; not least is the tendency to impose technical design criteria. These may not be entirely appropriate for acequia methods and may even undermine and weaken our commitment to local and more collective community-based approaches to problem solving. These technical designs are "Super-Sized" and inconsistent with the humbler scale of acequia form and function. Moreover, these efforts either deliberately or inadvertently inculcate a new modernist subjectivity by inducing *parciantes* to accept individual contracts; irrigators are also constantly invited to shift to drip irrigation, the use of gated pipe, and other techniques potentially at variance with sustaining acequia flood-irrigated practices.[49] These seemingly neutral designs can reduce our ability to act on the basis of shared norms of mutual reliance and Indigenous knowledge. It is seldom understood that collective mutual-aid interests are an alternative to the dominant individual rational actor model that dominates economic behavioral expectations in these NRCS programs. The soil conservation regime tries to define the behaviors acequia farmers must follow so they can be seen as acting effectively as land and soil managers—as subjects who can demonstrate the capacity for responsible behavior but "only to the extent that they have the managerial capacities to pursue economically 'rational' practices."[50]

As acequia farmers we continuously juxtapose ourselves against the imposition of this neoliberal form of the governmentality of soil conservation. Many in the acequia community continue to act on the basis of collective self-provisioning and Indigenous knowledge to meet our soil conservation needs. In 1995–97, Robert Curry and I logged a small set of "soil augur" surveys to corroborate local claims that acequia farms are soil reserves. Our augur survey found evidence at the Corpus A. Gallegos Ranches and three other sites of unusually deep topsoil horizons in excess of 1.8 meters, or about 6 feet. These are among the deepest in the highland parks of the entire southern Rocky Mountain biome. We found no evidence at all of hardpan; recall the Mexica concept of *tepetate*. We took this as indicative of continued use of sound local knowledge of soil regeneration practices, since the NRCS was doing little at the time.

As a community of traditional irrigators, we remain coinhabitants of a place where too many neighbors defect to neoliberal orientations and ignore evidence of the decline of perennial native plant communities and wildlife habitat. Too many of us fail to recognize the consequences of stubborn and indiscriminate plowing and monoculture habits reflective of a sodbuster mentality. At the TAI *almunyah* we are challenging farmers who defect to selfish modernist sensibilities by emphasizing Indigenous knowledge of soil conservation and regeneration in our own practices. Not all acequia farmers are that successful, but in our context the technology of gravity-driven flood irrigation, when combined with intensive permaculture practices, carries the possibility of sustained regenerative benefits from a "disturbance ecology" with deep roots in the soil knowledge of Indigenous peoples. The best acequia farmers are like beavers. We contribute to biological and landscape ecological diversity by following original instructions as coinhabitants and active shapers of a shared ecosystem we are allowed to transform but not ruin for use by others, including more-than-human beings. How we farm is the first step toward a politics of decolonial foods and derives from respect for the life of the soil itself as inherited from our Indigenous forbearers. In Nahuatl, *Teotlalyollotli* is the "sacred heart of the soil."[51] This grounds us with humility in service to the land. After all, *Sin suelo sano, no hay maíz.* Without healthy soil, there is no maize.

# CONCLUSION

Food is who we are, and how we grow, share, and consume our food informs us of the people we want to be, the places we wish to coinhabit, and the worlds we work to create. Decolonial food is not just a radical restructuring of our food system; it is the radical restructuring of the meaning of our food. It is the resurgence of a way of being human that celebrates ancestral practical knowledge while engaging in the acts of making and sharing food and expanding accountability across the food chain. We are not only dismantling hierarchies of power and privilege but working to bring dignity back into all of our food practices by providing pathways to healing through transformative, everyday lived experiences and direct action. Through our Indigenous and hybrid subjectivities, we are recovering the ancient traditions of our ancestors while simultaneously creating new expressions of agroecological, ethnoecological, and ethnomedical knowledge and adapting to the challenges of displacement and dispossession alongside climate change and other interwoven ecological, political, and cultural crises.

The goal of the coeditors was to create a space for dialogue by bringing together a group of scholars, farmers, gardeners, and activists who share a similar passion to remake our food practices and agrifood systems with Indigenous decolonial principles at heart. As we have seen, the Mexican-origin people, across our beautifully diverse ethnic spectrum, have long transformed ourselves, our relations, and our foods and foodways in ways that move beyond contemporary alternative food movements. Decolonization is a long process, and we recognize that this is only one step among many. One of the goals of theorists like Anzaldúa, Bonfíl Batalla, Fanon, Ñgugi, Sandoval, and Tuhiwai Smith—to name some of the thinkers that inspire us—is to come to terms with who we are by valuing our Indigenous ancestors and their knowledges, embracing and regrounding hybrid experiences, and determining how to move forward with our revolutionary subjectivities intact instead of remaining tagged within the governmentality regime of "identity politics." This collection is one such step addressing these concerns.

While the self-provisioning practices of Mexican-origin peoples are deeply rooted in centuries of place-based traditions and diverse ethnic cultures, these are by no means idle or static and fixed. As each contribution demonstrates, the living traditions of Mexican-origin people have always been grounded in the ability to adapt and the collective opportunities and decision-making processes that facilitate this openness to the adoption of new practices and knowledge. This is vital under the conditions of displacement and dispossession we face in the long-duration era of settler colonial and their affiliated capitalist nation-state empires.

This collection began by interrupting or fracturing the dominant food sovereignty paradigm. La Via Campesina has rightly come to represent the ideals and organizing intensity of the global food sovereignty movement, but in chapter 1, Devon Peña calls into question the very logics that set it apart, deeming these as still trapped within the logics of coloniality in neoliberal iterations and anthropocentrism. The organization's lack of self-reflexivity and use of an essentially anthropocentric concept of sovereignty limit our ability to bring freedom and dignity to the more-than-human world or to workers across all dimensions and levels of the food chain, including the two-thirds of the world living in a planet of urban slums.[1] By fragmenting these logics, Peña opens an Anzaldúan space of multiple, shifting, and transforming subjects. The rupture allows for new worlds to emerge from the in-between spaces of resistance and cooperation.

The chapters that follow all call upon Anzaldúa's "Coyolxauhqui imperative" because each one recognizes the ongoing process of making and unmaking. They also acknowledge that "there is never any resolution, just the process of healing."[2] As Peña argues in chapter 16, the sodbuster left a wound on the land and our bodies. Through collective action, people and places can begin to heal, and this involves relying on Indigenous knowledge of agroecology, ethnobotany, soil regeneration, water conservation, and our own traditions of mutual aid and direct participatory democracy; the iconic examples of the South Central Farmers, C2C, Familias Unidas por la Justicia, and the chico-making Culebra acequia farmers illustrate this in a compelling manner. It also seems important to note how the soil knowledge of Indigenous farmers prefigures the rise of modern-day biodynamic and permaculture farming methods and practices.

## Food, Place, and Decolonial Forms of Self-Organizing

There are multiple strategies people engage in to cultivate self-determining and coeval autonomous social relations. Many contributors to this volume reveal how Mexican-origin people modify, transform, and re-create Indigenous forms of self-organization. Rufina Juárez (chapter 2) shows how the multiethnic Mesoamerican farmers of South Central Farm were uprooted several times by neoliberal policies and enclosures; yet their struggle to create a self-determining community lives on. Juárez calls into question the "conflict food" nature of the capitalist industrial food system. Vandana Shiva has noted how women across the world are impacted differently by the industrial food chains of neoliberal capitalism but remain sources of healing knowledge, belief, and practice.[3] Silvia Patricia Solís, in chapter 4, echoes Juárez by demonstrating how women are impacted differently and further how they draw from their heritage and knowledge to heal the "colonial wound." As Luz Calvo and Catriona Rueda Esquibel argue in chapter 7, decolonizing our diets is not simply about a "return to the kitchen" along the lines, say, of the slow-food movement. By calling for the "liberation of the kitchen" and the "complete reconfiguration of gender," decolonizing our diets forces us to rethink how gender and gendered norms configure our food practices. The production, distribution, and consumption of food can be used as a tool of self-organization and mobilization when freed from the shackles of oppressive norms and structures, especially within our own movements, as Juárez also indicates. Similarly, Gabriel Valle illustrates in chapter 3 how a community can transform the value of food to represent the lifelines of convivial self-valorizing labor. The urban gardeners in San Jose share food from their kitchen gardens to strengthen social networks and improve cultural, social, and environmental well-being. This is no panacea, and many urban gardeners must still get emergency supplies at food pantries. Gardening strengthens a community's resilience, but it is not the only answer. Precarity increases vulnerability, including blocked access to heritage food items, but it also creates openings to new forms of autonomy. This should compel further community-based collaborative study of adaptive strategies and alterNative subjectivities involved in the remaking of our foodways as a form of collective struggle.

María Guillen Valdovinos, in chapter 9, echoes the ethnographic and theoretical findings of Valle, depicting her experiences with *el huerto familiar* (home kitchen garden) in helping to forge a "trans-border" sense of place and community. This resonates with Melissa K. Nelson, who argues that we are not only what we eat but "*where* we eat."[4] Guillen emphasizes the importance of home gardens for displaced people. Such gardens allow reconnection with the homeland while also creating a sense of place in new (home)lands. Growing food is a decolonial act when it provides people with a sense of who they are and where the come from. Guillen's family uses their home kitchen garden to create a sense of place, security, and well-being. Valle and Guillen both show how the act of producing and sharing heritage and healthy adopted foods involves an ability to re-create norms of reciprocity, respect, and trust and to act within constantly shifting open spaces and networks that are beyond the reach of the industrial capitalist food system and its global commodity chains.

While decolonization is a personal journey undertaken by each person, as a socially and culturally transformative practice it cannot be accomplished alone. Each step in the process requires deep embrace of the "Other." That dominant actors deploy technologies of govern-mentality and discursive regimes to undermine or block "escapes" from the system is one of those obstacles requiring collective energies. Neoliberal regimes are also seeking to recapture the spaces we have rendered or sustained as autonomous. In chapter 11, Tezozomoc and colleagues illuminate the ways the farmers respond to such adversity by transforming their struggles and changing tactics to meet the needs of the community through autonomous forms of organization that continuously draw from the deep well of Indigenous customary institutions of collective action and self-governance. The resilience of our diverse Mexican-origin diaspora community is revealed as one of the vital principles the South Central Farmers relied on to navigate around neoliberal enclosures and support the constant revaluing of the ancient wisdom of ancestors. They crafted new production scenarios and created new opportunities to remain autonomous as farmers, workers, and participants in Indigenous cooperatives. This involves innovative understanding of the actual forms of power/knowledge as these play out in urban-planning politics. The example of the planting of *Cipactli* (Tail of the Mother Earth Lizard) at the farm—which is the name of the *pochote* in Mixtec communities—becomes a subversive enunciation by displaced Indigenous farmers; it is part of their constituent will-to-power expressed as the symbolic politics of "naming" place for the reterritorializing of their own existence in autonomous spaces, which represent escapes from the neoliberal logics of enclosure, privatization, and hyperindividualism.

## Hybrid Subjectivities

Decolonial food projects are adaptive because they must be. They transform us because the material and political conditions of our lives and of the people and other relations in them compel us to embrace inventiveness so all may survive and flourish. Hybridity is in this sense a positive, affirming quality of revolutionary subjectivity and averts identity politics as the end point of our struggles. The knowledge of self-organizing practices described in each of these chapters helps us to better understand how our food practices and traditions are multileveled, multithreaded, multivalent, multiscaled, and always adapting and evolving. The practice of listening to our bodies, communities, and lands must be further cultivated to allow for the widening of our revolutionary subjectivity as Nepantleras/os.[5] This will allow us to navigate our way through these multiple worlds by means of the everyday lived practice of coevalness and conviviality and through the collaborative (re)making of places our alterNative foods and foodways create in common spaces and ideally through direct participatory democracy, as Pancho McFarland argues in chapter 14.

Hybridity is a key concept throughout this collection of essays. Our foodways heritage is often circumscribed by the legal status of Mexican-origin peoples who are forced to move between locations and across multiple shifting subjectivities. Consuelo Crow, in chapter 5, helps us better understand how the act of "crossing over" is a liminal state wherein travelers must move through a zone of exception undetected. In these "crossing times"—the equivalent of a refusal to obey the sovereign power—food and water become a matter of immediate survival, and things that are seemingly insignificant in other settings become vital to the precarious transborder traveler. Crow reveals the bare face of neoliberal governmentality as a technology that weaponizes the environment in the "war on immigrants"—via the deployment of a brutal and murderous state of exception that suspends these mobile bodies in a state of "rightlessness," of abject degradation as subjects suspended in an exclusion zone and wrongfully denied basic political and legal existence.

While "crossing over" may begin the journey for the traveler, finding home in *El Norte* is a never-ending process of continuous physical, social, and spiritual unrest and movement, even among people rooted for generations north of the border. In chapter 6, Lee Ann Epstein helps us to better understand how a constant state of "turned aroundness" may be confusing but can also be transformative. People make sense of the world in a variety of ways, and Mexican-origin communities are by no means homogenous in how we do this. We tend to embrace and enunciate a broad multiplicity of knowledges, beliefs, and practices that span locations and generations to reveal complex hybrid family genealogies. This is also how Calvo and Esquibel, in chapter 7, interpret the decolonial act of cooking and eating. For them, decolonial cooking is like the improvisation of jazz. This performance is an act of agency because the creative process of cooking allows us to take what we have and add a little of this and a little of that to create what we want but imbued with newer meanings grounded in the practical ecofeminist politics of care of the self in convivial acts residing beyond the neoliberal formula of egocentric self-care.

Just like adding spices to a dish, Rosalinda Guillen and C2C, in chapter 12, and Tomás Madrigal, in chapter 13, demonstrate how the damaging effects of neoliberal policies have not kept Mesoamerican diaspora and transborder workers from continuing to add and subtract what they need in order to foster well-being and dignity through organizing. This spirit of *rasquache* (makeshift inventiveness) enunciates ways of prospering on the margins because it involves finding common ground for colonized people the world over as Indigenous and oppressed communities. Everywhere, we see people using their ancestral knowledge bases to navigate the colonial environment, sometimes in strategic engagement with the juridical order and policing machinery of a state of *economic* exception that seeks to reduce us to bare life/habitance. To decolonize our foodways and bring dignity back into the practices of growing, cooking, serving, and eating food, we must transform ourselves in the wake of large- and small-scale inequalities. The racialization of Mexican-origin peoples, as is illustrated by the apprehension-detention-deportation industrial complex, is integral to the forces working to colonize our knowledge systems, bodies, and lands: Decolonization results when we transform our worlds by changing ourselves and retaining self-determination, dignity, and cultural integrity.

## The Subversive Uses of Knowledge

Agroecological practices of farming and associated traditional or ethnoecological knowledge (TEK) have deep roots in Mesoamerican and other Mexican diaspora communities. This knowledge-in-practice through food production, distribution, presentation, and consumption is a form of personal and cultural affirmation. Gregory A. Cajete, the Santa Clara Pueblo epistemologist, explains that native science is a "reflection of creative participation."[6] From this standpoint, TEK emerges from a group's intimate interaction with the environment *as our home*—a coinhabited space that becomes a place. John Mohawk reminds us that the universe is the "fountain of everything,"[7] and as human beings we gain knowledge through participation with the cosmos and must cultivate humility. Joseph Gallegos reminds us in chapter 8 of the virtue and multidimensionality of humility by showing how TEK is practical knowledge about how to live well in a place with the instructions handed down by ancestors. The humble act of producing chicos del horno affirms and enunciates a decolonial space by rupturing commodification and rejecting "imposter chicos" and transgenic corn products that are increasingly intruding on our land-based communities. The practices followed by Gallegos and his family embody the coevalness of the relationship between the Earth and the people through place-based knowledge realized in the convivial act of cooking. Chicos are the living spirit of ancestral ways passed down across generations and celebrating and rekindling our sociality.

In a similar way, in chapter 14, McFarland examines how "organic intellectuals" rise up from the ground in urban and rural spaces, culminating in a "revolution in the urban garden." These organic intellectuals have enunciated new performances to show that urban and rural environments can be spaces to celebrate people and places. While the oppressive nature of industrial capitalism has sought to silence these voices, the place-based organic intellectual cannot be silenced because s/he has the ability to navigate interstitial spaces and meet resistance with cooperation. Nowhere is there an exact or static definition of traditional knowledge, and Gallegos and McFarland both demonstrate how traditional knowledge is ancient *and* modern at once. It is the embodiment of many worlds and places, and this quality is one key to understanding the resilience of Mexican-origin and other colonized communities.

## Subversive Love: Autonomy, Food, and Healing Spaces

Chela Sandoval believes that "it is love that can access and guide our theoretical and political 'movidas' (revolutionary maneuvers) toward decolonized being."[8] Her point is that part of the colonizing process involves producing a relationship to the "Other." To counter the alienation and dehumanizing effects of colonization, we must "love" the "Other," and we believe this should include more-than-human beings. It is through love that we cross from one world to the other because it blurs the lines between "us" and "them." Transcending the logics of "Othering" opens pathways for collective healing. The works by Valle (chapter 3), Solís (chapter 4), Epstein (chapter 6), Calvo and Esquibel (chapter 7), Guillen (chapter 9), and Curry (chapter 10) all embrace food and foodways as part of healing practices and a collection of traditions, and they all illustrate how these can be similarly used to decolonize our own minds, bodies, communities, and lands.

These spaces of healing emerge in a variety of ways, and the accounts in these chapters offer myriad instances of the victories of self- and communal determination. Yet Mexican-origin peoples seeking to find spaces of healing still face many challenges, not the least of which is a lack of access to safe spaces to practice autonomy, understood here as an escape from the orbit of the neoliberal capitalist food system and its modes of surveillance, compliance, incitement, and compulsion. Countless motivated and committed people are seeking to bring dignity and justice back into our food system, but for many, the changes we are making and seeking will come too late. The current industrial capitalist food system is perhaps the second biggest human experiment ever launched without consent in the history of the planet, outranked only by conquest, colonialism, and the genocidal trajectories spun by European settler societies. With its goal being to insure profits rather than personal or collective well-being, the violence imposed every day by neoliberal capitalism across the entire food chain and our bodies and communities cannot be downplayed or wished away.

For many people, the agri-food system exists across such massive temporal and spatial scales that it becomes difficult to come to terms with our place in this construct. It is almost as if food and the food system have become a "hyperobject."[9] Decolonizing our food compels us to rethink our relation to place but also to consider larger-scaled and more long-duration processes. Only through humble acts in place can we begin to dismantle the oppressive nature of the globalized industrial capitalist food system while creating alterNative worlds. This volume details some of the myriad ways Mexican-origin peoples are growing, preparing, and consuming food as a common medium to cultivate new subjectivities of resistance and collaborate in the process of decolonizing our food and foodways and thus our bodies and communities.

Tezozomoc and colleagues, in chapter 11, and Peña, in chapter 1, both address a relevant lesson that has to do with the connection between the spaces and practices of healing and the political project of autonomy. The creation of spaces of healing requires the capacity for us to recognize and mobilize the use value of our own agency, and we must recognize how this derives from deep ethics grounded in TEK and related cultural artifacts and attributes centered on coevalness and conviviality. Tezozomoc and colleagues remind us that the "usual poststructuralist approach . . . has resulted in the celebration of rather fleeting ontological experiments," while the struggle at SCF focused on "heterotopic agency . . . mobilized *as a more permanent and permeable occupation of space*, one that can be a single real place juxtaposing diverse elements from multiple spaces." This is not just the remaking of heterotopic spaces but the enunciation of a decolonial ontology guiding placemaking itself. The gentle, loving art of place-making is part of the medicine that heals us, and this capacity is much older than the Euro-American modernist project of enclosure and so seems more likely to survive the looming decline of the neoliberal project, which heartlessly lacks respect for the senses of place and is therefore strategically vulnerable and phenomenologically weak. Or, to put this another way, neoliberal ideologies lack the lived experiences in which the subject can ground the possibility of cultivating higher-order moral interrelationships with all our relations.

If we reconsider this pithy argument in light of the work by Curry, in chapter 10, and Calvo and Esquibel, in chapter 7, then it becomes clearer that the decolonization of our food and foodways is a potent and viable pathway to healing precisely because it rejects the identity politics of "authenticity" and instead embraces the dynamic and creative fire of Indigenous labor's hybrid circumstances. This is especially vital when we are dealing with large numbers of displaced persons, families, and communities that have been forced to flee the violence of settler colonial dispossessions and neoliberal capitalist enclosures. This hits very close to home, since the liberation of the kitchen, as proposed by Curry and Calvo and Esquibel, involves a love of food and cooking that can only grow out of the *phenomenologically sustained* cultural and labor practices of our forbearers in convivial embrace of the Other—or as Gustavo Esteva and Mahdu Prakash explain, La Comida is not just food or nutrition (eating); it is a social practice that ties us to the full depth and range of food as a culinary art linked to cohabitation of the kitchen as an autonomous loving space.[10]

Peña offers a similar argument in chapter 1 when he states that "the autonomy perspectives . . . are guided by awareness that our movements do not seek permission from the state or corporate acquiescence in order for us to act in solidarity." The key to this is the principle of "relational accountability/solidarity," which Peña borrows from other Indigenous scholars, like Shawn Wilson,[11] to argue that this is a sustained praxis generative of theory rather than a pre-conceptualized theory applied to the interpretation of observed phenomena. Relational accountability/solidarity, Peña writes, is "*a method of resistance.*" The lesson drawn from the US food autonomy movement is that rural farmers and urban food justice activists use relational knowledge to build solidarity economies in contradistinction to the neoliberal capitalist predatory mode. As Peña states, "Build your own economy—one not separated from the political but converting politics into the art of cohabitation (in Arendt's sense) and dedicated to conviviality and cultural mentoring." As is well known, Arendt's principle of cohabitation is precisely about "love of the 'Other,'" and so this signifies a different but compatible path some of us are following in search of revolutionary subjectivity and healing, without falling into the trap of identity politics.[12]

## Twenty-First-Century Teocintle Narratives, Scholarship, and Activism

By proposing decolonial perspectives of alterNative food, environmental, and social justice struggles in the critical study of food, foodways, and social movements, the chapters in this collection are explicitly foregrounding Indigenous epistemologies. The study and practice of the foodways of Mesoamerican peoples reveal the resiliency of Indigenous traditions even if we, Chicanas/os and Mexicanas/os, do not always recognize it. Cultural historian and student of maize civilization Roberto Rodríguez recalls that one of his elder mentors, Julieta Villegas, acknowledged the loss of Indigenous identity and traditions among Chicanas/os. As an antidote to this tragic loss of the Indigenous soul, she advised her students: "Do not for one second doubt that you are Indigenous. If you ever do, eat a tortilla."[13]

The chapters in this volume serve a similar purpose as the parable retold by Rodríguez of how "eating a tortilla" has decolonial epistemological and ontological implications. Yet, eating a *non-GMO* corn tortilla is not always possible given the lack of labeling. But the act to make it so can help decolonize and re-indigenize our biopolitics and eventually our bodies. This requires ongoing analysis of our food and foodways to understand how to re-indigenize our minds and create more collective and organizationally sustainable practices. Our cultural heritage as Mexican-origin peoples includes a widely shared civilizational impulse centered on maíz, but this really involves the

"Three Sisters" (corn, bean, squash) and more since amaranth is also sacred to Indigenous peoples of Mesoamerica. We have to address the problems posed by the advent of transgenic (GMO) crops that threaten our sacred maize, as is so eloquently argued by Adelita Sanvicente Tello and Araceli Carreón in chapter 15.

The study of Mesoamerican foodways allows us to see ourselves differently and to understand our problems and triumphs from new vantage points. For example, the current attack on Indigenous peoples includes a wave of anti-immigrant hysteria in the United States that has led to increased militarization and surveillance of the border and the unjust apprehension and detention of Mexican and Central American displaced peoples, as recounted here by Guillen in chapter 12 and Madrigal in chapter 13. Sharing the narratives and histories of Indigenous foodways across *Abya Yala* (Turtle Island; the Americas), the contributors to this volume legitimize the demands and experiences of the people of the Mesoamerican diaspora since it becomes obvious that Indigenous Americans who have lived here for thousands of years—responsible for developing civilizations that extend throughout South and North America—cannot "rationally be considered illegal."[14] This is an ideological argument that we can hardly disagree with.

Several chapters in the this collection, including Crow's ethnoarchaeology of the detritus of the "migrant trail" in the Sonoran Desert in chapter 5, are concerned with revealing the governmentality strategies we must more effectively contest to challenge the *actually existing state of economic exception* that is driving our displaced members of the "corn civilizations" to risk their lives as transborder travelers subject to multiple environmental threats, including the presence of hostile, well-armed anti-immigrant vigilante groups and militias. In other words, the state of economic exception and its companion regime of "legality" do not have to be "rational." We need to recognize that the apparatus is inherently derived from settler colonial irrationality in the preponderance of the use of violence to further the aims of an ideology of disconnection.[15] This is what needs to be challenged more effectively, especially though cross-border transnational organizing campaigns that nurture alterNative solidarity economies and worker self-organizing projects across all levels and sectors of the food chain. A related dimension addressed in many of these chapters, one that moves us beyond a focus on purely symbolic or discourse politics and ideological self-critique, is the *study of direct-action activism*—our own and colleagues' myriad changing terrains of struggle, organizational forms, and strategies of resistance.

Several chapters focused on the health problems associated with colonized diets, which we deem to be an effect of structural violence resulting from the combination of exploitative wages that require multiple jobs and thus widespread working-class reliance on "food junkyards" and heavily processed foodstuffs. These structural contradictions are currently being confronted through the revaluing of our indigeneity, which is an important development. The new forms of epistemic disobedience must also engage with the process of Indigenous and other peoples' self-organizing *as workers* taking place across the food chain and most notably for us among Native Mesoamericans displaced from their farms in México and Central America, as is the case with the Triqui and Mixteca berry workers in Washington State, whose struggles are meticulously chronicled and discussed by Madrigal in chapter 13.

Projects embracing the decolonization of our diets are also focused on compelling, recovering, and reactivating a renewed appreciation for deep and first foods and the culinary and gastronomical practices and knowledge bases associated with the heritage cuisines of our ancestral and living Indigenous cultures. This will undoubtedly go a long way toward promoting healing in our hungry and malnourished communities. The thousands of years of Indigenous analyses of our environments have created a highly nutritious Mesoamerican diet. This diet is filled with ecologically grown chile, maíz, *quelites*, *frijol, calabacita, tomate, papalo, huazontle, menta*, and other staples that appear to heal us of the health problems caused by colonization and capitalism. Food autonomy is a struggle to dismantle the sources of racial, class, ethnonationalist, and patriarchal violence unleashed by the deployment of technologies like GMOs and CAFOs (confined animal feeding operations), which wreak havoc on our bodies and those of our more-than-human coinhabitants. The structural violence of poverty induces hunger and malnutrition, attendant effects of colonialism and capitalism. We actively and directly oppose this regime of slow death when we return to a diet based on Indigenous crops and heritage cuisines.[16]

A decolonial approach that protects local, slow, and deep food has great potential for positively affecting Mexican/Mexican American subjectivities, bodies, families, and communities. It can serve as a means to enunciate the struggle for material survival by furthering development of our Indigenous traditional environmental knowledge (TEK). We take this as an opportunity to mobilize these knowledge bases as one key to solving a multitude of problems, including global warming and climate chaos, mass extinctions, and other ecological catastrophes. Daniel R. Wildcat argues that the place-based knowledge of Indigenous peoples allows for a relationship between humans and all our relations that is guided by ecologically sound, sustainable, and social justice principles.[17] The biocentric Indigenous values of respect, harmony, and coevalness can help us confront and overcome the interlinked food, energy, climate, ecological, and cultural crises that we face. To facilitate this process of transformation, we need to enunciate cultural climate-change stories and ensure that community-based, collaborative, social-action research studies of our foods and foodways and other aspects of our TEK are more widely realized and shared.

Importantly, this volume affirms that decolonial knowledge is constituted and transmitted through daily-lived activities and experiences. TEK is a form of deep experiential knowledge, and it is *mobile*—a feature of Indigenous "transmotion." TEK challenges the separation of knowing from doing that is characteristic of Western scientific discourse. Indeed, Indigenous epistemology and ontology are seen as not separate domains. Wildcat instructs that "in a world where people accept the separation of knowing from doing, it is instructive to reflect on the value of knowledges retained and realized in activity itself."[18] Decolonial knowledges are situated, place-based, and created through careful study and respectful interaction with the more-than-human members of our ecosystems. Analyses of Mexicana/o and Chicana/o food practices in gardens, kitchens, fields, home gardens, restaurants, markets, and the forbidden landscapes of the Sonoran Desert reveal intense struggles over the maintenance of Indigenous foodways and their continued development, including the creative forms of hybridity discussed by Crow, Curry, Calvo and Esquibel, María Guillen, and Valle.

Decolonial knowledge found in the resurgences of our foodways is egalitarian, made to be coproduced and shared through an invitation to participate in the practice of preparing, presenting, and consuming *la comida*. The authors in this volume share an understanding of decolonization embracing recognition of gender equity and mutual respect among social groups, in relation to other species, and across generations.

Americanization and Christianization served as twin pillars of the ideological attack on Indigenous people, and this included efforts to erase and demolish the autochthonous social organization of gender relations. Gender inequality and violence resulted from hundreds of years of the imposition of white settler colonial gender norms as well as the dispossession of our ancestral lands. The flexibility and multidimensionality of gender roles and practices developed over generations by Indigenous trial and error were disrupted by invasion and subsequent genocide, trauma, and displacement of people from places, and this also diminished and in some cases destroyed Indigenous bioregional polities. The studies in this volume suggest that activist scholars become more mindful of the need for the restoration of heritage cuisines and traditional agroecological and ethnobotanical practices, certainly, but we also suggest there is an obligation to enable and enunciate "free-range subjectivities" that are unafraid, and indeed are compelled, to celebrate the diversity of human sexuality and the inherent sensuousness of species life aiming toward conviviality and cohabitation. It is not enough for us to understand the colonial disruption of our communities; we must also actively promote alterNative knowledge(s) developed by our grassroots food and social justice movements as these pertain to the multidimensionality of our egalitarian impulses, that is, across race, ethnic, class, gender, sexuality, and species differences.

The millenary maíz narratives of the Mesoamerican civilizations are an inherent and enduring part of the storytelling or oral tradition. Some scholars emphasize that storytelling aids in the development of a sense of self and the transmission of what it takes to become a member of a gastronomical culture and way of life. Roberto Cintli

Rodríguez writes that "*centeotzintli* [*teocintle* as used in this volume] narratives decenter colonization by restoring the centrality of maíz and Indigenous history, language, culture, and cosmovision to this continent."[19] This discursive resistance is one quality of the multi-faceted decolonization struggles for food autonomy that require, in practice, that we recognize how maize civilization finds resilience in a polyculture agroecology rather than a monoculture of corn, regardless of that grain's symbolic power and ritual and ceremonial centrality.

Sanvicente and Carreón further illustrate this point in their discussion in chapter 15 of the multiple other center-of-origin crops that coinhabit Indigenous corn milpas across Mesoamerica. In the Indigenous milpa, maize is part of a complex polyculture of crops and wild and intermediate relatives. Crops may include everything from amaranth to *zapote negro* (*Diospyros digyna*). In the milpa, the Three Sisters—corn, bean, and pumpkin/squash—form a holistic and integrated biodynamic relationship through the chemical language of plants, which Western scientists call allelopathy.[20] Maize may be the "cast director," but the complex shifting mosaic of plants and animals creates a whole that is greater than the sum of its parts.

Sanvicente and Carreón directly connect maíz culture to contemporary legal and political struggles for autonomous self-governance and self-defense in Indigenous communities. The struggle against Monsanto, Syngenta, and the other "Gene Giants"—the transnational corporations promoting transgenic corn, soy, canola, and cotton in México—are clearly and unequivocally defined as the mortal enemy. The authors quote Mixe Native scholar and activist Floriberto Díaz, who observes, "A community first establishes a series of relationships between people and space, and in the second instance, among the people together."

Essential to a more holistic understanding of the place of maize in Mesoamerican civilizations and our contemporary cultures, then, is attention to the ancient institutions of collective action and self-governance (*autonomía*) that derived from "Original Instructions"—the lessons offered by the self-willing land organism itself.[21] A key feature of these ancestral and still extant polities is their shared concept of property as relation rather than as possession, a point also brought up by Peña in chapter 1, and Tezozomoc and colleagues in chapter 11.

At the end of the chapter by Sanvicente and Carreón, in a passage that resonates with Chela Sandoval's instructive missive about "Othering," and for us entails smashing the smoking mirror altogether, they quote Díaz explaining this in eloquent terms:

> the sacredness of the land does not support division or possession; neither of the people over the land . . . or of a certain people over others. The integrity of all the natural elements and living things explains the primacy of community and family, as opposed to the individual, and respects the community's interests, and the corn is respected in itself, but also deserves the respect of the community and its family.

There are more than a thousand autonomous municipalities across México today, and these Indigenous communities are the heart and soul of the anti-GMO corn movement in México. None of them, significantly, has a word for *individual* in their many languages and dialects.

Indeed, the Zapatista-inspired *autonomía* movement has spread to numerous community-based movements and organizations working for environmental and food justice north of the border. While there is a certain continuity between ancient Mexica (Aztec) texts (codices) and the living traditions and cosmological philosophies of today, these autonomous Indigenous communities are demonstrating that the defense of maize today requires a dialogue and synthesis constituted by a much more diverse set of Indigenous knowledge traditions and ethics—for example, Hña Hñu, Mixtec, P'urhépecha, Triqui, Tzeltal, Zapotec, and so on—melding with the bodies of risk science and biosafety conservation evidence produced by allies among progressive environmental and agricultural scientists, lawyers, and other members of our subaltern civil society. These are our "postmodern" calmecac activists, as Tezozomoc and colleagues argue in chapter 11.

In the United States today, dislocated Mexicana/os and Chicanas/os are actively reclaiming and redefining territory as a home place. The South Central Farmers and hundreds of other farming and gardening groups in the United States demonstrate alterNative strategies of survival by reclaiming place and restoring even hostile landscapes through the revival of first foods, Native foodways, and TEK. Mexican@s have refashioned patches of industrial wasteland in places like South Los

Angeles and Pilsen and Roseland in Chicago. Remaking urban decay into verdant homes and convivial spaces, they are resisting the imposition of poverty, illness, and hunger under the vagaries of industrial capitalism by creating gardens, farms, and farmers' markets where they can continue to share maíz narratives and actual seeds, plants, and tubers. There is no more powerful and subversive a materially grounded and spiritually anchored discourse than the seed itself. This form of storytelling is compelled by the sharing, among trusted colleagues, of multigenerational seed stock from our sacred maize and its sisters.

Strategies of reterritorialization through these dimensions of our foodways take place at the individual, family, and community level. This volume covers a large cross section of the spaces of reterritorialization by the people of the Mesoamerican diaspora and those of us who waited for their arrivals. While the changes to place in urban areas reveal the extent of Indigenous cultural resilience in the United States, rural places are no less important. Studies of the Mesoamerican diaspora and reterritorialization in more locales across the United States will undoubtedly reveal the extent of alterNative food movements and strengthen communal ties across borders and within and between cities and the countryside. As Mexican "transmotion" continues to change, reflecting the imposition of capitalism and colonialism as well as the desires of Indigenous arrivants and their families, new maíz narratives and strategies for reterritorialization will develop. Maíz narratives are emerging from places of more recent destinations, such as the Deep South, New York City and the Northeast, the Pacific Northwest and up into Vancouver, British Columbia, and as far north as southeastern Alaska. These wait to be shared, understood, enunciated, and defended. We hope this volume will inspire new studies, stories, and food justice activism as part of burgeoning autonomía movements across the homelands of Indigenous origin places and our adopted diaspora landscapes, wrought of a diverse multiethnic transnational culture of Indigenous peoples who refuse to be ghosts of the primitive accumulation, or hapless wards and victims of colonial dispossession. Aquí estamos, y no nos vamos.

# NOTES

## Introduction

1. In 2003, the Mexican federal government adopted the "Ley General de Derechos Lingüísticos de Los Pueblos Indígenas" (Law of Linguistic Rights of Indigenous Peoples), which recognizes sixty-two indigenous languages as "national languages" with the same legal, educational, and political standing as Spanish in all territories in which they are spoken. The insidious politics of recognition have rendered these largely empty words, but Indigenous struggles for linguistic survival, tied directly to political autonomy, are resurgent on both sides of the border as is evident in Zapotec language radio programming in places as far north as Washington State.

2. Urban studies scholar Mike Davis (2001) elaborates on the half million or so Zapotec and other Indigenous peoples who created what is commonly called "Oaxacalifornia"—the Indigenous "transnational suburbs" detailed in his perceptive account of "Latino magical urbanism." On Mesoamerican Indigenous diaspora people and traveling foods and foodways in Alaska, see Komarnisky 2009; for the same in the context of Los Angeles and Seattle, see Mares and Peña (2010, 2011), who emphasize seed saving and exchange in their accounts.

3. Our usage of "Indigenous" follows the approach of the "Food Sovereignty Is Tribal Sovereignty" Facebook community, which offers this statement, "A Note on Membership":

> Following the wishes of the members of this group, it is intended
> to be a Native-only space, for those who have indigenous ancestry
> from North, Central or South America, the Arctic or the Pacific (i.e.,
> Native American/Indian/NDN, Kanaka Maoli/Native Hawaiian,
> First Nations, Métis, Inuit, Xicanx/Chicano/Latinx/indigenous to
> Mexico, Central and South America, Aboriginal, Maori, etc.).

The coeditors wish to offer a distinction (for ourselves) between Indigenous Mexicans who have retained their languages and tribal affiliations, and "non-tribal" Indigenous Chicana/os whose genealogies include undeclared or emerging affiliations with tribal nations north and south of the border but are still working on defining their decolonial ancestries. We have encouraged authors to adopt their own Indigenous "identity" and affiliation terms; these present a beautiful mosaic reflecting our diversity and shared commitments to radical decolonial activism. For us, the distinction highlights the unified strength within the diversity of our communities of origin while noting how legal status constitutes one key difference, and Chicana/os must work to abolish this, especially as this affects Indigenous but "out-of-status" persons in our communities.

4. Komarnisky 2009, 41. Mesoamericans in Alaska hail from Zapoteca, Chinanteca, Mixteca, and Mixe origins.

5. On transmotion, see Vizenor (1998), who considers the concept as indicating "native motion" and "an active presence." Vizenor sees this practice of "survivance" as a sui generis form of sovereignty, although some of the contributors to this volume would urge the adoption of the conceptual vocabulary of autonomy rather than sovereignty in this context.

6. The record on African forced arrival, labor exploitation, and eventual rebellion in colonial Mexico is extensive; see Davidson 1966; West Africans from Guinea-Bissau were among the rebels fleeing to the highlands of Veracruz and Puebla, eventually making their way to Guerrero and Oaxaca. Vinson (2000) used "Archivo General de La Nación" (AGN) archives for a detailed profile of Afro-Mexicans in Igualapa, Guerrero; A. B. Fisher (2006) examines relations between "Afromestizos" and Indigenous inhabitants of Guerrero and considers the effects of crippling epidemics and expanding enclosures under settler colonial incursions. This led to conflict and cooperation concurrent with increasing land and population pressures.

7. On the intersection of culture, capital, and empire in the transformation of the ecology of place in México and the US southwestern states, see Peña 2005c. Merchant (1989) first introduced the concept of "ecological revolutions" in the New England environmental history context.

8. For example, Pilcher (2005, 247) laments the industrialization of the Mexican tortilla as evidence of this decline and concludes by observing that "one of the modern world's great ironies is that only the wealthy can eat like peasants." Our collective view in this book is that food and foodways across México and the United States have been industrialized and are associated with increased prevalence of diet-related maladies and morbidity. But the resurgence of Indigenous food, foodways, and cuisines is also equally compelling and merits the attention we have given it.

9. The historical depth of the "fusion" of multiethnic and multinational cuisines in México has been observed for so long it has finally been more widely acknowledged. This is the focus of a recent National Public Radio program illustrating how Mexican cuisine "was doing fusion five hundred years ago"; see Godoy 2016.

10. See Krimsky and Gruber (2016) for a recent summary on scientific critiques of commercial agricultural biotechnology from scientists and activists affiliated with the Council for Responsible Genetics and their flagship journal, GeneWatch. Also see Swanson et al. 2014.

11. On the displacement of Mexican corn farmers since NAFTA, covering the period between 1994 and 2010, see Relinger 2010. The original estimate for 1994–2004 was provided by the 2004 Sierra Club report on the impacts of NAFTA; see Sierra Club 2004.

12. As of 2014 these include Monsanto, DuPont (Pioneer), Syngenta, Groupe Limagrain, Bayer CropScience, and Dow AgroSciences. These soon may be reduced to four, as ChemChina announced a bid for Syngenta in February 2016, and Monsanto is being pursued by Bayer CropScience since May 2016.

13. On the Latina/o health paradox, see Lara et al. 2005.

14. Agamben (2004, 42) defines the anthropological machine as "an ironic apparatus that verifies the absence of a nature proper to Homo, holding him suspended between a celestial and a terrestrial nature, between animal and human."

In chapter 1 Peña introduces an Indigenous standpoint epistemology as a strategy to overcome this apparent Eurocentric binary.

## 1. Autonomía and Food Sovereignty

An earlier version of this chapter was presented at the thirty-ninth annual conference of the National Association for Chicana and Chicano Studies, March 14–17, 2012, Chicago, panel on "Traveling Foodscapes: Displaced and Diaspora Peoples Remaking Urban Agriculture." I thank Silvia Patricia Solís and Tezozomoc for challenging my views on food sovereignty. I also gratefully acknowledge Gabriel R. Valle for vital assistance during revisions.

1. For example, Agarwal 2014; Bernstein 2014; Burnett and Murphy 2014; Desmarais and Wittman 2014; Edelman 2014; Grey and Patel 2014; Kloppenburg 2014; Martínez-Torres and Rosset 2014; McMichael 2014; and van der Ploeg 2014.

2. One exception is Grey and Patel (2014, 3), who note, "food sovereignty is (and should be) a far more radical anti-colonial project than is compatible with its origins in the Mexican state." By *decoloniality* I refer to the "delinking" of our knowledge production from the rationalities of colonial logic. I emphasize how delinking involves epistemic disobedience *and* the agency of Indigenous peoples' knowledge imbricated with a politics of self-enunciation rather than against the hegemonic representations; see Mignolo 2001, 2009.

3. Agrawal (2005) first used the concept of "environmentality" to examine the production of "nature" and "subjects" in the conduct of the management of natural resources.

4. See Cruishank 2014.

5. My approach in this chapter involves a critical reading of the work of Agamben (1998, 2004, 2005) inspired by Mark Rifkin (2009) on bare life and bare habitance.

6. M. Smith 2011, 106–7.

7. My views are grounded and constrained by experiences derived from three decades of collaborative ethnographic and ethnoecological social action research with urban and rural farmers and other food-chain workers along the West Coast of the United States and in the Southwest, where acequia farmers are asserting their autonomy; see Peña 1998, 2002, 2003a, 2003b, 2005a, 2005c; Hicks and Peña 2003; Mares and Peña 2010, 2011.

8. The concept of the "peasant" is problematic. An early valuable and relevant critique was advanced by Kearney (1996). The so-called peasant is more than a land-based farmer and many subsistence farmers today must work for wages as part of the rural proletariat/precariat.

9. Via Campesina 2001.

10. UN Office for the High Commissioner of Human Rights, Special Rapporteur on the Right to Food, http://goo.gl/boCTvP, accessed May 21, 2016.

11. Butler and Midzain Gobin 2015, 1.

12. I am introducing the term *alterNative* to designate an epistemology of resistance that combines the concepts of "alterity" and "indigeneity," that is, a de-centering, re-membering subject.

13. This applies to privileged landowners in places like Napa Valley or Tuscany. "Slow food" farmers own expensive tracts of land and cater to largely white and elite clientele; they inhabit a petite-bourgeoisie class location. India is said to have its own "little Tuscany" as per Trip Advisor (see photo at Trip Advisor of India's very own mini Tuscany, https://goo.gl/AhuAkM; accessed June 7, 2016). Conflicts between LVC affiliates in India and in other South Asian and Pacific Rim countries reflect divisions between smallholder subsistence and medium-sized trade-oriented producer cooperatives; see Borras, Edelman, and Kay 2008 and Borras 2008, 154–55.

14. Borras and Franco 2010, 35n22; emphasis added:

> we advance two gentle critiques . . . : *absence of class analysis and unnecessary localism*. It is important to note . . . that food sovereignty as originally advanced by Via Campesina has been, and remains to be, a very broad and flexible notion of an alternative founded on the fundamental principles of right, autonomy and sustainability. Once it gained traction . . . food sovereignty has been (re)interpreted and accorded various meanings by different interest groups. A particularly strong current among those (re)interpreting the meaning of food sovereignty pushes the alternative notion toward "unnecessary localism" and "socially undifferentiated local communities."

15. Esteva and Prakash (1998, 110–51) critique universal human rights as a "Trojan horse of recolonization." Alfred (2009, 3) criticizes the notion that Indigenous sovereignty will automatically imply a process of de-colonization: "In fact, without a cultural grounding, self-government becomes a kind of Trojan horse for capitalism, consumerism, and selfish individualism."

16. See, for example, Demarest 1995; Herlihy et al. 2008.

17. On relational knowledge and accountability, see Wilson 2009.

18. One popular school of thought follows Bourdieu's argument that "perception is being." For example, a recent study of farmworkers in Washington argues that the health care system and workplace reproduce race, class, and national-origin hierarchies. The "migrant body" is constructed as a subject through the perceptions held by farm owners, managers, clinicians, physicians, and the workers themselves who are seen as "internalizing" the symbolic violence of a "clinical subjectivation"; see Holmes 2013. This approach traps the subject in the discursive fields of the dominant settler colonial formation and occludes the myriad forms of agency pursued by "migrants" like the Triqui and Mixteca berry workers Holmes studied. From this standpoint, it is not "perception is being" but rather "relation is being." Mignolo's claims are relevant to this vital distinction being drawn here; see also Alcoff 2007.

19. See M. Taylor 2006; Coulthard 2014; M. Rifkin 2009; Wolfe 2013. I use *telluric* in the manner of Carl Schmitt (2004), who defines it as "the relationship to the land that makes the true partisan always a defender of local territory rather than an attacker or invader." Also see Pan 2013. Settler colonists outside Europe try to erase indigeneity by mythologizing themselves as original inhabitants of occupied Indigenous territories; the only true telluric partisans on Turtle Island are the Indigenous peoples.

20. I am thinking here of the work of political theorist Michael Taylor, who

deems rational choice theory to be "an ideology of disconnection"; see M. Taylor 2006.

21. Peña 2005a; Alfred 2009.

22. Peña 2005a, 141–42; I cited the work of Kimberly Tallbear, who explains how many tribal governments find it difficult to overcome the challenge of "applying cultural values and philosophies to contemporary governance, economic development and institution building." This is an enduring effect of the 1934 Indian Reorganization Act and disrupted the efficacy and evolution of self-governance while blocking access to "forbidden" Indigenous knowledge.

23. I depart from Deleuze here on the "plane of immanence" concept since Indigenous epistemic precepts view death as the continuity of the renewal of other life. On the philosophy of "Original Instructions," see Nelson 2008; Peña 2005a. Also see Bennett (2010) for a relevant discussion of "vibrant matter." Bennett asks: "Why advocate for the vitality of matter? Because my hunch is that the image of dead or thoroughly instrumentalized matter feeds human hubris and our earth-destroying fantasies of conquest and consumption."

24. Cleaver 1977, 1979. A study of records of fair trade–certified farmers indicates they receive higher prices for coffee, but there is "no evidence . . . workers received higher wages or benefited from certification." Valkila and Nygren as cited in Dragusanu, Giovannucci, and Nunn 2013, 28.

25. *Prosumption* refers to a unity of production and consumption and is especially evident in the process of producing food for autoconsumption in rural and urban food justice projects.

26. On urban agriculture as a strategy of working-class autonomy, see Valle 2015, 2016.

27. Quote taken from "Chiapas: The Southeast in Two Winds, a Storm, and a Prophecy" by Subcommander Marcos of the Zapatista National Liberation Army written in August of 1992 and released publicly in January 27, 1994; http://goo.gl/JaViNt, accessed May 21, 2016.

28. Via Campesina 2001; emphasis added.

29. On hierarchies of race, class, and gender, and discriminatory practices associated with these divisions in alternative, organic, and sustainable agriculture movements, see Guthman 2004; Peña 2002.

30. For a critical indigenized perspective on "ecosystem services" discourse, see Peña 2005a, 2005c. The *International Assessment for Agricultural Knowledge, Science and Technology for Development* (IAASTD) was a major UN Environmental Program (UNEP) study undertaken in 2005–7. The resulting report endorsed a shift toward support for agroecology and measured wariness toward biotechnology in the Global South; see UN Environmental Program 2009. This shift has been articulated by Indigenous communities for decades. Notably, the United States and Canada objected to proposals on biotechnology (e.g., intellectual property rights). Appendix 4 includes a statement that "China and US do not believe that this entire section is balanced and comprehensive." These objections are rooted in the position of China that IPR regimes may inhibit expansion of "sustainable" technologies. The United States fears that a failure to emphasize IPR management and modernization (qua adoption of the US patent system) constitutes an unacceptable attack on US prominence as global leader in transgenic crop biotechnologies.

31. The latest membership list of LVC (June 2013) is available at http://goo.gl /QSEZ92, accessed February 4, 2015. The list includes 164 organizations in seventy-three countries.

32. For some relevant critiques, see Guthman 2004, 2011; Peña 2002; VanLeuven 2007.

33. For example, see Rosset and Martínez-Torres 2012.

34. In other words, *teocintle*. In Oaxaca among Zapotec, Mixtec, and other Indigenous maize farmers there are concerted efforts to protect the habitat *of Zea diploperennis*, the wild relative known to the farmers there as the *Madre del Maíz* (Mother of Maize); see Peña 2005a, 46–48.

35. In February 2016, the Costilla County Land Use Planning Commission in San Luis, Colorado, approved by a vote of seven to zero an ordinance developed by the Sangre de Cristo Acequia Association, The Acequia Institute, and the Center for Food Safety that would establish a "Costilla County Center of Origin GMO-Free Protection Zone" designed to protect the local landrace known as maíz de concho (a white flint corn). For more on the struggle to protect center-of-origin crops in the United States, see Peña 2016a. Such efforts are also under way in tribal First Nations that have sought to ban GMO crops and animals within the boundaries of the tribal reservations and reserves (e.g., Yurok, Dineh). These struggles are constrained by the US refusal to ratify its signatory status to the Convention on Biological Diversity, Cartagena Biosafety protocols, or ILO 169.

36. For an extended environmental justice critique of dominant cost/benefit analytical models in US environmental regulation and policy, see Peña 2011c.

37. Confirmed repeatedly since the study by Quist and Chapela (2001); see Bellon and Berthoud (2006) for a more recent assessment of these threats.

38. Bellon and Berthoud (2006, 10) argue, "It is unlikely that the introduction of transgenes, just because they are transgenes, will automatically reduce the diversity of alleles in local maize populations or the morphological variants managed by small-scale farmers." This depends on a lot of factors, including the extent to which agroecological selection practices result in reversibility of transgene introgression. This is a matter of great uncertainty, and we would do well to resist further continued use of GE crops in center-of-origin zones. Equally grave as a threat to the survival of landraces is the abandonment of *ejidos* and small farms in México by more than 2.5 million maize farmers since the "Salinista" neoliberal reforms and NAFTA.

39. See Redclift 1987; Shiva 1988; Peña 1992, 1998, 2005c; Forsyth 2003; Agrawal 2005.

40. Walker and Salt 2006; in the context of Indigenous farmers, see Salmon 2012.

41. For more on the concept of *autosuficiencia alimentaria*, see Azpeitia Gómez 1987.

42. Again I refer the reader to Merchant (1989) for more on the concept of ecological revolutions and to Byrd (2011) for an indigenizing theoretical approach to this matter of the "clearing of space" for colonialism.

43. Agamben 2005. The effects of and resistance to the state of exception is discussed by Madrigal in chapter 13 for the case of Mesoamerican diaspora farmworkers in Washington State.

44. Agamben 2005, 2; emphasis added.

45. M. Rifkin 2009, 94. An example of the denial of biological flourishing or survival involves an Alabama state law that requires proof of citizenship or legal status before the activation of domestic water services; the law targets undocumented Mexican workers; see Peña 2011a.

46. Benhabib 2006, 23; emphasis added.

47. Boyer 2010, 330, 333; emphasis added.

48. Nabhan 2013.

49. On the assassination of Comandante Galiano, see Solidaridad Chiapas 2014.

50. LVC's only response to the Ayotzinapa massacre is a declaration posted December 1, 2014, to the Spanish language version of their web page; see Via Campesina 2014.

51. As previously discussed in Peña 2006a, 2006b; Mares and Peña 2010, 2011.

52. Davis 2006.

53. Aerni 2011, 27.

54. This and the previous quote are from UN Environmental Program 2009, 100.


## 2. Indigenous Women in the Food Sovereignty Movement

1. This chapter was originally prepared as a plenary address for the thirty-seventh annual conference of the National Association for Chicana and Chicano Studies, Seattle, October 7–10, 2010.

2. Editors' note: This refers to the aftermath of the beating of Rodney King by LAPD troopers. The video that captured that incident of wanton police brutality was among the very first ever to document such an event on a handheld device. The beating of King ushered a week of resistance on the streets between April 29–May 4, 1992. The mainstream media and politicos characterized these struggles as rioting, looting, and "race" riots. For a critical discussion of the media and popular framing of the events, see Giroux 1995.

3. Editors' note: The excerpt is from an interview with Mittal appearing in Danaher, Biggs, and Mark 2007.

4. Editors' note: This herb is *Porophyllum ruderale*, also known by the popular names *papaloquelite* and *yerba del venado* (deer weed). In English it is commonly known as the "butterfly plant." It is a pungent herb and is usually eaten raw on a sandwich or added to guacamole and salads. It is also used fresh with soups and stews, grilled meats, and beans, much like cilantro. Papalo is not cooked; it is only used raw or added at the last moment. For a detailed inventory of agrobiodiversity at South Central Farm, see Peña 2005b.

5. Editors' note: According to interviews conducted by coeditor Peña, another reason had to do with two men getting expelled for violating the "rules of the common," including the acquisition of multiple plots for commercial gain; this was against administrative rules agreed to by the general assembly through consensus. They proved to be among the disgruntled informants who lied to the journalist about a lack of democracy and theft of monies. This issue will be discussed in a forthcoming study by Juárez and Peña on the principles for enduring common property resources as outlined by Elinor Ostrom (1990).

## 3. Food Values

1. Altieri 1998; Gliessman 2007.
2. Davis 2006.
3. Buchmann 2009.
4. Bullard 2000.
5. Gottlieb and Joshi 2010, 5.
6. According to Harry Cleaver (1992, 129) , self-valorization is a "self-defining, self-determining process, which goes beyond the mere resistance to capitalist valorization to a positive project of self-constitution." For a more detailed discussion on self-valorization in the context of urban agriculture, see Valle 2015, 2016.
7. Mill 1998.
8. Munn 1986; Graeber 2001.
9. Gottlieb and Joshi 2010, 5.
10. Stédile and Martins de Carvalho 2011, 25.
11. I am using the critical political ecology framework championed by Forsyth (2003) that emphasizes the role of scientific and local knowledge in the framing of environmental discourses and the remaking of ecological politics and policies. This approach includes both local and global case studies and comparative approaches conducted at different temporal and spatial scales.
12. Gordon (1991) is among the first to make this argument in the context of political ecology; also see Peña 1992.
13. See, for example, Buck 1991.
14. L. T. Smith 1999, 142.
15. Guthman (2011) is among those who have criticized the "whiteness" of the alternative food movements; Mares and Peña (2010, 2011) have also presented critiques.
16. Buchmann 2009, 705.
17. Koc et al. 1999.
18. Moskow 1999, 79.
19. Ibid., 81.
20. Galluzzi, Eyzaguirre, and Negri 2010, 3635.
21. See Eyzaguirre and Linares 2004.
22. Mares and Peña 2010, 256.
23. Mares and Peña 2011, 199.
24. Ibid., 204.
25. Gulluzzi, Eyzaguirre, and Negri 2010, 3648.
26. Mares and Peña (2010, 2011) also make this argument for the case of the South Central Farm in Los Angeles; also see chapter 11, this volume.
27. Marx 1990, 167.
28. Henderson 2013, xxv.
29. Díaz del Castillo 1956, 269–70; this is a classic example of the "colonial gaze." For a *decolonial* environmental history of the Mexica (Aztec) and Mayan civilizations prior to and since the point of contact, see Peña 2005c. Chapter 16, this volume, provides a critique of the discourse on the ethnoedaphology and ethnopedology of the Mexica and other Indigenous peoples of Mesoamerica.
30. See Aguilar-Moreno 2006.
31. Losada et al. 1998, 58.

32. See Toensmeier and Bates 2013; Allen and Wilson 2012; Carpenter 2009; Thorp 2006; Lyson 2004.

33. On the gift economy, see Malinowski 1922 and Mauss 1990.

34. Scott 1977.

35. D. Díaz 2012, 41.

36. Munn 1973; 1986, 20; emphasis in the original.

37. Graeber 2001.

38. Marx 1993, 361.

39. Mares and Peña 2011.

40. Gullì 2005, 133.

41. Marx 1993, 611. Also, see Hardt and Negri 2009.

42. Montoya 2011, 109.

43. Ibid., 189.

44. Flores 2013, 11.

45. For a critical and historical analysis of food deserts and poor health outcomes, see McClintock 2011.

46. Ibid., 94.

47. Marx 1852, 1.

48. The principal point made by Gullì (2005) centers on the need for us to recognize that Marx distinguishes between "productive labor" (working for money or subsistence) and "living labor" (working for artistic creation). This distinction is a basis for rethinking value as a category of autonomous living labor capacity and is a concept Marx developed and used extensively in *Grundrisse*.

49. Pudup 2008, 1229.

50. Ibid., 1228.

51. Alkon and Mares 2012.

52. For more on the theory of racial formation, see Omi and Winant 1986.

53. McMichael 2010, 170.

54. Viertel 2011, 142.

55. Alkon and Mares 2012, 354.

56. Foucault as interpreted by Fornet-Betancourt et al. 1987, 118.

57. Hardt and Negri 2005, 83.

58. For a critical study of the different forms of citizenship and the "unauthorized practice of social ctizenship" among out-of-status workers, see Del Castillo 2007.

59. Cleaver 1992, 12.

60. Wilson 2008, 71.

61. Holt-Giménez and Shattuck 2011, 316.

62. Mares and Peña 2010, 256.

63. Guthman 2011, 264.

64. Agyeman and McLaren 2014, no page.

65. Ibid., no page.


## 4. *Del alivio y coraje la tuna nacera*

I wish to acknowledge Devon G. Peña for his various readings of this manuscript. The writings and contributions of Luz Calvo, Dolores Calderón, Frank Margonis,

Clayton Pierce, Graham Slater, and Andrea Garavito Martínez made this essay stronger. I especially wish to thank Juan José García, my partner, for his invaluable support, reading, and contributions to this essay.

1. The *nopal cardona* (*Opuntia streptacantha*) is native to the altiplano Potosino-Zacatecano, which is my grandmother's "*tierra natal*"; see López González and Rodríguez 1997. The *palma sabal* (*Sabal mexicana*) is native to deep South Texas and the northeastern Tamaulipas region. The *albahaca* (*Ocimum basilicum*) has various medicinal applications. My family uses it to treat migraines and as part of our *despojos*.

2. Pendleton Jiménez 2006.

3. After reading, among others, Cajete 1994; Deloria and Wildcat 2001; Grande 2004; LaDuke 1999; and Peña 1992.

4. Cajete 1999, 189.

5. Facundo Cabral (May 22, 1937–July 9, 2011) was an Argentine troubadour whose songs of protest have had a global reach. He is known for his song "No soy de aquí ni soy de allá."

6. Nandita Sharma (2014, 167) tells us, in her reading of Sylvia Wynter's "1492: A New World View," that

> part of the consequence of expanding capitalist social property rela-
> tions to the Americas (which began in earnest in the late sixteenth
> and early seventeenth centuries) was the territorialization of land,
> place, subjectivity, and belonging. Control over land and control
> over a sense of place and who "belonged" there and who did not
> were absolutely crucial to gaining control over people. This process
> of territorialization was significant in the ushering of a "new world
> order" of both constituting and partitioning putative "races," gen-
> ders, and later "nations" and later natives and "migrants."

I invoke this form of territorialization and its regime of terror by those who designate themselves inalienable powers. This terror produces historical traumas that shape our constructions of place and the ways we dwell. This echoes Peña's indigenized reframing of food sovereignty as a struggle against the reduction of Indigenous people to both "bare life" and "bare habitance" in chapter 1, this volume.

7. Pillow 2003.

8. Tuck and Yang 2012.

9. I would like to respond to Tuck and Yang (2012, 6) and how they delineate indigeneity:

> In order for the settlers to make a place their home, they must
> destroy and disappear the Indigenous peoples that live there.
> Indigenous peoples are those who have creation stories, not coloni-
> zation stories, about how we/they came to be in a particular place—
> indeed how we/they came to be a place.

I understand the need to put a stop to the metaphorical "*asignación*" of decolonial and Indigenous thought and practices. One way to do this is by interrogating theoretical arguments in light of the distinguishing and real-life consequences of actual decolonization struggles—such as those related to Indigenous sovereignty, land and water rights, or against work as *the* form of social control. There is also resistance against the administration of death as a form of neoliberal structural violence. But if distinctions are to be made on the basis of who has creation stories and who has colonization stories, then this will be taken up as a line drawn between what is Indigenous and what is colonizing. This will only revert us back to imposing the risk of dis-membering (forgetting and keeping separate) the experiences harmed precisely by these colonial binaries. More recently, Wolfe (2013, 258–57) makes a relevant contribution that resets the discussion somewhat, noting that

> the question is not . . . one of the ground for binarism, but one of where and when—if at all—the originary binarism became dis- solved or transcended. . . . The primal binarism that is maximally visible during the era of the frontier subsequently becomes less and less visible in settler-colonial discourse—which, as Lorenzo Veracini has pointed out, persistently seeks its own transcendence through declaring itself a thing of the past.

Devon G. Peña, in a personal communication to the author (May 22, 2016), advises that Wolfe's concept of "settler/colonist" as "originary binary" is not an argument for a naturalized postcolonial futurity. It is instead presented by Wolfe as the contested and evolving contexts of political projects over liberation from the administration of the life world of the colonized by the colonizer. Our aim is not to "essentialize" or "eternalize" the binary but to challenge and dissolve it through conscious political struggles for Indigenous autonomy and also against "recognition" by the settler colonial state, which only serves to legitimize that juridical and political order; on this final point, see M. Rifkin 2009; Wolfe 2013; and Coulthard 2014.

10. Moraga and Anzaldúa 1983, 23.

11. Alaimo and Hekman 2008.

12. Pendleton Jiménez 2006, 228.

13. See Santos 2014.

14. Walter Mignolo (2005, 8) draws from Franz Fanon, W. E. B. DuBois, Gloria Anzaldúa, and Ali Shai'ati to explain how "the wretched [as described by Franz Fanon] are defined by the colonial wound." More than just the physical and psychological violence of settler invasion and dispossession of Indigenous peoples, the colonial wound is instead

> the consequence of racism, the hegemonic discourse that ques- tions the humanity of all those who do not belong to the locus of enunciation (and the geopolitics of knowledge) of those who assign the standards of classification and assign to themselves the right to classify.

I align myself with Mignolo in viewing the colonial wound as a consequence of racism by those who have given themselves the inalienable right to classify others, but I would be more cautious in how we understand its manifestations. "Physical and/or psychological" fall within the binary rubrics designed by dominant schools of Western science. The world we know is messy and entangled, and the colonial wound we bear is a manifestation of place. As researchers and activists, we need to understand place to get a sense of the depth of our wounds and learn ways to heal them.

15. More than just knowledge, *saberes* include knowledge as well as wisdom and deeply held beliefs verified by shared historical and ethnic experiences. See Urrieta (2013) for an extended discussion of the importance of *saberes*.

16. Urrieta 2013, 320–21.

17. Fanon 2004, 50.

18. L. T. Smith 1999, 146.

19. Ibid., 146.

20. Lugones 2010, 743. I depart from Lugones in the use of "acting" within her interpretation of intimacies. I situate intimacies within the body and its material and intersubjective dimensionality. Therefore, before and if they are acted upon, they are felt. Pendleton Jiménez felt her transgression through her pain.

21. To understand the meaning of "fractures and limits," we can first recognize that coloniality and violence are constitutive of each other, and coloniality is a broken matrix of power with limits in how it is experienced and challenged by those whom it sets upon. Lugones (2010, 754) states that

> coloniality infiltrates every aspect of living through the circulation of power at the levels of the body, labor, law, imposition of tribute, and the introduction of property and land dispossession. . . . [However] its logic and efficacy are met by different concrete people whose bodies, selves in relation, and relations to the spirit world do not follow the logic of capital.

Lugones's (749) reading of Walter Mignolo is that he understands this as the "fractured locus" that involves multiple perceptions and inhabitations and includes

> the hierarchical dichotomy that constitutes the subjectification of the colonized. But the locus is fractured by the resistant presence, the active subjectivity [and intersubjectivities] of the colonized against the colonial invasion of self in community from the inhabitation of that self.

It is in these subject locations that Lugones (753) proposes we will find dwelling places of coalescence where we can learn from each other and, I would add, learn from the land. Peña believes we should see how Lugones's "fractured locus" is akin to Negri's argument of the principal lesson of Marx's *Grundrisse* residing in the idea that revolutionary subjectivity emerges from the intrinsic antagonistic quality of the value relation between labor and capital, which creates a fractured and alienated existence and can end only with the dissolution of the social control implicit in the negative dialectic between living labor and dead labor accumulated and circulated as capital; personal note to the author, June 20, 2016.

22. A search on the nomenclature and ethnobotany of *sangre de grado* rendered various Latin names, all belonging to the Euphorbiaceae family. *Croton draco* and *Croton lechler* are known for latex properties; *Jatropha dioica* and *Jatropha elbae* are known for root and bark infusion. These all have medicinal properties and can be toxic. Therefore, direct and experienced knowledge of the plant is a necessity for medicinal uses. Andrade-Cetto and Heinrich (2005) discuss classification and use in the treatment of diabetes.

23. I draw from a concept Peña (2005b, 6) identifies as "autotopography," which he characterizes as a process of "self-telling through place-making."

Peña speaks to the manner in which diaspora people and even entire communities are crossing borders carrying epistemologies of place. In the process of "re-emplacement" they transform encountered landscapes into something that feels, looks, smells, tastes, and acts like home. One could say this means we focus on spiritual healing rather than the colonial wound but never forget those who desire to plot on becoming our lords and masters.

24. It was after a long conversation with Peña that I was encouraged to see the values and possibilities of our food justice movements. He pushed me to see how my initial conceptualization of the body as place carried colonial tendencies. This will continue to be a source of ongoing reflection in my own epistemological thought and will influence my pursuit of further inquiry on decolonial and anti-colonial futurities. The current conceptualization of the body as place is of my own doing and sole responsibility.

25. Walker 1983, 238.

26. Ibid., 241.

27. Pendleton Jiménez 2006, 222.

28. Ibid., 222.

29. Ibid., 223.

30. Kirk 1999, 190.

31. Moraga 1993, 173.

32. Tuck and Mackenzie 2014, 56.


## 5. Tracing Food Packs and Tuna Cans on *La Línea*

1. Sundberg 2011, 318–19.

2. This statement is used in a dark, yet sarcastic manner by members of border control groups and agencies in reference to the dangers of crossing the desert.

3. Quoted in Sundberg 2011, 322.

4. Ibid., 323.

5. Chavez 2008, 2–3.

6. Ray 2010; Sundberg 2011.

7. Turner 2004, 106–7. Also see Clifford 1994.

8. Latour 1999, 175.

9. Cohen 1976, 9; also see Douglas 2002.

10. Barth 1998, 13.

11. Ibid., 1, 18.

12. *La Línea* translates to "The Line" and is a colloquial term for the U.S.-México border.

13. Mintz 1978, 69.

14. Craig, Vaughan, and Skinner 2001, 418; Ferguson, Price, and Parks 2010, 146–48; Regan 2010, 133.

15. M. Carney 2013, 32.

16. Ibid., 32, 34, 43.

17. Ibid., 32.

18. Durand and Massey 2007, 238; Cerruti and Massey 2001, 187–89.

19. Under Operation Gatekeeper, Operation Hold the Line, and Operation Safeguard.

20. Literally translated as "the other side," *el otro lado* is a colloquial term for the lands that lie north beyond the US-México border, and these can include Canada and Alaska as ultimate destinations.

21. De León 2012, 486.

22. See Nestlé 2014.

23. Electrolytes can be either bottles of Gatorade or Pedialyte or packets of electrolyte powder that is usually intended for babies and children and distributed in México by clinics and other health agencies.

24. De León 2012, 488.

25. Ferguson, Price, and Parks 2010, 23–24. Rhabdomyolysis is a condition where the muscle cells begin to break down and the protein from the muscle cells gets into the bloodstream and circulates through the body. During filtration by the kidneys, this muscle protein becomes trapped in the microscopic tubules of the kidneys and causes them to shut down.

26. Baboquivari Peak is a 7,730-foot peak visible from a variety of locations across southern Arizona. Travelers and coyotes use this natural feature as a location point to determine the four cardinal directions. Baboquivari Peak is also the site of Kitt Peak National Observatory. As the center of the Tohono O'odham cosmology and the home of the creator, I'itoi, it is the most sacred place to the Tohono O'odham people. According to tribal legend, I'itoi resides in a cave below the base of the mountain.

27. The Department of Homeland Security estimated that by 2008 almost half of unauthorized border crossings occurred at the México-Arizona border, primarily in the area of Nogales, Sasabé, and the Tohono O'odham Indian Reservation. One especially critical analysis (D. Martínez et al. 2014, 270) of the DHS report summarizes the geography of death as border control strategy:

> A survey of Mexican deportees conducted by the Migrant Border Crossing Study of the University of Arizona and The George Washington University found that the average number of days spent crossing in the Tucson sector increased from 2.3 days in 2008 to 3.3 days in 2011. Recent accounts of people traveling with UBCs [Undocumented Border Crossers] when they were last seen alive also suggest that it is not uncommon for people to spend four or five days crossing in southern Arizona. In the early 2000s a popular crossing route near the city of Nogales following the Santa Cruz River towards Tucson would typically take less than two nights. However, increased enforcement and a permanent checkpoint on Interstate-19 have pushed people west on a longer route towards the Tohono O'odham Indian Reservation through more remote and rough terrain.

28. High-density polyethylene (HDPE), or polyethylene high-density (PEHD), is made from petroleum. With a high strength-to-density ratio, it is used in the production of lightweight but strong plastic bottles.

29. Nestlé 2014.

30. Making reference to these sites in no way poses any greater threat to travelers or the supplies provided for them than leaving them out in the desert. One humanitarian told me that "there are no secrets out here."

31. Canisters contain preassembled food packs as well as hand warmers, fresh

socks, sweatpants, sweatshirts, blankets, and, if available, electrolyte drinks; also see De León 2013.

32. Translates to "The North" and is used colloquially to refer to the United States.

33. One example: While pinto beans are common to parts of northern México, the preferred varieties in the south (e.g., Chiapas) involve black beans.

34. African honeybees were introduced to Brazil during the 1950s. In 1957, twenty-six swarms of hybridized bees escaped quarantine and arrived in North America in 1985. By 1994, Africanized bees had colonized 15 percent of the native honeybee population in the Tucson, Arizona, region.

35. Ankri and Mirelman 1999, 1.

36. Travelers can experience a range of profound injuries, including massive blisters, broken bones, sprains, embedded cactus spines, yeast infections, and cracked skin.

37. *Coyote* is a colloquial term used to refer to smugglers for hire to guide travelers through the desert.

38. A name chosen by NMD to present their group as being akin to that of similarly named political prisoner groups or Latino gangs.

39. A video clandestinely recorded by NMD reveals the destruction of goods and materials left for travelers in the desert. View the YouTube clip at https://goo.gl /so9mEa; accessed May 21, 2016. See also "Border Patrol took blankets intended for immigrants crossing border, group says (VIDEO)," *Huffington Post*, January 18, 2013, http://goo.gl/oiBlCa; accessed May 27, 2016.

40. Tucson-based Humane Borders maintains maps of reported deaths of migrants found in southern Arizona, including the Tohono O'odham Indian Reservation. Go to http://goo.gl/BfhIgs; accessed May 26, 2016.

41. Rauer and Pancera 2006.

42. Travelers reportedly use mayonnaise jars to their hide cash from coyotes and *bajadores*, the gang members who patrol and control crossing points at the border to prey on the transborder travelers, whom they routinely rob, beat up, and sexually assault, often for money and goods and in acts designed to control trafficking. This term literally translates as "the unloader." Women and girls begin a regimen of birth control pills before crossing due to the high rate of sexual assault as the unnegotiated price of crossing into the United States.

43. Agamben 1998.

44. Vizenor 1990, 162.


## 6. Norteada/o en el barrio

1. PONS Online Dictionary, s.v. "nortear," http://en.pons.com/translate/ spanish-english/nortearse; accessed June 2, 2016.

2. Real Academia Española, s.v. "nortear": 1. tr. Dirigirse hacia el norte, especialmente por mar."

3. See Gómez de Silva 2001. Also see the discussion board exchange at Saberia.com, http://goo.gl/4fnGVF, accessed May 21, 2016:

En México y algunos otros países de Sudamérica se utiliza la expresión 'estar norteado' como sinónimo de estar perdido y/o desorientado, de no saber hacia dónde se va, de no tener muy claros cuáles son los objetivos en la vida o en una parcela de la misma, de no encontrar el camino. Es lo que en España se conoce como 'perder el Norte' . . . .El dicho 'estar norteado' tiene su origen en los navegantes que surcaban las aguas del Atlántico en el siglo XVI. Aquellos, a medida que se acercaban al Ecuador, comprobaban como la estrella polar se aproximaba al horizonte, ocultándose hasta desaparecer debajo del mar. Lógicamente, los marinos se desorientaban, o lo que es lo mismo 'perdían el norte' . . . . Tal es la popularidad de la expresión que nos ocupa que existe incluso una película mexicana titulada 'Norteado'. Dirigida por Rigoberto Pérezcano, esta cinta narra la historia de Andrés, un hombre que llega a Tijuana para tratar de cruzar la frontera con EEUU. Tras varios intentos fallidos, el protagonista descubre que la vida en su nueva ciudad de adopción no es ni mucho menos sencilla . . . . Volviendo al Norte -también llamado septentrión-, recordar que es lo opuesto al Sur; uno de los cuatro puntos cardinales . . . e inequívoco punto de referencia para viajeros y navegantes.

Also see the discussion at a Cervantes studies site, http://cvc.cervantes.es/foros /leer_asunto1.asp?vCodigo=39295, accessed May 21, 2016.

Queridos contertulios: En México, en el habla popular (coloquial) se emplea el vocablo «nortear» como equivalente de "desorientar". Mireille, lo de "estar sin norte" definitivamente se entendería como "estar desorientado"; por lo menos en México . . . Hace unos años me di cuenta que el DRAE sí toma cuenta de «norteado» y lo asigna como mexicanismo y vulgarismo. . . . Aunque soy originalmente de México, no conocía esta palabra hasta muchos años después, ya cuando tenía varios años aquí en EE.UU. Saludos, C.G. 1971–1981 México D.F. 1981–Presente EE.UU. California

4. Anzaldúa 2007, 99.

5. Anzaldúa 1999; Delgado Bernal 2001.

6. Perhaps a future discussion might align the concept of estar norteada/o with the economic and northward pull of the braceros, a group of Mexican laborers pulled to the United States for paid farm labor during the industrialization of agriculture, and even speak to those braceros who made the United States home despite its intent to push them "back" to México; see Montejano 1987.

7. Pérez 1999, 77.

8. Mares and Peña, 2011, 199.

9. Ibid., 208.

10. Mignolo 2001; Quijano 2007; Pérez 1999.

11. Tuck and Yang 2012, 1.

12. L. T. Smith 2004, 28.

13. See Pérez 1999; Sandoval 2000; Anzaldúa 2002, 2007, 2009; Saldívar-Hull 2000.

14. Pérez 1999; Anzaldúa 1999, 2002; Sandoval 2000.

15. Pérez 1999, 6.

16. Ibid., 108.

17. Anzaldúa 2002, 547. Editors note: Coyolxauhqui is the Mexica deity associated with the Milky Way and star beings. She embodies re-membering, and Anzaldúa images her as a Moon Goddess who is dis-membered, "split and torn apart," and who then "pulls herself together again."

18. In addition to Anzaldúa (2002), other Indigenous epistemologies claim that the classical Westernized ideology separated the spirit/mind/body as three unconnected parts, but in Indigenous ways of thinking, bodymindspirit are inextricably connected; see Meyer 2003.

19. Sandoval 2000, 36.

20. Ibid., 151.

21. Peña 2005c, 71.

22. Calvo and Esquibel 2015.

23. Champion and Collins 2013; Carrera, Xiang, and Tucker 2007.

24. Calvo and Esquibel 2015.

25. Ibid., 17.

26. Anzaldúa 2007.

27. Anzaldúa 2007, 99. Translation: I'm turned around/disoriented by all of the voices that speak to me simultaneously.

28. Keating 2006, 8.

29. Vartabedian 2006.

30. On the concept of linguistic landscape, see Sayer 2010; Shohamy and Gorter 2009.

31. USDA 2015.

32. At a panel session of the fortieth annual conference of the National Association for Chicana and Chicano Studies (NACCS) in 2013 in San Antonio, coeditor Peña raised this issue and suggested that the term *food desert* can be seen as part of the vernacular of the colonizer's gaze and obscures the diversity of life desert ecosystem peoples have known and used to thrive, despite the apparent paucity of food sources as seen through the settler colonial lens.

33. As claimed by Schroeder 1993.

34. A recent example extends into Chiapas, where soft drinks have replaced the indigenous corn-meal drink, and rates of diabetes have increased exponentially; see Kearns 2014.

35. Tuck and Yang 2012, 5.

## 7. Tortilleras, testimonios, y recetas

1. Calvo and Esquibel 2015.

2. Pilcher 1998.

3. This artwork can be viewed on the artist's home page; http://goo.gl/dbnOcX. Accessed May 24, 2016.

4. Pilcher 1998, 10.

5. Ibid., 17.

6. Hidalgo de la Riva 2004, 144–45.

7. Irene Barraza Sánchez in Rebolledo, Gonzalez-Berry, and Marquez 1988, 17.

8. The original was first prepared around 1557 by Fray Bernardino de

Sahagún as *Historia general de las cosas de Nueva España: el Códice Florentino* and is available online from the World Digital Library; https://www.wdl.org/en /item/10096/; accessed February 27, 2017. The first and most notable English translation, cited here, is Anderson and Dibble 1961.

9. See, generally, A. L. Morales 2001.

10. John et al. 2005, 2907.

11. Ibid., 2911.

12. Jones et al. 1997.

13. Ibid., 527.

14. Jones et al. 1997.

15. Anzaldúa 2002, 312.

16. G. J. Sánchez 1995, 102.

17. LaDuke 1999, 15.

18. Ibid., 18–19.

19. Ibid., 19.

20. Ibid., 20.

21. Ibid.

22. LaDuke 2005, 70.

23. Ibid., 194–95.

24. Ibid., 200.

25. Ibid., 200–201.

26. Ibid., 201.

27. Ibid.

28. O'Brien 2008.

29. Nabhan in Tohono O'odham Community Action 2010, 97.

30. O'Dea et al. 1984.

31. Ibid., 598.

32. Shintani et al. 1991.

33. White Earth Land Recovery Project, Defeat Diabetes Program 2010.

34. Galindo 2013.

35. Anzaldúa 2002, 312.

## 9. Travels of a Diaspora Community

1. Following Peña (2005c), I refer to traditional ecological/environmental knowledge as a particular form of place-based knowledge of the plant and animal species, landforms, watercourses, and other biophysical properties of the environment in a given place.

2. A. B. Fisher (2006) provides a rare look at movement and resettlement involving what he terms "Afromestizo" and Indigenous partnerships and cohabitation. He reconstructs the historical development of Cacalotepeque and views it as the "resurrection" of an essentially pre-Hispanic polity. He notes that "the ties its members formed with neighboring people reflect the complex relationship between indigenous and black colonial subjects in Mexico." He attributes the myriad roles of the Afromestizos in the area's Indigenous communities to the combination of "the severe demographic collapse of the sixteenth and seventeenth centuries" and of

"earlier waves of epidemic disease and state congregación or resettlement campaigns." Fisher further notes that the settlement of Spanish colonialists "took longer to develop in the mid-Balsas River Depression compared to much of central and western Mexico." The first-generation encomienda estates were soon abandoned, and the "decline of both the colony's encomendero elite and mining sector served as a death knell for most private properties in the region."

3. For studies of the admixture of populations in Guerrero, see Bonilla et al. (2005) and Juárez-Cedillo et al. (2008), who confirm the presence of genome sequences from the peoples of Guinea-Bissau.

4. Ramsay 2004, 447.

5. I am critiquing the concept of *mestizaje* as it is based on the racial theory of José Vasconcelos, who envisioned the creation of the perfect or fifth race as a result of amalgamation (or miscegenation). Vasconcelos believed that a synthesis (really whitening) of all known races of the world would reign supreme in the future and therefore logically determine national identity, thus solving ethnic, racial, and social inequalities in México. This has been a rationale throughout Latin America to justify the erasure of people of African descent and other ethnic communities. See Alonso 2004.

6. For more on Cihuatéotl, see the commentary by the late Mexican novelist and social critic Carlos Fuentes (2012, 43–44), who writes this about an important Mexican sculptural representation of Cihuatéotl at the Museum of Xalapa:

> quien podría ser una diosa, o simplemente el alma de una mujer que murió al dar a luz. . . . Sin embargo, la más suntuosa de estas mujeres, quien al morir puede reclamar la divinidad, es la Cihuatéotl sentada, infinitamente apacible, incluso serena, pues se sienta en el trono de la eternidad, coronada por un mono ciego y sostenida por cuatro rayos de luz a sus espaldas.

> (she could be a goddess, or simply the soul of a woman who died in childbirth (upon giving the light). . . . However, the most sumptuous of these women is Cihuatéotl sitting, who when dying can claim divinity, is infinitely peaceful, even serene, as she sits on the throne of eternity, crowned by a blind monkey and supported by four beams of light.)

As eloquent as the account is, Fuentes remains silent about the continued meaning and tribute inspired by Cihuatéotl, as the maker and giver of medicine, among Indigenous women who still celebrate and observe the divinity of the female body in life and in death.

7. These two words—"her-story" and "his-story"—are an act of *mujerista* epistemic and linguistic disobedience. The etymology of the word *history* establishes the root source in the Greek word ἴστωρ "hìstōr" (wise man; witness). The Ionian Greek ἱστορία "hìstoria" refers to the stories told by men. The Greek word σοφία translates as "orators." All are related and centered in the idea of men as the storytellers; see Müller 1864, 296–97. In the ways of the Indigenous and Afromexicana/o cultural traditions of my family, women and men share the esteemed role of storyteller.

8. Mares and Peña 2010, 246.

9. Peña 2005a, 138.

10. Mares and Peña 2010, 364.

11. Peña 2005a.

12. Berkes and LaRochelle 2003, 362.

13. Mares and Peña 2011, 201.

14. LaDuke 2005, 191.

15. Gottlieb 2009, 8.

16. Peña 2005a, 132.

17. Ibid., 134.

18. A. Smith 2005, 69.

19. Ibid., 74.

20. Ibid., 117.

21. Esteva 2003, 5.

22. Mares and Peña 2011; also see Peña, chapter 1, this volume.

## 10. Food, Class, Ethnicity, and Race in the Classroom

1. See, for example, the varied work of scholars such as Baca Zinn (1979); Baca Zinn, Eitzen, and Wells (2015); Moraga (1983); Zavella (1987, 1991); and Pesquera and Segura (1993).

2. Gamio's study was first published in 1931; for the class I consulted Gamio (1972), Esquivel (1992), and Martin (1992).

3. Esteva and Prakash 1998, 69–75.

## 11. Fragmentary Food Flows

1. The authors wish to acknowledge the collaboration of the dedicated organic scholar activist Devon G. Peña, who has collaborated with SCF for more than a decade now and has made crucial contributions to the development of our political work and the theoretical arguments presented in this chapter.

2. Originally in Peña 2005b; also see Mares and Peña 2010, 2011.

3. Johnson and Hunn 2012; Hondagneu-Sotelo 2014.

4. Peña 2005b, 2006a, and 2006b.

5. Mares and Peña 2011.

6. See the undated *Wikipedia* entry for South Central Farm, http://goo.gl /bjubUZ; accessed May 24, 2016.

7. A. Martínez 2006.

8. Development of the prospectus for a community farm trust acquisition of the fourteen acres involved Alina Bodke, consulting with the Annenberg Foundation; Devon G. Peña of The Acequia Institute and professor at the University of Washington; and Tezozomoc of SCF. In a June 2005 private report to the group, Peña presented an outline for a management plan articulating how the farmers already managed the land in a manner complying with the eight principles of enduring common property resource (CPR) management as outlined by Elinor

Ostrom (1990) and her students and colleagues. On the $16 million offer fashioned by Annenberg, see *Wikipedia* 2017; the primary source cited in endnote 9 of the *Wikipedia* entry for South Central Farm is in the KTLA TV station archive, which is currently inactive.

9. hecubus 2006.

10. Peña 2006a.

11. Johnson and Hunn 2012, 288.

12. Deleuze and Guattari 1972.

13. Giddens 1990.

14. Negri 2008.

15. Foucault 1984, 9.

16. Lazzarato 2014, no page number.

17. Ibid.

18. Maidan 2014, paragraph 5; no page number.

19. See Pels 1998, especially the discussion of the "primacy puzzle."

20. Ibid., 15, 22–32.

21. Hardt and Negri 2009, 6–7.

22. Ibid., 9.

23. Peña, *Environmental and Food Justice Blog* (3 January 2011), http://goo.gl /7EnUKC; accessed May 21, 2016.

24. Lash and Friedman 1992; see also Zupancic 2005.

25. Eagleton 1991, 1. This is a proposal often made by critical theorists of the Frankfurt school.

26. Also see McMillan 2012. Of course we allude here to the central exchange between Morpheus and Neo in the motion picture *The Matrix*.

27. Martineau and Ritskes 2014.

28. Mares and Peña 2011.

29. Martineau and Ritskes 2014, 2; emphasis added.

30. Holman 2014.

31. Davis 2001.

32. On the concept of *vecindad*, see Peña (2011b) and Esteva (1996), who defines *vecindad* as more than just a neighborhood because it implies the existence of spaces and relations of conviviality.

33. Devon G. Peña introduced us to the concept of the "phenomenological weakness" of neoliberal capitalism in email correspondence with the author, May 11, 2015.

34. Lundborg and Vaughan-Williams 2011, 367.

## 12. Growing Justice in the Fields

1. As quoted on the U.S. Food Sovereignty Alliance website, http://goo.gl /SGWX70, accessed May 24, 2016.

2. Shreck, Getz, and Feenstra 2005.

3. The Declaration of Nyéléni was formulated at the first global forum on food sovereignty held in Mali in 2007. The text is available online at http://goo.gl /Eco4GO; accessed May 25, 2016. Also see chapter 1, this volume, for further discussion.

## 13. "We Are Human!"

1. The Border Protection, Anti-terrorism, and Illegal Immigration Control Act of 2005 (H.R. 4437) was a bill in the 109th US Congress that would have, among other actions deemed by many as unconstitutional, "criminalized" the act of "unauthorized entry," an act that is currently considered a civil, rather than a criminal, violation. It is also known as the "Sensenbrenner Bill," after Congressman Jim Sensenbrenner (R-WI).

2. Rios 2006.

3. Robinson and Santos 2014.

4. The Border Security, Economic Opportunity, and Immigration Modernization Act (S. 744) is a proposal first floated by Senator Charles E. Schumer (D-NY). For a critical analysis of the bill, see Schey 2016.

5. Community Alliance for Global Justice 2012, 17.

6. Ibid., 19; this is taken from the June 2001 Via Campesina declaration.

7. Stewart 2014, no page.

8. Before his death, Karl Marx wrote in a draft of a letter to Vera Zazulich that he acknowledged the capacity of the Russian peasantry to engage in a different type of economic practice, one that might not require the brutal collectivization and industrialization that occurred under Stalin in the 1930s. In 1928, José Carlos Mariátegui also foresaw the move by Indigenous people in South America to engage in a different type of economy.

9. Wolf 1982, 381; Luxemburg 1951 [1913].

10. Kautsky 1988 [1899].

11. Wolf and Mintz 1957; Katz 1974.

12. Goldsmidt 1947; McWilliams 1949; P. Taylor 1954.

13. Griffith and Kissam 1995; Krissman 1996; Meillassoux 1975; Á. Palerm 1980; J. V. Palerm and Urquiola 1993.

14. Galarza 1964; Gamboa 1990.

15. Alamillo 2006; Camarillo 1979; Garcia 2001; Grandin 2009; J. V. Palerm 1991, 1998, 2000.

16. Wolf 1982, 381; see also Haley 2009, 12.

17. Griffith and Kissam 1995, 14.

18. Montejano 1977, 1.

19. Ibid., 1.

20. L. H. Fisher 1953; Goldsmidt 1947; McWilliams 1939; Zaragoza 2007.

21. L. H. Fisher 1953; Kautsky 1988 [1899]; J. V. Palerm 1991, 1998, 2000.

22. Kautsky 1988 [1899]; Meillassoux 1975; Á. Palerm 1980; J. V. Palerm and Urquiola 1993.

23. L. H. Fisher 1953.

24. Omi and Winant 1986, 56; Molina 2006; Ngai 2004; Gilmore 2007, 27–28.

25. Chacon and Davis 2006, 139.

26. TVW 2013, no page.

27. Telles and Ortíz 2009.

28. Ramírez, McDevitt, and Farrell 2000, 3; Lovrich et al., 2003, 15.

29. The quote is taken from the author's field notes.

30. Holmes 2013.

31. Rosalinda Guillen had also been asked to intervene in the 2008 work stop-

page, but had decided not to become involved after realizing that the farmworkers wanted someone to resolve the matter for them, instead of solving it themselves, because they were only familiar with the process established by Holmes and his colleagues in 2004.

32. In 2012, Rhett Searcy was accused in a complaint to the California Agricultural Labor Review Board by the United Farm Workers against Montalvo Farms in Ventura County of using a taser on union organizers during a labor dispute; see Kern 2012.

33. This excerpt is taken from the unpublished primary document (print copy) of the original NWDC Resistance signature statement.

34. For a record of that decade of organizing for immigrant rights in Greater Seattle through 2012, see Murray 2012.

35. See the Washington Compact website for a list of these special interest groups, http://washingtoncompact.com/supporters; accessed December 9, 2014.

36. Stoeve 2014.

37. Ibid.

38. These excerpts are all based on the author's English translation of Colectiva de Detenidos's hunger strike demands by NWDC Resistance.

39. Subcomandante Insurgente Marcos, EZLN communique, March 17, 1995.

40. To read more on the concept of solidarity economies, see Madrigal 2014b.

41. Devon Peña, personal communication, June 3, 2015.

42. Felimón Pineda addressed these words to five grower-lobby spokesmen at his first meeting as a delegate to Governor Inslee's Farm Work Group on October 16, 2014.

43. As quoted in Dominick 2014.

44. McKinley 2014.

45. For more on Juan Chavez Alonso, see Madrigal 2013.

46. Dunbar-Ortíz 2014.

47. Cuevas 2005; Madrigal 2014a.

## 14. Organic Intellectuals and Direct Action Fifty Years Past Chicago's "War on Poverty"

1. An earlier version of this chapter was originally prepared for the fiftieth anniversary of the "War on Poverty" at Chicago State University; February 11, 2014.

2. Gramsci 1971, 113.

3. See McFarland 2008, 2013, especially part I, chapters 1–2 in McFarland 2013.

4. The movie of the same name was released in 1973.

5. Churchill and Vander Wall 2001, 146.

6. See McFarland (2008, 2013) for a critical discussion of hypermasculinity in the hip-hop culture of Black and Chicano men.

7. See Gallagher (2009, 2011) for details of food deserts in Chicago. Also, see posts prepared by the author for the *Environmental and Food Justice Blog* (http://ejfood.blogspot.com) in a series of essays examining how activists are using the food justice movement in Chicago to combat poverty and lack of access to fresh, culturally appropriate food.

8. Literally a "goat-head"—conveying the sense of a person characterized

by stubborn determination. All translations are by the author unless otherwise indicated.

9. Among neoliberal intellectuals, Friedrich von Hayek, who won the Nobel Prize in economics for his theory of information, knowledge, and prices, is distinctly honest in the admission that freedom (of the market) requires inequality, and any attempt by government to address poverty and inequality is an attack on freedom. See Hayek 1978. Also, see the discussion of Hayek in Peña (2008), who asks:

> Did Hayek deserve the Nobel Prize for thinking through these ontological assumptions about the constitution of power and freedom in the form of the inviolate rational individual actor seeking maximum knowledge of prices to plan a better world for optimum self-conservation and self-satisfaction?

10. Gibson-Graham 2006.

11. See Peña 1997, 2005a.

12. See Gomberg (2006) for detailed examination of shared labor and the concept of contributive justice.

13. Carney and Rosomoff 2009.

14. For a discussion of the Black American population as a colonized population, see Waberi (2009), and for a discussion of the Chican@ population as colonized, see McFarland (2013).

15. For an early path-defining study of Chicana/o environmental ethics, see Peña and Martinez 1998.


## 15. *Sin maíz, no hay país*

1. The authors wish to gratefully acknowledge the superb translation by coeditor Devon G. Peña, who also compiled page references for quotes from various authors and in particular the key work of Mixe scholar Floriberto Díaz. The translation is based on the authors' chapter, "La disputa por el maíz: comunalidad vs transgénicos en México" pp. 493-526 in: El maíz en peligro ante los transgénicos: Un análisis integral sobre el caso de México, edited by Elena R. Álvarez-Buylla Alma Piñeyro Nelson. Mexico City: UNAM.

2. Bonfil Batalla 1990, 32–34.

3. López 2003, 29.

4. Florescano 1994, 37.

5. Solares 2007, 346–47.

6. Benz 1997.

7. Bernal 1979.

8. Galinat 1995.

9. Ibid., 3.

10. George Beadle, as quoted in Flannery 1973, 287.

11. Editors' note: Nixtamalization is the process of soaking and boiling the maize kernels in lime (or food quality lye) to separate the pericarp (husk) from the germ. This process makes the vital nutrients, which include amino acids, available for absorption through the human body's digestive and circulatory systems.

12. Solares 2007, 258.

13. Shiva 2003, no page numbers.
14. Massieu and Lechuga 2002, 301.
15. Sánchez et al. 2004.
16. Food and Agriculture Organization 2010, 9; translation by coeditor Peña.
17. J. Rifkin 1999.
18. Bartra 2001.
19. Ayala, Schwentesius, Almaguer, et al. 2012, 91–92.
20. V. Martínez 2007.
21. Perrea 2009.
22. Espinoza and Turrent 2007, no page; quote is from notes taken at the lecture presentation.
23. Expansión 2005.
24. V. Martínez 2007.
25. Norandi 2007.
26. Espinosa and Turrent 2007.
27. As cited in V. Martínez 2007.
28. Espinosa et al. 2008.
29. CEDRSSA 2007.
30. *Imagen Agropecuaria* 2009, 1.
31. Ibid.
32. Massieu and San Vincente 2006.
33. Union of Concerned Scientists 2009.
34. Salamanca 2010.
35. Benbrook 2003.
36. Restrepo 2009. The coeditors of the volume wish to note that the research on the toxicity and carcinogenic risks of glyphosate is quite extensive of late. Moreover, in February 2016 the International Agency for Research on Cancer (IARC) classified the herbicide as a human carcinogen after reviewing all of the peer-reviewed science available. This has set off an intense debate over the inadequacies of the use of proprietary industry studies by regulatory agencies in the United States and Europe in setting their risk standards.
37. Marielle and Peralta 2007, 59.
38. Ruíz, 2006.
39. Covantes 1999.
40. Ruíz 2006.
41. Larson and Chauvet 2002.
42. For more on the social and environmental impacts of GMOs in Mexico, as seen through the vantage points of Mexican scientists, visit the home page of the Unión de Científicos Comprometidos con la Sociedad, http://www.uccs.mx.
43. Diario Oficial de la Federación 2005, http://goo.gl/IdQca9, accessed May 26, 2016.
44. Bonfil Batalla 1990, 32.
45. Common property resources, or CPRs, as proposed by Ostrom (2000).
46. Sánchez et al. 2004, 275.
47. See, generally, Ostrom 1990.
48. Shiva 2002, no page numbers.
49. F. Díaz 2007, 37.
50. Ibid., 136.

51. Berlanga and Concheiro 2010; also cited in Lopez 2012.
52. Boege 2008, 20.
53. Serratos-Hernández 2009, 4.
54. Kato et al. 2009, 5.
55. F. Díaz 2007, 65.
56. Nadal 2002, 26.
57. See generally, Kato et al. 2009.
58. Ortega 2003.
59. F. Díaz 2007, 53.
60. Sánchez et al. 2004, 46.
61. F. Díaz 2007, 54.

## 16. Sodbusters and the "Native Gaze"

1. This paper was originally inspired by invitation of the University of Washington Department of Anthropology Spring 2008 Colloquium, "Epistemologies of Anthropological Research," convened on May 23, 2008. The author thanks Elaine H. Peña, Ann Anagnost, Mario Montaño, and Gabe Valle for comments on earlier drafts.

2. For a history of sodbusting in the Upper Midwest between 1820–60, see Schob 1973.

3. See Marx 1990; especially chapters 23–24.

4. On the Indigenous ethics of "original instructions," see Nelson 2008.

5. Lai, Reicosky, and Hanson 2007.

6. Nelson 2008; Kimmerer 2015. In 2008, the American Prairie Partners recounted the underlying scientific wisdom of the joke:

> The story was taken to [illustrate] the Indian's ignorance, but in fact when the native grasses are turned under and the soil aerated, the organic matter decomposes faster. This creates a flush of nutrients available to cultivated crops, but when the crops are harvested, the nutrients are removed with the harvest, and the soil continues to be depleted year after year. Today's dependence on chemical fertilizers is evidence that perhaps there was more wisdom in that old Indian's statement than was recognized at the time. Certainly in terms of recovering the lost prairie, his statement was true.

This statement was originally on the Web (http://goo.gl/Qi6rCA) and accessed March 29, 2008. The link was inactive on January 12, 2015. The statement is now reposted to the Native American Seed website, at http://goo.gl/u5eSZp; accessed May 26, 2016. This illustrates the persistently "viral" nature of the sodbuster's joke.

7. Some scholars propose that we should combine soil science with anthropology to search for congruence between Indigenous and Western scientific soil classification and conservation; see Krough and Paarup-Laursen 1997. For a global multidisciplinary survey of studies on the "soil knowledge of local people," see Barrera-Bassols and Zinck 2003.

8. Bocco and Pulido 2003, 200.

9. Peña 1998, 2005c. An illustrious example of Indigenous landscape mosaics is the Ohlone oak savannas of California; see Anderson 2007.

10. Nabhan 2011. In Rio Culebra of Colorado, one study confirms the occurrence of a reversion to wild intermediary "tunicate" forms in the in-bred parent lines of the local landrace, maíz de concho, a white flint corn; see Peña 2016b and Green Fire Times 2016.

11. The *frijol tapado* (covered beans) system is among the many Indigenous mulching and fallow practices verified as regenerative of soil fertility. It is also proven to be as productive as more conventional mono-crop systems using fertilizers; see Fiedler and Kareiv 1998, 433. Also cf. Rosemeyer et al. 2000; Thurston et al. 1994.

12. For further discussion of autonomy in Indigenous philosophies of power/knowledge and governance, see Alfred 1999, 2009.

13. Patrick Henry as quoted in Helms 1991, 24.

14. Middlebury College has Cato's manuscript with an extended commentary; http://goo.gl/EE3Iju; accessed May 27, 2016.

15. Williams and Ortíz-Solorio 1981.

16. For a recent and rare discussion of agricultural instruction in calmecacs, see Colín 2014.

17. Williams and Ortíz-Solorio 1981. A more general treatment of "natural resource management" and Indigenous knowledge in Latin America is Pichón, Uquillas, and Frechione 1999.

18. Warkentin 2006.

19. Handiwerk 2008.

20. The Tepetlaoztoc Codex is in the British Museum; some elements can be examined on the British Museum's blog home page; https://goo.gl/LEkJOo; accessed May 27, 2016.

21. Peña 2005c; Geo-Mexico (2011) insightfully notes:

> The Spaniards did not maintain the Aztec civil works, deforested the surrounding hillsides, and started filling Lake Texcoco. This contributed to major flooding in 1555, 1580, 1604. The city was actually underwater . . . continuously . . . from 1629–1634. During this period the Spaniards invested in several flood control efforts, but they were not successful. In 1788 they started construction of a massive canal to connect the basin to rivers north of the city flowing to the Gulf of Mexico. The open canal, which was up to 30 m (100 ft) deep in places, provided flood relief, but did not completely solve the problem, and flooding continued.

22. Maffee (2014, 101–2) directs our attention to the fact that in the 1960s Miguel León-Portilla first identified the root word of *Neltiliztli* as derived from the same radical as *Tla-nel-huatl*, "root," and which in turn leads to *nelhuayotl*, "base" or "foundation." León-Portilla insists this concept of truth is best rendered as meaning "a well-grounded stability, well-foundedness, and well-rootedness." See León-Portilla 1990, 8.

23. *Tepetate* (hardpan) is a compacted clay strata caused by overtillage of the wrong soil profile; it can also refer to a naturally occurring property of certain soil compositions characterized by a dense clay lens underneath shallow topsoils of various sandy loam varieties.

24. Barrera-Bassols, Zink, and van Ranst 2009.

25. For critiques of the slash-and-burn stereotype, see Fedick 1996; Peña 2005c.

26. Fedick 1996.

27. González, Ventura, and Castellanos 2006, 1.

28. Ibid.; emphasis in the original.

29. A more recent study of the Tepetlaoztoc Codex concurs with my view of the calmecacs as institutions of agroecological research and learning; Williams and Pierce (2014, 160) note:

> No information has come to light about the procedures followed to produce the population and land registers for Tepetlaoztoc. Did a single individual possess all of the requisite knowledge . . . or were tasks divided among a multidisciplinary team of several specialists: a census taker, a surveyor with assistants to take measurements, a mathematician who calculated areas, a soil specialist who classed the soils, and a painter who recorded the information? . . . The codex data show that metrology, mathematics, and pedology were foundational sciences of specialists serving as civil servants.

30. An encouraging sign: Zapatista village-based *Caracoles* (Snail Shells) include schools where students learn by doing by following Indigenous agroecology and Mayan soil knowledge. For more evidence of the continuity of soil TEK, see Barrera-Bassols, Zink, and van Ranst 2009.

31. On the Maya managed mosaic, see Fedick 1996; Peña 2005a; on the surprisingly low level of soil erosion during the height of the city-states in the southern Maya Lowlands of Belize, Guatemala, Honduras, and Mexico, see Beach et al. 2006.

32. Simon 1997; Peña 2005c.

33. For a critique of so-called classical norms of "distanced normalizing observation" that remains fresh and relevant today, see Rosaldo 1989.

34. The one-million-acre Sangre de Cristo Land Grant was among the last issued by the Mexican government. The Culebra Mountain Tract, or "La Sierra," which pertains to our watershed, comprises some eighty thousand acres of montane, subalpine, and alpine forests. Communal use rights granted in the nineteenth century and confirmed as a result of a decades-long land rights struggle were restored by a 2002 Colorado Supreme Court ruling. As a matter of disclosure, TAI has successor rights to these common lands. On the lawsuit to restore land grant rights to this montane area, see the published opinions for *Lobato v. Taylor*, 70 P.3d 1152 (Colo. 2003); *Lobato v. Taylor*, 71 P.3d 938 (Colo. App. 2002); *Rael v. Taylor*, 876 P.2d 1210 (Colo. 1994). The American Civil Liberties Union-Colorado website offers primary sources including amicus briefs and a complete collection of court opinions. These may be accessed at http://goo.gl/pq6eZn; accessed May 26, 2016. For an unmatched general history of land grant litigation in New Mexico, see Ebright 1994. For more on Colorado land grant, see Stoller 1980; on the history of the Sangre de Cristo Land Grant, see Mondragón-Valdéz 2006.

35. Hicks and Peña 2003.

36. The Ute people—now living south and west of Durango on a reservation—gave this land rights case a blessing as resistance by Indigenous peoples, according to Frank White, Ute-Chicano elder, in a personal communication to the author,

June 1988. I have not found a record of an official tribal statement on this Chicana/o land rights lawsuit.

37. The irony amazes since the colonel did not know how acequias were rooted in antiquity when "al Andalus" was the heart of a multicultural Islamic Spain (711–1492 CE).

38. I did not use the term *race*. I had asked a simple question: "What is it about the local culture around here that makes for good irrigation?" He launched into a diatribe about the superiority of mechanical sprinklers and corrugated pipe. I had expected him to evoke racist stereotypes and misconceptions about acequias. Interestingly, in 2013, corrugated pipe was introduced to our local acequia systems in a hybrid adaptation that still relies on gravity for energy but keeps the *Acequia Madre* (Mother Ditch) and "*linderos*" (lateral ditches) intact.

39. For example, see Weber 1991.

40. We do use a small reservoir on the highest ground and a system of pipes and valves to divert water using gravity from the SLPD to an outtake for the fields farthest from the *Acequia Madre*.

41. A classic and widely cited study of soil erosion is Blaikie and Brookfield (1987). A more recent study on the economics of irrigation systems by Texas A&M University researchers acknowledges that, paired with furrows, center-pivot irrigation systems "can cause some environmental problems, such as soil erosion, sediment transport, loss of crop nutrients, deep percolation of water, and movement of dissolved chemicals into groundwater"; see Amosson et al. 2011, 1.

42. Research by my colleague Robert D. Curry demonstrated how the art of acequia flood irrigation, if managed with skill and attentiveness, is actually regenerative of topsoil horizons. Curry, a hydrologist and watershed scientist, found the acequias of the Culebra watershed are actually producers of topsoil; he called them "soil banks." See Curry's chapter and discussion in Peña, Martinez, and Curry (forthcoming); also, see Fernald, Baker, and Guldan (2007) for a study of the hydrologic, riparian, and agroecosystem functions of acequia irrigation systems; see Hicks and Peña (2003) for the results of an earlier study conducted in 1999–2002.

43. The term *naturalized exotic* is used here to refer to a plant or animal species introduced from outside a given ecosystem and adapted to local conditions without displacing native species on a scale that threatens the continued reproductive viability of the native organisms.

44. See Crosby (1986), who termed the British Empire the "Empire of the Dandelion" in reference to one of the most pervasive plants accompanying the settler colonists across the planet.

45. I learned the word *almunyah* from the late Estevan Arellano, who explained it during a visit to his family's permaculture farm in Embudo, New Mexico. He explained the almunyah as a private farmer-led agricultural teaching and research farm during the heyday of the Islamic period in Andalusia. TAI adopted this name to refer to our acequia farm as *Almunyah de las Dos Acequias* (Research Farm of the Two Communal Irrigation Ditches) to honor these deep antecedent roots. The almunyah is also analogous to a calmecac, and our work is tied to ongoing community-directed research projects involving the participation and leadership of local youth and acequia farmers collaborating with university-based scholars, like Adrianna Nieto of Metropolitan State University in Denver and

Lynn Sikkink of Western State Colorado University. Since 2007, TAI has also hosted classes from Ft. Lewis College (Durango), Colorado College, and numerous other higher-education institutions. We also host a semiannual Acequia Agroecology and Permaculture Field School from the University of Washington.

46. Anonymous weed technician, local NRCS, conversation with the author, March 3, 2007.

47. It took several federal lawsuits, like *Garcia v. Venneman*, 224 F.R.D. 8 (D.D.C. 2004), for the USDA to start addressing decades of discrimination against Latina/o farmers. The EQIP outreach is one example of new programs designed to address patterns of racism and neglect; progressive Chicana/os and white men and women in our local NRCS offices are leading these long-overdue investments in the provision of technical and financial assistance to acequia farmers.

48. Mike McGowan, personal note to the author, June 2006.

49. One confidential source informed me that about 80 percent of the Costilla County acequia farmers with individual EQIP contracts are in noncompliance with any given number of technical requirements. As a result, they have not received their full funding despite expenditure of their own resources. The technical issues seem arbitrarily constructed and overzealously applied. For example, several farmers have complained that they did not get paid by the EQIP program because they had placed cross-wire (hatches) at a distance that exceeded the technical requirement by six inches. There is no evidence the scale of the "misplacement" of the hatches causes any significant problems with the effectiveness of the fencing.

50. Higgins and Lockie 2002, 419ff.

51. I wish to thank Tezozomoc for introducing me to this Nahuatl word in May 2016.

## Conclusion

1. Cf. Davis 2006.
2. Anzaldúa 2007, 122.
3. Shiva 1988.
4. Nelson 2008, 180.
5. For essential originary discussions of "nepantla," see Anzaldúa 2007; Keating 2006.
6. Cajete 2000, 19.
7. Mohawk 2008, 52.
8. Sandoval 2000, 141.
9. On the concept of the hyperobject, see Morton 2013.
10. Esteva and Prakash 1998; also see Valle in chapter 3, this volume.
11. Wilson 2009.
12. For discussion of identity and subjectivity in new Chicana/o social movements, see Peña 2012.
13. Rodríguez 2014, xxii.
14. Ibid., 177.
15. M. Taylor 2006.
16. Berlant 2011.

17. See, generally, Wildcat 2009.

18. Ibid., 16.

19. Rodríguez 2014, 9.

20. Allelopathy can be understood as the process involving chemical signals that plants use to "communicate" with other plants and organisms.

21. Nelson 2008; Peña 2017.

# BIBLIOGRAPHY

Aerni, Philipp. 2011. Food sovereignty and its discontents. *African Technology Development Forum (ATDF) Journal* 8, no. 1-2: 23–40. https://goo.gl/1Df8HF. Accessed February 23, 2017.

Agamben, Giorgio. 1998. *Homo sacer: Sovereign power and bare life.* Stanford, Calif.: Stanford University Press.

———. 2004. *The open: Man and animal.* Stanford, Calif.: Stanford University Press.

———. 2005. *State of exception.* Chicago: University of Chicago Press.

Agarwal, Bina. 2014. Food sovereignty, food security and democratic choice: Critical contradictions, difficult conciliations. *Journal of Peasant Studies* 41, no. 6: 1247–68.

Agrawal, Arjun. 2005. *Environmentality: Technologies of government and the making of subjects.* Durham, N.C.: Duke University Press.

*Agueda Martinez: Our people, our country.* Directed by Esperanza Vasquez. 1977. Produced by Moctesuma Esparza. http://goo.gl/LZK9Vn. Accessed June 2, 2016.

Aguilar-Moreno, Manuel. 2006. *Handbook of life in the Aztec world.* New York: Facts on File.

Agyeman, Julian, and Duncan McLaren. 2014. "Smart cities" should mean "sharing cities." *Time,* September 29. http://goo.gl/QkVhUj. Accessed May 27, 2016.

Alamillo, José M. 2006. *Making lemonade out of lemons: Mexican American Labor and leisure in a California town 1880–1960.* Urbana: University of Illinois Press.

Alcoff, Linda. 2007. Mignolo's epistemology of coloniality. *CR: The New Centennial Review* 7, no. 3: 79–101.

Alfred, Taiaikke. 1999. *Peace, power, righteousness: An indigenous manifesto.* New York: Oxford University Press.

———. 2009. *Wasáse: Indigenous pathways of action and freedom.* Toronto: University of Toronto Press.

Alkon, Alison Hope, and Teresa Marie Mares. 2012. Food sovereignty in US food movements: Radical visions and neoliberal constraints. *Agriculture and Human Values* 29, no. 3: 347–59.

Allen, Will, and Charles Wilson. 2012. *The good food revolution: Growing healthy food, people, and communities.* New York: Gotham Books.

Alonso, Ana María. 2004. Conforming disconformity: "Mestizaje," hybridity, and the aesthetics of Mexican nationalism. *Cultural Anthropology* 19, no. 4: 459–90.

Altieri, Miguel. 1998. *The potential of agroecology to combat hunger in the developing world.* Washington, DC: International Food Policy Research Institute.

Amosson, Steve, et al. 2011. *Economics of irrigation systems.* AgriLIFE Extension Report B-6113 (October). College Station: Texas A&M University.

Anderson, Arthur J. O., and Charles Dibble. 1961. *Florentine codex: Book 10: The people. (The Florentine codex: A general history of the things of New Spain by Fray Bernardino de Sahagún),* 2nd rev. ed. Salt Lake City: University of Utah Press.

Anderson, M. Kat. 2007. *Indigenous uses, management, and restoration of oaks of the far western United States.* Technical Note No. 2. National Plant Data Center. Natural Resources Conservation Service, USDA. http://goo.gl/oz9NwZ. Accessed May 27, 2016.

Andrade-Cetto, Adolfo, and Michael Heinrich. 2005. Mexican plants with hypoglycemic effect used in the treatment of diabetes. *Journal of Ethnopharmacology* 99, no. 3: 325–48.

Ankri, Serge, and David Mirelman. 1999. Antimicrobial properties of allicin from garlic. *Microbes and Infection* 1, no. 2: 125–29.

Anzaldúa, Gloria. 1999. *Borderlands/La frontera: The new mestiza.* 2nd ed. San Francisco: Aunt Lute Books.

———. 2002. Now let us shift. In *This Bridge We Call Home: Radical Visions for Transformation,* edited by Gloria Anzaldúa and AnaLouise Keating, 40–57. New York: Kitchen Table Press/Women of Color Press.

———. 2007. *Borderlands/La frontera: The new mestiza.* 3rd ed. San Francisco: Aunt Lute.

———. 2009. Let us be the healing of the wound. In *The Gloria Anzaldúa Reader,* edited by AnaLouise Keating, 303–17. Durham, N.C.: Duke University Press.

Ayala Garay, Alma V., Rita Schwentesius Rindermann, Gustavo Almaguer Vargas, Sergio Roberto Márquez Berber, Benjamin Carrera Chávez, and José Luis Jolalpa Barrera. 2012. *Competitividad del sector agropecuario en México: implicaciones y retos.* Mexico: Instituto Nacional de Investigaciones Forestales, Agrícolas y Pecuarias (INIFAP). https://goo.gl/ofOHJo. Accessed March 12, 2017.

Azpeitia Gómez, Hugo. 1987. La autosuficiencia alimentaria en México. *Nueva Antropología* 9, no. 32: 129–50. http://goo.gl/lwgqlj. Accessed February 12, 2015.

Baca Zinn, Maxine. 1979. Chicano family research: Conceptual distortions and alternative directions. *Journal of Ethnic Studies* 7, no. 3: 59–71.

Baca Zinn, Maxine, D. Stanley Eitzen, and Barbara Wells. 2015. *Diversity in families.* 10th ed. New York: Pearson.

Barrera-Bassols, N., and J. A. Zinck. 2003. Ethnopedology: A worldwide view on the soil knowledge of local people. *Geoderma* 111: 171–95. https://goo.gl/XxD1YP. Accessed June 6, 2016.

Barrera-Bassols, N., J. A. Zinck, and E. van Ranst. 2009. Participatory soil survey: Experience in working with a Mesoamerican indigenous community. *Soil Use and Management* 25, no. 1: 43–56.

Barth, Fredrik, ed. 1998. *Ethnic groups and boundaries: The social organization of culture difference.* Long Grove, Ill.: Waveland Press.

Bartra, Armando. 2001. La renta de la vida. *Cuadernos Agrarios: Biopirateía y bioprotección* 21: 19–23.

Beach, Timothy, Sheyrl Luzzadder-Beach, Nicholas Dunning, Jon Hageman, and Jon Lohse. 2002. Upland agriculture in the Mayan lowlands: Ancient Maya soil conservation in Northwestern Belize. *Geographical Review* 92, no. 3: 372–97.

Beach, Timothy, Nicholas Dunning, Sheyrl Luzzadder-Beach, Duncan E. Cook, and Jon Lohse. 2006. Impacts of the ancient Maya on soils and soil erosion in the central Maya Lowlands. *Catena* 65: 166–78.

Bellon, Mauricio, and Julien Berthaud. 2006. Traditional Mexican agricultural systems and the potential impacts of transgenic varieties of maize diversity. *Agriculture and Human Values* 23: 3–14.

Benbrook, Charles. 2003. Impacts of genetically engineered crops on pesticide use in the United States: The first eight years. *Biotech Infonet Technical Paper*, no. 6. http://goo.gl/MlI50G. Accessed June 6, 2015.

Benhabib, Seyla. 2006. *Another cosmopolitanism*. Oxford: Oxford University Press.

Bennett, Jane. 2010. *Vibrant matter: A political ecology of things*. Durham, N.C.: Duke University Press.

Benz, B. F. 1997. Diversidad y distribucion prehispanica del maíz mexicano. *Revista de Arquelogia Mexicana* 5, no. 25: 16–23.

Berlanga, Héctor Robles, and Luciano Concheiro Bórquez. 2010. Balance de los territorios agrarios y perspectiva de una reforma agraria en México. In *Disputas territoriales. Actores sociales, instituciones y apropiación del mundo rural*, edited by Carlos Rodríguez Wallenius et al., 333–35. México: Universidad Autónoma Metropolitana-Unidad Xochimilco.

Berlant, Lauren Gail. 2011. *Cruel optimism*. Durham, N.C.: Duke University Press.

Bernal, John. 1979. *La ciencia en la historia*. México: Nueva Imagen.

Bernstein, Henry. 2014. Food sovereignty via the "peasant way": A skeptical view. *Journal of Peasant Studies*, 41, no. 6: 1031–63.

Blaikie, Piers, and Harold Brookfield, eds. 1987. *Land degradation and society*. New York: Methuen.

Boaventura de Sousa, Santos. 2014. *Epistemologies of the South: Justice against epistemicide*. Boulder, Colo.: Paradigm Publishers.

Bocco, Gerardo, and Juan Pulido. 2003. Geomorphological and landscape wisdom: Using local knowledge to manage slopes. In *Contemporary meanings in physical geography: From what to why?*, edited by Stephen Trudgill and André Roy, 199–210. New York: Routledge.

Boege, E. 2008. *El patrimonio biocultural de los pueblos indígenas de México. Hacia la conservación in situ de la biodiversidad u agrobiodiversidad en los territorios indígenas*. México: INAH-CONDEPI. http://goo.gl/ZeC9fY. Accessed May 27, 2016.

Bonfil Batalla, Guillermo. 1990. *México profundo: Una civilización negada*. México: Grijalbo/Consejo Nacional para la Cultura y Las Artes.

Bonilla, Carolina, Gerardo Gutiérrez, Esteban J. Parra, Christopher Kline, and Mark D. Shriver. 2005. Admixture analysis of a rural population of the state of Guerrero, Mexico. *American Journal of Physical Anthropology* 128, no. 4: 861–69.

Borras, June, Marc Edelman, and Cristóbal Kay, eds. 2008. *Transnational agrarian movements confronting globalization*. New York: Wiley-Blackwell.

Borras, Saturnino. 2008. *Competing views and strategies on agrarian reform*. International perspectives, vol. 1. Manila, Philippines: Ateneo de Manila University Press.

Borras, Saturnino, and Jennifer Franco. 2010. Towards a broader view of the politics of global land grab: Rethinking land issues, reframing resistance. ICAS Working Paper Series No. 001 (May). https://goo.gl/nWh9eT. Accessed June 7, 2016.

Boyer, Jefferson. 2010. Food security, food sovereignty, and local challenges for transnational agrarian movements: The Honduras case. *Journal of Peasant Studies* 37, no. 2: 319–51.

Buchmann, Christine. 2009. Cuban home gardens and their role in social-ecological resilience. *Human Ecology* 37, no. 6: 705–21.

Buck, Pem Davidson. 1991. Colonized anthropology: Cargo-cult discourse. In

*Decolonizing anthropology: Moving further toward an anthropology for liberation*, 24–41. Washington, DC: Association of Black Anthropologists, American Anthropological Association.

Bullard, Robert D. 2000. *Dumping in Dixie: Race, class, and environmental quality.* Boulder, Colo.: Westview Press.

Burnett, Kim, and Sophia Murphy. 2014. What place for international trade in food sovereignty? *Journal of Peasant Studies* 41, no. 6: 1065–84.

Butler, Benjamin, and Liam Midzain Gobin. 2015. Recognition, rights, and the indigenous subject of settler societies: The Special Rapporteurs visit to Canada. Paper presented at the 2015 Indigenous Studies Association. Available at Academia.edu, https://goo.gl/GcEsek. Accessed June 2, 2016.

Byrd, Jodi A. 2011. *Transit of empire: Indigenous critiques of colonialism.* Minneapolis: University of Minnesota Press.

Cajete, Gregory. 1944. *Look to the mountain: An ecology of indigenous education.* Durango, Colo.: Kivaki Press. https://goo.gl/gbZU4l. Accessed February 23, 2017.

———. 1999. Reclaiming biophilia: Lessons from indigenous people. In *Ecological education in action: On weaving education, culture, and the environment,* edited by Gregory A. Smith and Dilafruz R. Williams, 189–206. Albany: State University of New York Press.

———. 2000. *Native science: Natural laws of interdependence.* Santa Fe, N.Mex.: Clear Light Publishers.

Calvo, Luz, and Catriona R. Esquibel. 2015. *Decolonize your diet: Plant-based Mexican-American recipes for health and healing.* Vancouver, B.C.: Arsenal Pulp Press.

Camarillo, Albert. 1979. *Chicanos in a changing society: From Mexican pueblos to American barrios in Santa Barbara and southern California, 1848–1930.* Cambridge, Mass.: Harvard University Press.

Carney, Judith A., and Richard Nicholas Rosomoff. 2009. *In the shadow of slavery: Africa's botanical legacy in the Atlantic world.* Berkeley: University of California Press.

Carney, Megan A. 2013. Border meals: Detention center feeding practices, migrant subjectivity, and questions on trauma. *Gastronomica* 13, no. 4: 32–46.

Carpenter, Novella. 2009. *Farm city: The education of an urban farmer.* New York: Penguin Press.

Carrera, P. M., G. Xiang, and K. L. Tucker. 2007. A study of dietary patterns in the Mexican-American population and their association with obesity. *Journal of the American Dietetic Association* 107, no. 10: 1735–42.

CEDRSSA. 2007. *Maíz indicadores básicos.* H. Cámara de diputados, Centro des Estudios para el Desarrollo Rural y la Soberanía Alimentaria, Dirreción de Evaluación de Políticas Públicas Rurales. LX Legislatura. México, D.F.

Cerrutti, Marcela, and Douglas S. Massey. 2001. On the auspices of female migration from México to the United States. *Demography* 38, no. 2: 187–200.

Chacón, Justin Akers, and Mike Davis. 2006. *No one is illegal: Fighting violence and state repression on the U.S.-México border.* Chicago: Haymarket Books.

Champion, Jane D., and Jennifer L. Collins. 2013. Retrospective chart review for obesity and associated interventions among rural Mexican-American adolescents accessing healthcare services. *Journal of the American Association of Nurse Practitioners* 25, no. 11: 604–10. https://goo.gl/R4HC43. Accessed February 28, 2017.

Chavez, Leo Ralph. 2008. *The Latino threat: Constructing immigrants, citizens, and the nation*. Stanford: Stanford University Press.

*Chicana*. Directed by Sylvia Morales. 1979. Hollywood, Calif.: Ruiz Productions. 23 min.

Churchill, Ward, and Jim Vander Wall. 2001. *Agents of repression: The FBI's secret wars against the Black Panther Party and the American Indian Movement*. Boston: South End Press.

CIBIOGEM. 2011. *Pruebas de campo (de 1988 al 13 de junio de 2005) y permisos de liberación al ambiente (del 14 de junio de 2005 a 2009), de Organismos Genéticamente Modificados aprobados por México, por cultivo y comparación entre aprobados conforme la Ley Federal de Sanidad Vegetal (LFSV) y La Ley de Bioseguridad de Organismos Genéticamente Modificados (LGOGM). Sistema Nacional de Información Disponible para Consulta*. http://goo.gl/sMq57v. Accessed June 2, 2016.

Cleaver, Harry. 1977. Food, famine, and the international crisis. *Zerowork Political Materials Flyer*. http://zerowork.org/CleaverFoodFamine.html. Accessed February 23, 2017.

———. 1979. *Reading capital politically*. Austin: University of Texas Press.

———. 1992. The inversion of class perspective in Marxian theory. In *Theory and Practice*, Open Marxism, vol. 2, edited by Werner Bonefeld, Richard Gunn, and Kosamas Psychopedis, 106–44. London: Pluto Press. https://goo.gl/INTZVG. Accessed June 2, 2016.

Clifford, James. 1994. Diasporas. *Cultural Anthropology* 9, no. 3: 302–38.

Cohen, Abner. 1976. *Two-dimensional man: An essay on the anthropology of power and symbolism in complex society*. Berkeley: University of California Press.

Colín, Ernesto Tlahuitollini. 2014. *Tlacahuapahualiztli* (The art of educating a person). In *Indigenous education through dance and ceremony*, 65–78. New York: Palgrave Macmillan.

Community Alliance for Global Justice, Food Justice Project. 2012. *Our food, our right: Recipes for food justice*. Seattle: CAGJ.

Concheiro, L. 1995. *Estructura agraria y mercado de tierras en México*. México: UAM-Xochimilco and Food and Agriculture Organization.

*Corridos: Tales of passion and revolution*. Directed by Luis Valdez. 1987. KQED (Recorded April 1, 1987). Accessed through El Teatro Campesino Collection, Hemispheric Institute for Performance and Politics, NYU Libraries. http://goo.gl/xfIDhv. Accessed June 2, 2016.

Coulthard, Glen S. 2014. *Red skin, white masks: Rejecting the colonial politics of recognition*. Minneapolis: University of Minnesota Press.

Covantes, L. 1999. Cuando el destino nos alcanzó. *Revista Este País* (July). México.

Craig, James R., David J. Vaughan, and Brian J. Skinner. 2001. *Resources of the earth: Origin, use, and environmental impact*. 3rd ed. Upper Saddle River, N.J.: Prentice Hall.

Crosby, Alfred. 1986. *Ecological imperialism: The biological expansion of Europe, 900–1900*. Cambridge: Cambridge University Press.

Cruishank, Julie. 2014. *Do glaciers listen? Local knowledge, colonial encounters, and social imagination*. Vancouver, B.C.: UBC Press.

Cuevas, Maria. 2005. "As close to God as one can get": Rosalinda Guillen, a Mexicana farmworker organizer in Washington State. In *Memory, community, and activism: Mexican migration and labor in the Pacific Northwest*, edited by Jerry Garcia and Gilberto Garcia, 277–307. East Lansing, Mich.: Julian Samora Research Center.

Danaher, Kevin, Shannon Biggs, and Jason Mark, eds. *Building the green economy: Success stories from the grassroots*. Sausalito, Calif.: PoliPointPress, 2007. http://goo.gl/dAH6yZ. Accessed March 19, 2013.

Davidson, David M. 1966. Negro slave control and resistance in Colonial Mexico, 1519–1650. *Hispanic American Historical Review* 46, no. 3: 235–53.

Davis, Mike. 2001. *Magical urbanism: Latinos reinvent the U.S. big city*. London: Verso.

———. 2006. *Planet of slums*. London; New York: Verso.

Del Castillo, Adelaida. 2007. Illegal status and social citizenship: Thoughts on Mexican immigrants in a postnational world. In *Women and migration in the U.S.-México borderlands: A reader*, edited by Denise A Segura and Patricia Zavella, 92–105. Durham, N.C.: Duke University Press.

De León, Jason. 2012. "Better to be hot than caught": Excavating the conflicting roles of migrant material culture. *American Anthropologist* 114, no. 3: 477–95.

———. 2013. Undocumented migration, use-wear, and materiality of habitual suffering in the Sonoran Desert. *Journal of Material Culture* 18, no. 4: 321–45.

Deleuze, Gilles, and Félix Guattari. 1972. *Anti-Oedipus: Capitalism and schizophrenia*. Translated by Robert Hurley et al. London and New York: Continuum Books.

Delgado Bernal, Dolores. 2001. Learning and living pedagogies of the home: The mestiza consciousness of Chicana students. *International Journal of Qualitative Studies in Education* 14, no. 5: 623–39.

Deloria, Vine, Jr., and Daniel R. Wildcat. 2001. *Power and place: Indian education in America*. Golden, Colo.: Fulcrum Publishing.

Demarest, Geoffrey. 1995. Geopolitics and urban armed conflict in Latin America. Foreign Military Studies Office, Fort Leavenworth, Kans. This article was previously published in *Small Wars and Insurgencies* 6, no. 1 (Spring).

Desmarais, Annette Aurélie, and Hannah Wittman. 2014. Farmers, foodies and First Nations: Getting to food sovereignty in Canada. *Journal of Peasant Studies* 41, no. 6: 1153–73.

Diario Oficial de la Federación (DOF). 2005. *Ley de Bioseguridad de Organismos Genéticamente Modificados* (LBOGM). March 18. México. http://goo.gl/YzV7jY. Accessed June 7, 2016.

Díaz, David R. 2012. Barrios and planning ideology: The failure of suburbia and the ·dialectics of new urbanism. In *Latino urbanism: The politics of planning, policy, and redevelopment*, edited by David R. Díaz and Rodolfo D. Torres, 21–46. New York: New York University.

Díaz, Floriberto. 2007. *Comunalidad, energía viva del pensamiento mixe: Escrito*. México: UNAM/Programa Universitario México Nación Multicultural. https://goo.gl/4JFTYu. Accessed June 7, 2016.

Díaz del Castillo, Bernal. 1956. *The discovery and conquest of México, 1517–1521*. New York: Farrar, Straus, and Cudahy.

Dominick, Brian. 2014. Marcos is dead. Long live Marcos! *Medium*, May 26, 2014. https://goo.gl/RDKfgr. Accessed May 25, 2016.

Douglas, Mary. 2002. *Purity and danger: An analysis of concepts of pollution and taboo*. London: Routledge.

Dragusanu, Raluca, Daniele Giovannucci, and Nathan Nunn. 2013. The economics of fair trade. Paper prepared for the *Journal of Economic Perspectives*. http://goo.gl/bQ2GN2. Accessed June 1, 2016.

Dunbar-Ortíz, Roxanne. 2014. *An indigenous peoples history of the United States*. Boston: Beacon Press.

Durand, Jorge, and Douglas S. Massey, eds. 2007. Crossing the border: Research from the Mexican Migration Project. *Economic Development and Cultural Change* 56, no. 1: 236–41.

Eagleton, Terry. 1991. *Ideology: An introduction*. London: Verso.

Ebright, Malcolm. 1994. Land grants and lawsuits in northern New Mexico. Albuquerque: University of New Mexico Press.

Edelman, Marc. 2014. Food sovereignty: Forgotten genealogies and future regulatory challenges. *Journal of Peasant Studies* 41, no. 6: 959–78.

Elac, John Chala. 1961. *Employment of Mexican workers in United States agriculture 1900–1960; A binational economic analysis*. PhD diss. University of California Los Angeles.

Espinosa, Alejandro, and Antonio Turrent Fernández. 2007. ¿Que pasa con el maíz y la tortilla? Paper presented to CIEESTAM, Universidad Autónoma de Chapingo.

Espinosa, Alejandro, et al. 2008. El potencial de las variedades nativas y mejoradas del maíz. *Revista Ciencias* 92–93: 160.

Esquivel, Laura. 1992. *Like water for chocolate: A novel in monthly installments, with recipes, romances, and home remedies*. Translated by Carol Christensen and Thomas Christensen. New York: Doubleday.

Esteva, Gustavo. 1996. Tepito: No thanks, First World. Context Institute. Originally published in *Reclaiming Politics*, Fall/Winter 1991. http://goo.gl/eIcpSv. Accessed May 23, 2016.

———. 2003. A flower in the hands of the people. Translated by Mark Fried. *New Internationalist* 360 (September). http://goo.gl/YobMoz. Accessed June 3, 2015.

Esteva, Gustavo, and Mahdu S. Prakash. 1998. *Grassroots post-modernism: Remaking the soil of cultures*. London: Zed Books.

Expansión. 2005. Cómo el maíz transgénico de Monsanto cambiará al campo. *Revista Expansión* 924 (14 September).

Eyzaguirre, Pablo, and Olga Linares. 2004. Introduction to *Home garden and agro-biodiversity*. Washington, DC: Smithsonian Books.

Fanon, Frantz. 2004. *The wretched of the earth*. Translated by Richard Philcox. New York: Grove Press.

Fedick, Scott L. 1996. *Managed mosaic: Ancient Maya agriculture and resource use*. Berkeley: University of California Press.

Ferguson, Kathryn, Norma A. Price, and Ted Parks. 2010. *Crossing with the virgin: Stories from the migrant trail*. Tucson: University of Arizona Press.

Ferlay, J., I. Soerjomataram, M. Ervik, R. Dikshit, S. Eser, C. Mathers, M. Rebelo, D. M. Parkin, D. Forman, and F. Bray. 2015. Cancer incidence and mortality worldwide: Sources, methods and major patterns in GLOBOCAN. *International Journal of Cancer* 136, no. 5 (March): E359–E386.

Fernald, Alexander G., Terrell T. Baker, and Steven J. Guldan. 2007. Hydrologic, riparian, and agroecosystem functions of traditional acequia irrigation systems. *Journal of Sustainable Agriculture* 30: 3: 147–71.

Fiedler, Peggy L., and Peter M. Kareiv, eds. 1998. *Conservation biology: For the coming decade.* New York: Chapman and Hall.

Fisher, Andrew B. 2006. Creating and contesting community: Indians and Afromestizos in the late-colonial Tierra Caliente of Guerrero, Mexico. *Journal of Colonialism and Colonial History* 7, no. 1.

Fisher, Lloyd H. 1953. *The harvest labor market in California.* Cambridge, Mass.: Harvard University Press.

Flannery, Kent V. 1973. The origins of agriculture. *Annual Review of Anthropology* 2: 271–310.

Flores, Yvette Gisele. 2013. *Chicana and Chicano mental health: Alma, mente, y corazon.* Tucson: University of Arizona Press.

Florescano, Enrique. 1994. *Memoria mexicana.* México: Fondo de Cultura Económica.

Food and Agriculture Organization (FAO). 2010. *El Segundo informe sobre el estado de los recursos fitogenéticos en el mundo para la alimentación y la agricultura.* http://www.fao.org/docrep/014/i1500s/i1500s.pdf. Accessed May 25, 2016. English version, *The second report on the state of the world's plant genetic resources.* http://fao.org/agriculture/seed/sow2/. Accessed May 25, 2016.

Fornet-Betancourt, Raúl, Helmut Becker, Alfredo Gomez-Müller, and J. D. Gauthier. 1987. The ethic of care for the self as a practice of freedom: An interview with Michel Foucault on January 20, 1984. *Philosophy and Social Criticism* 12, no. 2–3: 112–31.

Forsyth, Tim. 2003. *Critical political ecology: The politics of environmental science.* New York: Routledge.

Foucault, Michel. 1984. Of other spaces, heterotopias. *Architecture/Mouvement/ Continuité* 5: 46–49. Translated by Jay Miskowiec. http://web.mit.edu/allanmc/ www/foucault1.pdf. Accessed March 10, 2017.

Fuentes, Carlos. 2012. Desplazarse en el tiempo y en el espacio. *La palabra y el hombre* 21, no. 2 (verano): 41–45. https://goo.gl/78dM6X. Accessed May 24, 2016.

Galarza, Ernesto. 1964. *Merchants of labor: The Mexican bracero story.* Santa Barbara, Calif.: McNally and Loftin.

Galinat, Walton C. 1995. El orígen del maíz: El grano de la humanidad. *Economic Botany* 49: 3–12.

Galindo, María. 2013. *No se puede descolonizar sin despatriarcalizar: teoría y propuesta de la despatriarcalización.* La Paz, Bolivia: Mujeres Creando.

Gallagher, Mari. 2009. *Chicago 2009 food desert progress report.* Chicago: Mari Gallagher Consulting Group. http://goo.gl/DF7zMP. Accessed June 7, 2016.

———. 2011. *The Chicago 2011 food desert progress report.* Chicago: Mari Gallagher Consulting Group. http://goo.gl/DF7zMP. Accessed June 7, 2016.

Galluzzi, Gea, Pablo Eyzaguirre, and Valeria Negri. 2010. Home gardens: Neglected hotspots of agro-biodiversity and cultural diversity. *Biodiversity and Conservation* 19, no. 13: 3635–54.

Gamboa, Erasmo. 1990. *Mexican labor and World War II: Braceros in the Pacific Northwest, 1942–1947.* Austin: University of Texas Press.

Gamio, Manuel. 1972. *The life story of the Mexican immigrant: Autobiographic documents*. New York: Dover Publications.

Garcia, Matt. 2001. *A world of its own: Race, labor, and citrus in the making of Greater Los Angeles, 1900–1970*. Chapel Hill: University of North Carolina Press.

Geo-Mexico. 2011. Attempts to provide drainage for Mexico City date back to Aztec times. April 26. http://goo.gl/Rqy4W7. Accessed June 16, 2016.

Gibson-Graham, J. K. 2006. *The end of capitalism (as we know it): A feminist critique of political economy*. Minneapolis: University of Minnesota Press.

Giddens, Anthony. 1990. *The consequences of modernity*. Cambridge: Polity.

Gilmore, Ruth Wilson. 2007. *Golden gulag: Prisons, surplus, crisis, and opposition in globalizing California*. Berkeley: University of California Press.

Giroux, Henry A. 1995. The politics of insurgent multiculturalism in the era of the Los Angeles Uprising. In *Critical multiculturalism: Uncommon voices in a common struggle*, edited by Barry Kanpol and Peter McLaren, 107–24. Westport, Conn.: Greenwood.

Gliessman, Stephen R. 2007. *Agroecology: The ecology of sustainable food systems*. Boca Raton, La.: CRC Press.

Godoy, María. 2016. How Mexican cuisine was doing fusion 500 years ago. National Public Radio. http://goo.gl/OvJ3Wi. Accessed May 24, 2016.

Goldschmidt, Walter. 1947. *As you sow*. New York: Harcourt, Brace.

Gomberg, Paul. 2006. *How to make opportunity equal: Race and contributive justice*. Boston: Wiley-Blackwell.

Gómez de Silva, Guido. 2001. *Diccionario breve de mexicanismos*. México: Academia Mexicana, Fondo de Cultura Económica. https://goo.gl/olrsQJ. Accessed February 27, 2017.

González, Javier, Eusebio Ventura, and Javier Castellanos. 2006. History and development of soil science in México. In *Abstracts of the 18th World Congress of Soil Science*, July 9–15, 2006, Philadelphia. CD-ROM.

Gordon, Edmund T. 1991. Anthropology and liberation. In *Decolonizing anthropology: Moving further toward an anthropology for liberation*, 149–67. Washington, DC: Association of Black Anthropologists, American Anthropological Association.

Gottlieb, Robert. 2009. Where we live, work, play . . . and eat: Expanding the environmental justice agenda. *Environmental Justice* 2, no. 1: 7–8.

Gottlieb, Robert, and Anupama Joshi. 2010. *Food justice*. Cambridge, Mass.: MIT Press.

Graeber, David. 2001. *Toward an anthropological theory of value: The false coin of our own dreams*. New York: Palgrave.

Gramsci, Antonio. 1971. Prison notebooks: The intellectuals. In *An anthology of Western Marxism*, edited by Roger S. Gottlieb, 113–19. New York: Oxford University Press.

Grande, Sandy. 2004. *Red pedagogy: Native American social and political thought*. Lanham, Md.: Rowman and Littlefield.

Grandin, Greg. 2009. *Fordlandia: The rise and fall of Henry Ford's forgotten jungle city*. New York: Metropolitan Books.

Green Fire Times. 2016. Saving New Mexico's rare seeds. Special Issue 8, no. 5 (May). http://goo.gl/ojOWGQ. Accessed May 26, 2016.

Grey, Sam, and Raj Patel. 2014. Food sovereignty as decolonization: Some contributions from indigenous movements to food system and development politics. *Agriculture and Human Values*; published online October 5, 2014. DOI 10.1007/s10460-014-9548-9.

Griffith, David, and Ed Kissam. 1995. *Working poor: Farmworkers in the United States.* Philadelphia: Temple University Press.

Gulli, Bruno. 2005. *Labor of fire: The ontology of labor between economy and culture.* Philadelphia: Temple University Press.

Guthman, Julie. 2004. *Agrarian dreams: The paradox of organic farming in California.* Berkeley: University of California Press.

———. 2011. "If they only knew": The unbearable whiteness of alternative food. In *Cultivating food justice: Race, class, and sustainability,* edited by Alison Hope Alkon and Julian Agyeman, 263–81. Cambridge, Mass.: MIT Press.

Haley, Brian D. 2009. *Reimagining the immigrant: The accommodation of Mexican immigrants in rural America.* New York: Palgrave Macmillan.

Handiwerk, Brian. 2008. Aztec math decoded, reveals woes of ancient tax time. *National Geographic News,* April 3. http://goo.gl/DZG44Z. Accessed June 7, 2016.

Hardt, Michael, and Antonio Negri. 2005. *Multitude: War and democracy in the age of empire.* New York: Penguin Books.

———. 2009. *Commonwealth.* Cambridge, Mass.: Belknap/Harvard University Press.

Hayek, Friedrich A. von. 1978. *The constitution of liberty.* Chicago: University of Chicago Press.

hecubus. 2006. On the side of Horowitz. *We have seen better days* (blog). June 14. https://goo.gl/hhDFKO. Accessed March 7, 2017.

Helms, Douglas. 1991. Two centuries of soil conservation. *OAH Magazine of History* 4. Reprinted in *Readings in the history of the Soil Conservation Service,* edited by Douglas Helms. Washington, DC: Soil Conservation Service, Economics and Social Science Division.

Henderson, George L. 2013. *Value in Marx: The persistence of value in a more-than-capitalist world.* Minneapolis: University of Minnesota Press.

Herlihy, Peter H., Jerome E. Dobson, Miguel Aguilar Robledo, Derek A. Smith, John H. Kelly, and Aida Ramos Viera. 2008. A digital geography of indigenous México: Prototype for the American Geographical Society's Bowman Expeditions. *Geographical Review* 98, no. 3: 395–415.

Hernández, Kelly Lytle. 2010. *Migra! A History of the US Border Patrol.* Berkeley: University of California Press.

Hicks, Gregory A., and Devon G. Peña. 2003. Community acequias in Colorado's Rio Culebra Watershed: A customary commons in the domain of prior appropriation. *University of Colorado Law Review* 74, no. 2: 387–486.

Hidalgo de la Riva, Teresa. 2004. *Mujerista moviemaking: Chicana filmmakers Sylvia Morales and Lourdes Portillo.* Los Angeles: University of California Press.

Higgins, Vaughan, and Stewart Lockie. 2002. Re-discovering the social: Neo-liberal and hybrid practices of governing in rural natural resource management. *Journal of Rural Studies* 18, no. 4: 419–28.

Holman, Jordan. 2014. Urban farm bill could help transform South LA's empty lots. *Annenberg Media Center Intersections: South LA* (October 22). http://goo.gl/eehSUZ. Accessed June 7, 2016.

Holmes, Seth. 2013. *Fresh fruit, broken bodies. Migrant farmworkers in the United States*. Berkeley: University of California Press.

Holt-Giménez, Eric, and Annie Shattuck. 2011. Food movements unite! Making a new food system possible. In *Food movements unite! Strategies to transform our food systems*, edited by Eric Holt-Giménez, 315–23. Oakland, Calif.: Food First Books.

Hondagneu-Sotelo, Pierrette. 2014. *Paradise transplanted: Migration and the making of California gardens*. Berkeley: University of California Press.

Imagen Agropecuaria. 2009. Planea agroindustria que el maíz transgénico absorba el 80% del mercado de semilla convencional. *Imagen agropeciaria*, June 1. http://goo.gl/PRBaB8. Accessed May 27, 2016.

Jackson, Wes. 1980. *New roots for agriculture*. San Francisco: Friends of the Earth and The Land Institute.

John, Esther M., Amanda I. Phipps, Adam Davis, and Jocelyn Koo. 2005. Migration history, acculturation, and breast cancer risk in Hispanic women. *Cancer Epidemiology, Biomarkers and Prevention* 14, no. 12: 2905–13.

Johnson, Leslie M., and Eugene S. Hunn. 2012. Landscape ethnoecology: Reflections. In *Landscape ethnoecology: Concepts of biotic and physical space*, edited by Leslie M. Johnson and Eugene S. Hunn, 279–97. New York: Berghahn Books.

Jones, L. A., R. Gonzalez, P. C. Pillow, S. A. Gomez-Garza, C. J. Foreman, J. A. Chilton, A. Linares, J. Yick, M. Badrei, and R. A. Hajek. 1997. Dietary fiber, Hispanics, and breast cancer risk? *Annals of the New York Academy of Sciences* 837 (December): 524–36.

Jordan, June. 2002. *Some of us did not die: New and selected essays of June Jordan*. New York: Basic/Civitas Books.

Juárez-Cedillo, Teresa, Joaquín Zuñiga, Victor Acuña-Alonzo, Nonanzit Pérez-Hernández, José Manuel Rodríguez-Pérez, Rodrigo Barquera, Guillermo J. Gallardo, Rosalinda Sánchez-Arenas, María del Carmen García-Peña, Julio Granados, Gilberto Vargas-Alarcón. 2008. Genetic admixture and diversity estimations in the Mexican Mestizo population from Mexico City using 15 STR polymorphic markers. *Forensic Science International: Genetics* 2, no. 3: e37–e39.

Kato, Takeo, et al. 2009. *Origen y diversificación del maíz: una revisión analítica*. Universidad Nacional Autónoma de México, Comisión Nacional para el Conocimiento y Uso de la Biodiversidad (CONABIO). México. http://goo.gl/JWLzCJ. Accessed May 27, 2016.

Katz, Friedrich. 1974. Labor conditions on haciendas in Porfirian Mexico: Some trends and tendencies. *Hispanic American Historical Review* 54, no. 1: 1–47.

Kautsky, Karl. 1988 (1899). *The agrarian question*. Winchester, Mass: Zwan Publications.

Kearney, Michael. 1996. *Reconceptualizing the peasantry: Anthropology in global perspective*. Boulder, Colo.: Westview Press.

Kearns, Rick. 2014. The soft drink invasion of indigenous Chiapas; Increased diabetes death. *Indian Country Today Media Network*. Originally posted June 1, 2014. http://goo.gl/PpCKH9. Accessed June 2, 2016.

Keating, AnaLouise. 2006. From borderlands and new Mestizas to Nepantlas and Nepantleras: Anzaldúan theories for social change. *Human Architecture: Journal of the Sociology of Self-Knowledge* 4, no. 3: 5–16.

Kern, Rebekah. 2012. Labor Board alleges vicious anti-unionism. *Courthouse News Services,* May 22. http://goo.gl/zEaCx4. Accessed June 2, 2016.

Kimmerer, Robin. 2015. *Braiding sweetgrass: Indigenous wisdom, scientific knowledge and the teachings of plants.* Minneapolis: Milkweed Editions.

Kirk, Gwyn. 1999. Ecofeminism and Chicano environmental struggles: Bridges across gender and race. In *Chicano culture, ecology, politics: Subversive kin,* edited by Devon G. Peña, 177–200. Tucson: University of Arizona Press.

Kloppenburg, Jack. 2014. Re-purposing the master's tools: The open source seed initiative and the struggle for seed sovereignty. *Journal of Peasant Studies* 41, no. 6: 1225–46.

Koc, Mustafa, Rod MacRae, Luc Mougeot, and Jennifer Welsh, eds. 1999. *For hunger-proof cities: Sustainable urban food systems.* Ottawa: International Development Research Centre.

Komarnisky, Sara V. 2009. Suitcases full of mole: Traveling food and the connections between México and Alaska. *Alaska Journal of Anthropology* 7, no. 1: 41–56.

Krimsky, Sheldon, and Jeremy Gruber, eds. 2016. *The GMO deception: What you need to know about the food, corporations, and government agencies putting our families and our environment at risk.* New York: Skyhorse Publishing.

Krissman, Fred. 1996. California agribusiness and Mexican farm workers (1942–1992): A bi-national agricultural system of production/reproduction. Ph.D. diss., University of California Santa Barbara.

Krogh, Lars, and Bjarke Paarup-Laursen. 1997. Indigenous soil knowledge among the Fulani of northern Burkina Faso: Linking soil science and anthropology in analysis of natural resource management. *GeoJournal* 43, no. 2: 189–97.

LaDuke, Winona. 1999. *All our relations: Natives struggles for land and life.* Cambridge, Mass.: South End Press.

———. 2005. *Recovering the sacred: The power of naming and claiming.* Cambridge, Mass.: South End Press.

Lai, R., D. C. Reicosky, and J. D. Hanson. 2007. Evolution of the plow over 10,000 years and the rationale for no-till farming. *Soil and Tillage Research* 93, no. 1: 1–12.

Lara, Marelena, Cristina Gamboa, M. Iya Kahramanian, Leo S. Morales, and David E. Hayes Bautista. 2005. Acculturation and Latino health in the United States: A review of the literature and its sociopolitical context. *Annual Reviews of Public Health* 26, no. 3: 367–97.

Larson, Jorge, and Michelle Chauvet. 2002. Understanding complex biology and community values: Communication and participation. In *Maize and biodiversity: The effects of transgenic maize in Mexico,* edited by Secretariat of the Commission for Environmental Cooperation of North America. http://goo.gl/S7cQud. Accessed May 26, 2016.

Lash, Scott, and Jonathan Friedman. 1992. Subjectivity and modernity's other. Introduction to *Modernity and identity,* edited by Scott Lash and Jonathan Friedman, 1–30. Malden, Mass.: Blackwell Publishers.

Latour, Bruno. 1999. *Pandora's hope: Essays on the reality of science studies.* Cambridge, Mass.: Harvard University Press.

Lazzarato, Maurizio. 2014. *Signs and machines: Capitalism and the production of subjectivity.* Los Angeles: Semiotext(e).

León-Portilla, Miguel. 1990 (1963). *Aztec thought and culture: A study of the ancient Nahuatl mind.* Norman: University of Oklahoma Press.

López, Alfredo Austin. 2003. Cuatro mitos mesoaméricanos del maíz, In *Sin maíz no hay país*, edited by Gustavo Esteva and Catherine Marielle, 29–35. México: Consejo Nacional para las Culturas y las Artes. https://goo.gl/dXFFAE. Accessed June 2, 2016.

López Bárcenas, Francisco. 2012. Pueblos indígenas y megaproyectos: las nuevas rutas del despojo. *Contralínea*, October 22. https://goo.gl/qlCRJ2. Accessed March 12, 2017.

López González, Juan José, and Jesús Manuel Fuentes Rodríguez. 1997. Industrialización de la tuna Cardona (*Opuntia streptacantha*). *Journal of the Professional Association for Cactus Development* 2: 169–75.

Lorde, Audre. 1997. *The cancer journals*. Special ed. San Francisco: Aunt Lute Books.

Losada, H., H. Martinez, J. Vieyra, R. Pewaling, and J. Cortes. 1998. Urban agriculture in the metropolitan zone of México City: Changes over time in urban, suburban and peri-urban areas. *Environment and Urbanization* 10, no. 2: 37–54.

Lovrich, N., M. Gaffney, C. Mosher, M. Pickerill, and M. R. Smith, 2003. *Washington State Patrol traffic stop data analysis project report*. Pullman: Washington State University. https://goo.gl/u4Cv00. Accessed March 12, 2017.

Lugones, María. 2010. Toward a decolonial feminism. *Hypatia* 25, no. 4: 742–60.

Lundborg, Tom, and Nick Vaughn-Williams. 2011. Resilience, critical infrastructure, and molecular security: The excess of "life" in biopolitics. *International Political Sociology* 5, no. 4: 367–83.

Luxemburg, Rosa. 1951. *The accumulation of capital*. New York: Monthly Review Press.

Lyson, Thomas A. 2004. *Civic agriculture: Reconnecting farm, food, and community*. Medford, Mass: Tufts University Press; Lebanon, N.H.: University Press of New England.

Madrigal, Tomás. 2013. We make the road by walking, part II: Autonomy. *Kárãni* (blog), August 24. https://goo.gl/ZInPoq. Accessed December 21, 2014.

———. 2014a. "Nuestro trabajo es la vida": Spiritual transformations in organizing for farmworker justice. *The Inbreaking* 1: no 4 (Winter). https://seattlecatholicworker.files.wordpress.com/2014/12/inbreakingwinter14final.pdf. Accessed June 2, 2016.

———. 2014b. Building non-violent agricultural economies. *Kárãni* (blog), May 19. http://goo.gl/aybAoi. Accessed December 21, 2014.

Maffee, James. 2014. *Aztec philosophy: Understanding a world in motion*. Boulder, Colo.: University Press of Colorado.

Maidan, Michael. 2014. Review of *Signs and machines: Capitalism and the production of subjectivity* by Maurizio Lazzarato. *Marx and Philosophy Review of Books*. October 17.

Malinowski, Bronislaw. 1922. *Argonauts of the Western Pacific*. London: G. Routledge and Sons.

Marazzi, Christian. 2008. *Capital and language: From the new economy to the war economy*. Cambridge, Mass.: Semiotext(e) and MIT Press.

Mares, Teresa, and Devon G. Peña. 2010. Urban agriculture in the making of insurgent spaces in Los Angeles and Seattle. In *Insurgent public space*, edited by Jeffrey Hou, 241–54. New York: Routledge.

———. 2011. Environmental and food justice: Toward slow, local, and deep food systems. In *Cultivating food justice: Race, class, and sustainability*, edited by Alison Alkon and Julian Agyeman, 197–219. Cambridge, Mass.: MIT Press.

Marielle, Catherine, and Lizy Peralta. 2007. *La contaminación transgénica del maíz en México: Las luchas civiles en defensa del maíz y la soberanía alimentaria.* México: Grupo de Estudios Ambientales, A.C. http://goo.gl/kqQQM6. Accessed June 2, 2016.

Martin, Patricia Preciado. 1992. *Songs my mother sang to me: An oral history of Mexican American Women.* Tucson: University of Arizona Press.

Martineau, Jarrett, and Eric Ritskes. 2014. Fugitive indigeneity: Reclaiming the terrain of decolonial struggle through indigenous art. *Decolonization: Indigeneity, Education and Society* 3, no. 1: 1–12.

Martínez, Al. 2006. Celeb protesters, where's the do-re-mi? *Los Angeles Times*, June 2. http://goo.gl/F88ozo. Accessed May 12, 2016.

Martínez, Daniel, Robin C. Reineke, Raquel Rubio-Goldsmith, and Bruce O. Parks. 2014. Structural violence and migrant deaths in southern Arizona: Data from the Pima County Office of the Medical Examiner, 1990–2013. *Journal on Migration and Human Security.* 2: 4: 257–86.

Martínez, V. 2007. Crean dependencia en semillas. *Periódico Reforma*, 20 March.

Martínez-Torres, María Elena, and Peter M. Rosset. 2014. *Diálogo de saberes* in La Vía Campesina: Food sovereignty and agroecology. *Journal of Peasant Studies* 41, no. 6: 979–97.

Marx, Karl. 1852 [1995]. The Eighteenth Brumaire of Louis Bonaparte. https://www.marxists.org/archive/marx/works/1852/18th-brumaire/ch01.htm. Accessed June 3, 2016.

———. 1976. *Capital.* London: Penguin Group.

———. 1990. *Capital: A Critique of Political Economy.* Translated by Ben Fowkes. London; New York: Penguin Books/New Left Books.

———. 1993. *The Grundrisse.* New York: Penguin Classics.

Massieu, Yolanda T., and Jésus Lechuga Montenegro. 2002. El maíz en México: Biodiversidad y cambios en el consumo. *Análisis Económico* 36: 281–303. Universidad Autonóma Metropolitana-Azcapotzalco. http://goo.gl/WXwJoV. Accessed June 2, 2016.

Massieu, Yolanda T., and Adelita San Vicente Tello. 2006. El proceso de aprobación de la ley de bioseguridad: Política a la Mexicana e interés nacional. *El Cotidiano* 136: 39–51. http://goo.gl/3oG3rq. Accessed June 2, 2016.

Mauss, Marcel. 1990. *The gift: The form and reason for exchange in archaic societies.* Translated by W. D Halls. New York: W. W. Norton.

McClintock, Nathan. 2011. From industrial garden to food desert: Demarcated devaluation in the flatlands of Oakland, California. In *Cultivating food justice: Race, class, and sustainability*, edited by Alison Hope Alkon and Julian Agyeman. Cambridge, Mass.: MIT Press.

McFarland, Pancho. 2008. *Chicano rap: Gender and violence in the postindustrial barrio.* Austin: University of Texas Press.

———. 2013. *The Chican@ hip hop nation: Politics of a new millennial mestizaje.* East Lansing: Michigan State University Press.

McKinley, Luke. 2014. "Campesino." Video. December 21. http://vimeo.com/94948238.

McMichael, Philip. 2010. Food sovereignty in movement: Addressing the triple crisis. In *Food sovereignty: Reconnecting food, nature and community*, edited by Hannah Wittman, Annette Aurelie Desmarais, and Nettie Wiebe, 168–85. Halifax, NS: Fernwood Publishing.

———. 2014. Historicizing food sovereignty. *Journal of Peasant Studies* 41, no. 6: 933–57.

McMillan, Tracie. 2012. Food's class warfare: Do poor people eat badly because of limited options or personal preference? *Slate*, June 27. http://goo.gl/JpVfzn. Accessed June 2, 2016.

McWilliams, Carey. 1949. *California: The great exception.* Westport, Conn.: Greenwood Press.

Meillassoux, Claude. 1975. *Maidens, meal, and money: Capitalism and the domestic community.* Cambridge: Cambridge University Press.

Merchant, Carolyn. 1989. *Ecological revolutions: Nature, gender, and science in New England.* Chapel Hill: University of North Carolina Press.

Meyer, M. A. 2003. Hawaiian hermeneutics and the triangulation of meaning: Gross, subtle, casual. *Social Justice* 30, no. 4: 54–63.

Mignolo, Walter D. 2001. Coloniality of power and subalternity. In *The Latin American Subaltern Studies Reader*, edited by Ileana Rodriguez, 424–44. Durham, N.C.: Duke University Press.

———. 2005. *The idea of Latin America.* Malden: Blackwell Publishing.

———. 2009. Epistemic disobedience, independent thought and de-colonial freedom. *Theory, Culture and Society* 26, no. 7-8: 1–23.

Mill, John Stuart (Roger Crisp). 1998. *Utilitarianism.* Oxford: Oxford University Press.

Mintz, Sydney. 1979. Time, sugar and sweetness. *Marxist Perspectives* 2, no. 4: 56–73.

Mohawk, John. 2008. Clear thinking: A positive solitary view of nature. In *Original instructions: Indigenous teachings for a sustainable future*, edited by Melissa K. Nelson, 48–52. Rochester, N.Y.: Bear and Company.

Molina, Natalie. 2006. *Fit to be citizens? Public health and race in Los Angeles, 1879-1939.* Los Angeles: University of California Press.

Mondragón-Valdéz, María. 2006. Challenging domination: Local resistance on the Sangre de Cristo Land Grant. PhD diss. Department of American Studies, University of New Mexico.

Montejano, David. 1977. Race, labor repression, and capitalist agriculture: Notes from south Texas, 1920-1930. *Institute for the Study of Social Change Working Paper Series* 102, 1–54. Berkeley: University of California Berkeley.

———. 1987. *Anglos and Mexicans in the making of Texas, 1836-1986.* Austin: University of Texas Press.

Montoya, Michael J. 2011. *Making the Mexican diabetic: Race, science, and the genetics of inequality.* Berkeley: University of California Press.

Moraga, Cherrie. 1983. *Loving in the war years: Lo que nunca paso por sus labios.* Boston: South End Press.

———. 1993. *The last generation.* Boston: South End Press.

Moraga, Cherrie, and Gloria Anzaldúa, eds. 1983. *This bridge called my back: Writings by radical women of color.* 1st ed. New York: Kitchen Table: Women of Color Press.

Morales, Aurora Levins. 2001. *Remedios: Stories of earth and iron from the history of Puertorriqueñas.* Boston: South End Press.

Morton, Timothy. 2013. *Hyperobjects: Philosophy and ecology after the end of the world*. Minneapolis: University of Minnesota Press.

Moskow, Angela. 1999. The contribution of urban agriculture to gardeners, their households, and surrounding communities: The case of Havana, Cuba. In *For hunger-proof cities: Sustainable urban food systems*, edited by Mustafa Koc, Rod MacRae, Luc Mougeot, and Jennifer Welsh, 77–83. Ottawa: International Development Research Centre.

Müller, Max. 1864. *Lectures on the science of language: Delivered at the Royal Institution of Great Britain*. London: Longman, Green, Longman, Roberts and Green.

Munn, Nancy. 1973. Symbolism in a ritual context: Aspects of symbolic action. In *Handbook of social and cultural anthropology*, edited by John J. Honigmann and Alexander Alland, 579–612. Chicago: Rand McNally Co.

———. 1986. *The fame of Gawa: A symbolic study of value transformation in a Massim (Papua New Guinea) society*. Cambridge: Cambridge University Press.

Murray, Star Angelina. 2012. Legality, survival, and action: Immigrant and refugee organizing in the Pacific Northwest. Master's thesis, University of Washington, Tacoma Campus, Department of Interdisciplinary Arts and Sciences.

Nabhan, Gary P. 2011. *Where our food comes from: Retracing Nikolay Vavilov's quest to end famine*. Washington, DC: Island Press.

———. 2013. *Food, genes, and culture: Eating right for your origins*. Washington, DC: Island Press.

Nadal, Alejandro. 2002. *Corn in NAFTA: Eight years after*. México: El Colegio de México. http://goo.gl/vsDnIx. Accessed June 2, 2016.

Negri, Antonio. 2008. Sovereignty: That divine ministry of the affairs of earthly life. http://www.jcrt.org/archives/09.1/Negri.pdf. Accessed February 2, 2015.

Nelson, Melissa K. 2008. *Original instructions: Indigenous teachings for a sustainable future*. Rochester, N.Y.: Bear and Company.

Nestlé, Inc. 2014. Inaugura Nestlé primera fábrica CERO AGUA en el mundo, para disminuir en 15% su consumo anual de agua en México. Posted October 22. https://goo.gl/C8cwR4. Accessed June 2, 2016.

Ngai, Mae M. 2004. *Impossible subjects: Illegal aliens and the making of modern America*. Princeton, N.J.: Princeton University Press.

Norandi, Mariana. 2007. Transnacionales controlan 90% del mercado de semillas: CNC. *La Jornada* (April 2). http://goo.gl/B8xpNZ. Accessed June 2, 2016.

Obama, Barack. 2013. President Obama's 2013 State of the Union. *The White House* (blog). https://goo.gl/sVGwni. Accessed June 2, 2016.

O'Brien, Susie. 2008. Survival strategies for global times: The Desert Walk for Biodiversity, Health and Heritage. Interventions: International Journal of Post-Colonial Studies 9, no. 1: 83-98.

O'Dea, Kerin, P. A. Jewell, A. Whiten, S. A. Altmann, S. S. Strickland, and O. T. Oftedal. 1991. Traditional diet and food preferences of Australian aboriginal hunter-gatherers [and discussion]. *Philosophical Transactions of the Royal Society of London B: Biological Sciences* 334, no. 1270: 233–41.

Omi, Michael, and Howard Winant. 1986. *Racial formation in the United States: From the 1960s to the 1980s*. New York: Routledge and Kegan Paul.

Ortega, Rafael. 2003. La diversidad del maíz en México. In *Sin maíz no hay país*, edited by Gustavo Esteva and Catherine Marielle, 123–52. México: Consejo

Nacional para las Culturas y las Artes. 15 June 2015. https://goo.gl/YjC8Sn. Accessed June 5, 2017.

Ostrom, Elinor. 1990. *El gobierno de los bienes comunes: La evolución de las instituciones de acción colectiva*. México: Fondo de Cultura Económica/CRIM-UNAM.

———. 2000. Collective action and the evolution of social norms. *Journal of Economic Perspectives* 14, no. 3: 137–58.

Palerm, Ángel. 1980. *Antropologia y marxismo*. México, D.F.: Editorial Nueva Imagen.

Palerm, Juan Vicente. 1991. Farm labor needs and farmworkers in California 1970 to 1989. *California Agricultural Studies* 91-2. Sacramento: California Employment Development Department, Labor Market Information Division.

———. 1998. The expansion of California agriculture and the rise of peasant worker communities. In *Immigration: A civil rights issue for the Americas*, edited by Susanne Jones and Suzie Dod Thomas, 221–50. Wilmington, Del.: Scholarly Resources.

———. 2000. Farmworkers putting down roots in Central Valley communities. *California Agriculture*, 54, no. 1. Berkeley: Division of Agriculture and Natural Resources, University of California.

Palerm, Juan Vicente, and José Ignacio Urquiola. 1993. A binational system of agricultural production: The case of the Mexican Bajío and California. In *México and the United States: Neighbors in crisis*, edited by Daniel G. Aldrich Jr. and Lorenzo Meyer, 311–67. San Bernardino: Borgo Press.

Pan, David T. 2013. Carl Schmitt's theory of the partisan and the stability of the nation-state. *Telosscope*, January 2. Paper was presented at Telos in Europe: The L'Aquila Conference. September 7–9, 2012, L'Aquila, Italy. http://goo.gl/IXk5yb. Accessed June 11, 2016.

Parker, Pat. 1978. *Movement in black: The collected poetry of Pat Parker, 1961–1978*. Oakland, Calif.: Diana Press.

Pels, Dick. 1998. *Property and power in social theory: A study in intellectual rivalry*. New York: Routledge Press.

Peña, Devon G. 1992. The "brown" and the "green": Chicanos and environmental politics in the Upper Rio Grande. *Capitalism, Nature, Socialism*, 3, no. 1: 79–103.

———. 1997. *The terror of the machine: Technology, work, gender, and ecology on the U.S.-México border*. Austin: CMAS Books/University of Texas Press.

———, ed. 1998. *Chicano culture, ecology, politics: Subversive kin*. Tucson: University of Arizona Press.

———. 2002. Environmental justice and sustainable agriculture: Linking social and ecological sides of sustainability. *Occasional Paper Series*, Second National People of Color Environmental Leadership Summit, Washington, DC, October 23–27. Acequia Institute Research Reports. Link No. 1. http://goo.gl/ULatDG. Accessed June 1, 2016.

Peña, Devon G. 2003a. Identity, place, and communities of resistance. In *Just sustainabilities: Development in an unequal world*, edited by Julian Ageyman, Robert D. Bullard, and Bob Evans, 146–67. London: Earthscan; Cambridge, Mass.: MIT Press.

———. 2003b. The watershed commonwealth of the Upper Rio Grande. In *Natural assets: Democratizing environmental ownership*, edited by James Boyce and Barry Shelley, 169–85. Washington, DC: Island Press.

———. 2005a. Autonomy, equity, and environmental justice. In *Power, justice, and the environment: A critical appraisal of the environmental justice movement*, edited by David Pellow and Robert Brule, 131–52. Cambridge, Mass.: MIT Press; London: Earthscan.

———. 2005b. Farmers feeding families: Agroecology in South Central L.A. Lecture presented to the Environmental Science, Policy and Management Colloquium. University California, Berkeley, October 10. Acequia Institute Research Reports. PDF Link No. 2. http://goo.gl/WRil4B. Accessed June 21, 2016.

———. 2005c. *Mexican Americans and the environment: Tierra y vida*. Tucson: University of Arizona Press.

———. 2005d. Preliminary list of botanical species grown at South Central Community Garden. Acequia Institute Research Reports. DOC Link No. 7. http://goo.gl/WRil4B. Accessed June 21, 2016.

———. 2006a. Place, identity and social justice in the city: The story of an indigenous diaspora. The Second Annual Samuel E. Kelly Distinguished Lecture. Ethnic Cultural Theatre, University of Washington, April 18. Available from the reports page at www.acequiainstitute.org. Also available from the University of Washington public television station, UWTV.

———. 2006b. Putting knowledge in its place: Epistemologies of place-making in a time of globalization. Plenary address prepared for the Place Matters Conference, Diversity Research Institute, University of Washington. Urban Horticulture Center, October 27.

———. 2008. Anti-economics of EJ: Toward a political economy of justice? *Environmental and Food Justice*, May 26. http://goo.gl/D1BACa. Accessed May 25, 2016.

———. 2011a. Alabama's state of exception: Cut the water off to those sons of . . . . *History and Politics of Mexican Immigration* (blog), October 10. https://goo.gl/EGPjmC. Accessed June 5, 2017.

———. 2011b. Postliberalism, autonomy, and the (re)making of cities. Unpublished manuscript at Academia.edu home page. https://goo.gl/Uf4N6V. Accessed June 2, 2016.

———. 2012. Not identity, subjectivity: The Mesoamerican diaspora and the re-invention of revolutionary subjectivity. Paper presented at the fortieth annual conference of the National Association for Chicana and Chicano Studies, Palmer House Hilton, Chicago. Panel: Holy Sh!t: Moral Panics, Biopower, and the Scatta-Politics of Race. March 15–17.

———. 2016a. Costilla County introduces ordinance establishing a GMO-free zone to protect traditional farmers' varieties. *Environmental and Food Justice* (blog). http://goo.gl/uR2j5L. Accessed May 21, 2016.

———. 2016b. Deep seeds and first foods: Centers of origin, diversification of maíz in the Río Arriba, and the survival of heritage cuisines. *Environmental and Food Justice* (blog) http://goo.gl/cGotxm. Accessed May 21, 2016.

———. 2017. The hummingbird and the red cap. In *Wildness: Relations of people and place*, edited by Gavin Van Horn and John Haufensdoerffer, 88–99. Chicago: University of Chicago Press.

———. Forthcoming. *The last common: Endangered lands and disappeared people in the politics of place.* Tucson: University of Arizona Press.

Peña, Devon G., and Ruben O. Martinez. 1998. The capitalist tool, the lawless, and the violent: A critique of recent Southwestern environmental history. In *Chicano culture, ecology, politics: Subversive kin,* edited by Devon G. Peña, 146–57, Tucson: University of Arizona Press.

Peña, Devon G., Ruben O. Martinez, and Julia Curry Rodríguez, eds. Forthcoming. *Voces de agua y tierra: Environment, culture, and acequia farms in the Rio Arriba bioregion, 1598–1998.* Tucson: University of Arizona Press.

Pendleton Jiménez, Karleen. 2006. "Start with the land": Groundwork for Chicana pedagogy. In *Chicana/Latina education in everyday life: Feminista perspectives on pedagogy and epistemology,* edited by Dolores Delgado Bernal, C. Alejandra Elenes, Francisca E. Godinez, and Sofia Villenas, 219–30. Albany: State University of New York Press.

Pérez, Emma. 1999. *The decolonial imaginary: Writing Chicanas into history.* Bloomington: Indiana University Press.

Perrea, Ernesto. 2009. Mercado de semillas, negocio que germina y crece. *Imagen Agropecuaria* March 23. https://goo.gl/W6pZtR. Accessed March 17, 2017.

Pesquera, Beatriz M., and Denise A. Segura. 1993. There is no going back: Chicanas and feminism. In *Chicana Critical Issues,* 95–115. MALCS Editorial Board. Berkeley, Calif.: Third Woman Press,.

Pichón, Francisco J., Jorge E. Uquillas, and John Frechione, eds. 1999. *Traditional and modern natural resource management in Latin America.* Pittsburg: University of Pittsburgh Press.

Pilcher, Jeffrey M. 1998. *Que vivan los tamales: Food and the making of Mexican identity.* Albuquerque: University of New Mexico Press.

———. 2005. Industrial tortillas and folkloric Pepsi: The nutritional consequences of hybrid cuisines in Mexico. In *The Cultural Politics of Food and Eating: A Reader,* edited by James L. Watson and Melissa L. Caldwell, 235–49. New York: Wiley.

Pillow, Wanda. 2003. Confession, catharsis, or cure? Rethinking the uses of reflexivity as methodological power in qualitative research. *International Journal of Qualitative Studies in Education* 16, no. 2: 175–96.

Pudup, Mary Beth. 2008. It takes a garden: Cultivating citizen-subjects in organized garden projects. *Geoforum* 39, no. 3: 1228–40.

Quijano, Anibel. 2007. Coloniality and modernity/rationality. *Cultural Studies* 21, no. 2-3: 168–78.

Quist, David, and Ignacio Chapela. 2001. Transgenic DNA introgressed into traditional maize landraces in Oaxaca, México. *Nature* 414, no. 11: 541–43.

Ramirez, Deborah, Jack McDevitt, and Andy Farrell, A. 2000. *A resource guide on racial profiling data collection systems.* Washington, D.C.: U.S. Department of Justice. https://goo.gl/k4sBE3. Accessed March 12, 2017.

Ramsay, Paulette A. 2004. History, violence, and self-glorification in Afro-Mexican corridos from Costa Chica de Guerrero. *Bulletin of Latin American Research* 23, no. 4: 446–64.

Rauer, Stephanie, and Rinaldo Pancera. 2006. Testimony of a former coyote on the U.S.-México border. Voice recorder-video. Consuelo Crow Personal Archive.

Ray, Sarah J. 2010. Endangering the desert: Immigration, the environment, and security in the Arizona–Mexico borderland. *Interdisciplinary Studies in Literature and Environment* 17. no. 4: 709–34.

Rebolledo, Tey Diana, Erlinda Gonzales-Berry, and Teresa Márquez, eds. 1988. *Las mujeres hablan: An anthology of Nuevo Mexicana writers.* Albuquerque: University of New Mexico Press.

Redclift, Michael. 1987. *Sustainable development: Exploring the contradictions.* New York: Routledge.

Regan, Margaret. 2010. *The death of Josseline: Immigration stories from the Arizona borderlands.* Boston: Beacon Press.

Relinger, Rick. 2010. NAFTA and US corn subsidies: Explaining the displacement of Mexico's corn farmers. *Prospect: Journal of International Affairs at UCSD*, April 19. https://goo.gl/jNnzX3. Accessed May 24, 2016.

Restrepo, Ivan. 2009. Condena en La Haya contra el uso de glifosato. *La Jornada*, March 18. Print edition.

Rifkin, Jeremy. 1999. *El siglo de la biotechnología: El comercio genético y el nacimiento de un mundo feliz.* Barcelona: Crítica-Macrombo. Spanish translation of *The Biotech Century: Harvesting the Gene and Remaking the World.*

Rifkin, Mark. 2009. Indigenizing Agamben: Rethinking sovereignty in light of the "peculiar" status of Native peoples. *Cultural Critique* 73, no. 3: 88–124.

Rios, Victor M. 2006. The hyper-criminalization of Black and Latino male youth in the era of mass incarceration, *Souls* 8, no. 2: 40–54.

Robinson, William I., and Xuan Santos. 2014. Global capitalism, immigrant labor, and the struggle for justice. *Class, Race and Corporate Power* 2, no. 3: 1–16. http://goo.gl/hf8FIM. Accessed June 2, 2016.

Rodríguez, Roberto. 2014. *Our sacred maíz is our mother: Indigeneity and belonging in the Americas.* Tucson: University of Arizona Press.

Rosaldo, Renato. 1989. *Culture and truth: The remaking of social analysis.* Boston: Beacon Press.

Rosemeyer, M., N. Viaene, H. Swartz, and J. Kettler. 2000. The effect of slash/mulch and alleycropping bean production systems on soil microbiota in the tropics. *Applied Soil Ecology* 15: 49–59. https://goo.gl/8MwblC. Accessed June 2, 2016.

Rosset, Peter, and Maria E. Martínez-Torres. 2012. Rural social movements and agroecology: Context, theory, and process. *Ecology and Society* 17, no. 3: 17–39. https://goo.gl/UTLk2T. Accessed February 23, 2017.

Ruíz, A. 2006. *Respuesta en Europa y America ante los transgénicos (1994–2006).* Ensayo Para Obtener Grado de Maestria en Antropología Física, Escuela Nacional de Antropología y Historia (ENAH), México.

SAGARPA (Secretaria de Agricultura, Ganadería, Desarrollo Rural, Pesca y Alimentación). 2008. *Aviso por el que se ha dado a conocer el listado de información relativo a las solicitudes presentadas y títulos de obtentor de variedades vegetales expedidos, durante el period del mes de enero de 2004 al mes de junio de 2008.* http://goo.gl/BkSRDQ. Accessed June 2, 2016.

Salamanca, Fabrice. 2010. Transgénicos ¿es sensato darles la espalda? *Este País* 235 (November). http://goo.gl/DKrGNo. Accessed June 2, 2016.

Saldívar-Hull, Sonia. 2000. *Feminism on the border: Chicana gender politics and literature.* Berkeley: University of California Press.

Salmon, Enrique. 2012. *Eating the landscape: American Indian stories of food, identity and resilience.* Tucson: University of Arizona Press.

Sánchez, David, Isabel V. Lucena Cid, and Norman J. Solórzano Alfaro, eds. 2004. *Nuevos colonialismos del capital: Propiedad intelectual, biodiversidad y derechos de los pueblos.* Barcelona: Icaria Editorial.

Sánchez, George J. 1995. *Becoming Mexican American: Ethnicity, culture, and identity in Chicano Los Angeles, 1900–1945.* New York: Oxford.

Sandoval, Chela. 2000. *Methodology of the oppressed.* Minneapolis: University of Minnesota Press.

Sayer, P. 2010. Using the linguistic landscape as a pedagogical resource. *English Language Teaching Journal* 64, no. 2: 143–54.

Schey, Peter. 2016. Analysis of Senate Bill 744 pathway to legalization and citizenship. http://goo.gl/CcFKuw. Accessed May 26, 2016.

Schmitt, Carl. 2004 (1963). *The theory of the partisan: A commentary/remark on the concept of the political.* Lansing: Michigan State University Press.

Schob, David E. 1973. Sodbusting of the upper Midwestern frontier, 1820–60. *Agricultural History* 47, no. 1: 47–56.

Schroeder, Fred E. H., ed. 1993. *Front yard America: The evolution and meanings of a vernacular domestic landscape.* Cold Spring Harbor Monographs. Madison: Popular Press/University of Wisconsin Press.

Scott, James C. 1977. *The moral economy of the peasant: Rebellion and subsistence in Southeast Asia.* New Haven, Conn.: Yale University Press.

Serratos-Hernández, José Antonio. 2009. *El origen y diversidad del maíz en el continente americano.* México: Greenpeace. http://goo.gl/oota5Z. English translation available at http://goo.gl/G5c4Q6. Accessed May 27, 2016.

Sharma, Nandita. 2015. Strategic anti-essentialism: Decolonizing decolonization. In *Sylvia Wynter: On being human as praxis,* edited by Katherine McKittrick, 164–82. Durham, N.C.: Duke University Press.

Shintani, T. T., C. K. Hughes, S. Beckham, and H. K. O'Connor. 1991. Obesity and cardiovascular risk intervention through the ad libitum feeding of traditional Hawaiian diet. *American Journal of Clinical Nutrition* 53, no. 6: 1647S–51S.

Shiva, Vandana. 1988. *Staying alive: Women, ecology, and development.* London: Zed Books.

———. 2002. Democracy of commons: Biodiversity, water, and air. In *Governance of commons and livelihood security,* edited by Himadri Simha and Anant Kumar. Jharkhand, India: Xavier Institute of Social Service. https://goo.gl/UhyNah. Accessed March 12, 2017.

———. 2003. *Cosecha robada: El Secuestro del suministro mundial de alimentos.* Translation of *Stolen Harvest: The Hijacking of the Global Food Supply.* Buenos Aires: Editorial Paidós.

Shohamy, E., and D. Gorter. 2009. *Linguistic landscape: Expanding the scenery.* New York: Routledge.

Shreck, Aimee, Christy Getz, and Gail Feenstra. 2005. Farmworkers in organic agriculture: Towards a broader notion of sustainable agriculture. *Newsletter of the University of California Sustainable Agriculture Research and Education Program* 17, no. 1: 1–3. Davis, Calif.: UC Division of Agriculture and Natural Resources.

Sierra Club. 2004. NAFTA's impact on Mexico. Sierra Club Vault. https://goo.gl/VRMIGq. Accessed June 5, 2017.

Simon, Joel. 1997. *Endangered México: An environment on the edge*. San Francisco: Sierra Club Books.

Smith, Andrea. 2005. Rape of the land. In *Conquest: Sexual violence and American Indian genocide*, 55–79. Cambridge, Mass.. South End Press.

Smith, Linda Tuhiwai. 1999. *Decolonizing methodologies: Research and indigenous peoples*. London/New York: Zed Books.

———. 2004. *Decolonizing methodologies: Research and indigenous peoples*. 2nd. ed. London: Zed Books.

Smith, Mick. 2011. *Against ecological sovereignty: Ethics, biopolitics, and saving the natural world*. Minneapolis: University of Minnesota Press.

Solares, Blanca. 2007. *Madre terrible: La diosa en la religión del México antiguo*. México: UNAM and Barcelona: Anthropos Editorial. https://goo.gl/lgGrvp. Accessed May 27, 2016.

Solidaridad Chiapas. 2014. RvsR: El paramilitarismo "campesino y democrático" contra las comunidades Zapatistas. August 22. https://goo.gl/abuQIj. Accessed February 3, 2015.

Stédile, João Pedro, and Horácio Martins de Carvalho. 2011. People need food sovereignty. In *Food movements unite! Strategies to transform our food system*, edited by Eric Holt-Giménez and Annie Shattuck, 21–34. Oakland, Calif.: Food First Books.

Stewart, Ashley. 2014. Sakuma speaks up: Farm labor pariah defends his family's business. The Seattle Globalist, Feb 10. https://goo.gl/V6LKRG. Accessed June 9, 2017.

Stoeve, Rachael. 2014. They started by blockading a bus full of detainees—and went on to shake up the immigration debate. *Yes! Magazine*, April 16. https://goo.gl/57a3kn. Accessed June 2, 2016.

Stoller, Marianne. 1980. Grants of desperation, lands of speculation: Mexican period land grants in Colorado. *Journal of the West* 19: 22–39.

Sundberg, Juanita. 2011. Diabolic caminos in the desert and cat fights on the Río: A posthumanist political ecology of boundary enforcement in the United States–México borderlands. *Annals of the Association of American Geographers* 101, no. 2: 318–36.

Swanson, Nancy L., Andre Leu, Jon Abrahamson, and Bradley Wallet. 2014. Genetically engineered crops, glyphosate and the deterioration of health in the United States of America. *Journal of Organic Systems* 9, no. 2: 6–37.

Tallbear, Kimberly. 2001. Racializing tribal identity and the implications for political and cultural development. Paper presented at the Indigenous Peoples and Racism Conference, Sydney, Australia.

Taylor, Michael. 2006. *Rationality and the ideology of disconnection*. Cambridge: Cambridge University Press.

Taylor, Paul S. 1954. Plantation laborer before the Civil War. *Agricultural History* 28, no. 1: 1–21.

Telles, Edward, and Vilma Ortíz. 2009. *Generations of exclusion: Mexican Americans, assimilation, and race*. New York: Russell Sage Foundation.

Thorp, Laurie. 2006. *The pull of the earth: Participatory ethnography in the school garden*. Lanham, Md.: AltaMira Press.

Thurston, David H., Margaret Smith, George Abawi, and Steve Kearl, eds. 1994. *Tapado: Slash/mulch: How farmers use it and what researchers know about it.* Ithaca, N.Y.: Cornell International Institute for Food, Agriculture, and Development (CIIFAD).

Toensmeier, Eric, and Jonathan Bates. 2013. *Paradise lot: Two plant geeks, one-tenth of an acre, and the making of an edible garden oasis in the city.* White River Junction, Vt.: Chelsea Green Publishing.

Tohono O'odham Community Action, Mary Paganelli Votto, and Frances Sallie Manuel. 2010. *From I'itoi's garden: Tohono O'odham food traditions.* Tohono O'odham Community Action (TOCA)/Blurb.

Tuck, Eve, and Marcia McKenzie. 2014. *Place in research: Theory, methodology, and methods.* New York: Routledge and Francis Group.

Tuck, Eve, and K. Wayne Yang. 2012. Decolonization is not a metaphor. *Decolonization: Indigeneity, Education and Society* 1, no. 1: 1–23. https://goo.gl/jY7ot4. Accessed June 2, 2016.

Turner, Victor. 2004. Liminality and communitas. In *The performance studies reader*, edited by Henry Bial, 89–97. London: Routledge.

TVW. 2013. *House Labor and Workforce Development Committee*, Work Session: Discussion of agricultural labor issues and possible policy changes. Washington State Public Affairs Network. November 14. http://goo.gl/Hy5wwb. Accessed June 2, 2016.

Union of Concerned Scientists (UCS). 2009. *Failure to yield: Biotechnology's broken promises.* Issue Briefing. http://goo.gl/AXKEfp. Accessed June 2, 2016.

United Nations Environmental Program (UNEP). 2009. Agriculture at a Crossroads: International Assessment of Agricultural Knowledge, Science and Technology for Development—Global Report. Washington, DC: Island Press. https://goo.gl/6gujT2. Accessed June 7, 2016.

Urrieta, Luis. 2013. Familia and comunidad-based saberes: Learning in an indigenous heritage community. *Anthropology and Education Quarterly* 44, no. 3: 320–35.

USDA (US Department of Agriculture). 2015. Food desert locator tool. http://www.ers.usda.gov/data/fooddesert. Accessed June 2, 2016.

Valle, Gabriel R. 2015. Gardens of sabotage: Food, the speed of capitalism, and the value of work. *Aztlán: A Journal of Chicano Studies* 40, no. 1: 63–87.

———. 2016. Cultivating subjectivities: The class politics of convivial labor in the interstitial spaces of neoliberal neglect. PhD diss., Department of Anthropology, University of Washington, Seattle.

van der Ploeg, Jan Douwe. 2014. Peasant-driven agricultural growth and food sovereignty. *Journal of Peasant Studies* 41, no. 6: 999–1030.

VanLeuven, Sheena. 2007. Environmental justice and agriculture: A case study of conventional and organic farming in México and an exploration of the greater relationship linking environmental justice, organic farming, and environmental sustainability. Unpublished research competency paper (master's thesis), Environmental Studies Program, University of Chicago.

Vartabedian, Ralph. 2006. Cancer stalks a "Toxic Triangle": Scientists disagree about the risks of TCE. But residents near a former air base are dead certain. *Los Angeles Times*, March 30. https://goo.gl/MjGo5o. Accessed February 28, 2017.

Via Campesina. 2001. Our world is not for sale: Priority to people's food sovereignty. https://goo.gl/z6PvR6. Accessed June 2, 2016.

———. 2014. La Via Campesina Internacional se solidariza con Ayotzinapa y se une al reclamo global de justicia y de la presentación con vida de los 43 estudiantes desaparecedios por el estado mexicano. http://goo.gl/Mkwg3z. Accessed June 2, 2016.

Viertel, Josh. 2011. Beyond voting with your fork: From enlightened eating to movement building. In *Food movements unite! Strategies to transform our food systems*, edited by Eric Holt-Giménez and Annie Shattuck, 137–47. Oakland, Calif.: Food First Books.

Vinson, Ben. 2000. The racial profile of a rural Mexican province in "Costa Chica": Igualapa in 1791. *The Americas* 57, no. 2: 269–82.

Vizenor, Gerald. 1990. *Interior landscapes: Autobiographical myths and metaphors.* 1st ed. Minneapolis: University of Minnesota Press.

———. 1998. *Fugitive poses: Native American scenes of absence and presence.* Lincoln: University of Nebraska Press.

Waberi, Abdourahman A. 2009. *In the United States of Africa.* Translated by David Ball and Nicole Ball. Lincoln: Bison Books/University of Nebraska Press.

Walker, Alice. 1983. *In search of our mothers' gardens.* New York: Harcourt.

Walker, Brian, and David Salt. 2006. *Resilience thinking: How can landscapes and communities absorb disturbance and maintain function?* Washington, DC: Island Press.

Warkentin, Benno P., ed. 2006. *Footprints in the soil: People and ideas in soil history.* Amsterdam: Elsevier Science.

Weber, Kenneth. 1991. Necessary but insufficient: Land, water, and economic development in Hispanic southern Colorado. *Journal of Ethnic Studies* 19: 127–42.

White Earth Land Recovery Project. Defeat Diabetes Program. 2010. *Anishinaabe traditional food pyramid.* Anishinaabe Center, Callaway, Minn. http://www.recipesforlifewithdrbeth.com/nativeamericanfoodspyramid.php.

*Wikipedia.* 2016. Bracero Program. Wikimedia Foundation. https://goo.gl/QCAiWE. Accessed May 26, 2016.

———. 2017. South Central Farm. Wikimedia Foundation. https://goo.gl/JaaxaK. Modified January 15, 2017. Accessed March 7, 2017.

Wildcat, Daniel R. 2009. *Red alert! Saving the planet with indigenous knowledge.* Speaker's Corner Books. Golden, CO: Fulcrum.

Williams, Barbara J., and Carlos A. Ortíz-Solorio. 1981. Middle American folk soil taxonomy. *Annals of the Association of American Geographers* 71, no. 3: 335–58.

Williams, Barbara J., and Janice K. Pierce. 2014. Evidence of Alcohua science in pictorial land records. In *Texcoco: Prehispanic and colonial perspectives*, edited by Jongsoo Lee and Galen Brokow, 147–64. Boulder: University of Colorado Press.

Wilson, Shawn. 2008. *Research is ceremony: Indigenous research methods.* Black Point, Nova Scotia: Fernwood Publishing.

Wolf, Eric R. 1982. *Europe and the people without history.* Berkeley: University of California Press.

Wolf, Eric R., and Sydney W. Mintz. 1957. Haciendas and plantations in Middle America and the Antilles. *Social and Economic Studies* 6, no. 3: 380–411.

Wolfe, Patrick. 2013. Recuperating binarism: A heretical introduction. *Settler Colonial Studies* 3, no. 3–4: 257–79.

Woolcott, Ina. Huichol Indians and their rituals "future of Earth." *Shamanic Journey*. Shamanicjourney.com, 8 Feb. 2015. http://goo.gl/RHUjj4. Accessed June 2, 2016.

Zaragoza, Tony. 2007. Apple capital: Growers, labor and technology in the origin and development of the Washington state apple industry, 1890–1930. Ph.D. diss., Washington State University.

Zavella, Patricia. 1987. *Women's work and Chicano families: Cannery workers of the Santa Clara Valley*. Ithaca, N.Y.: Cornell University Press.

———. 1991. Beyond the double day: Work and family in working-class women's lives. *Feminist Studies* 16, no. 1: 53–67.

*Zoot Suit*. Directed by Luis Valdez. 1981. Universal Pictures. 103 minutes.

Zupancic, Tomaz. 2005. Understanding students' artworks according to the Scot Lash's post-modern regime of signification. http://goo.gl/rjFvGX. Accessed June 2, 2016.

# CONTRIBUTORS

**Luz Calvo** is a professor of ethnic studies at California State University, East Bay. Calvo received a PhD in History of Consciousness from the University of California, Santa Cruz, and has published in the areas of Chicana visual culture, critical race theory, and Chicana feminist analysis. Calvo is coauthor of *Decolonize Your Diet: Plant-Based Mexican-American Recipes for Health and Healing* (Arsenal Pulp Press, 2015) and curates "Decolonize Your Diet" social media accounts on Facebook, Instagram, Pinterest, and Twitter.

**Araceli Carreón** graduated from the Universidad Autónoma Metropolitana in Mexico City with a degree in social communication. She has been active in environmental movements in Mexico through Greenpeace and other organizations and is currently the coordinator of public policy for Bicitekas A.C., a civic organization dedicated to the expanded use of bicycles as a form of urban transportation. Carreón is also a major advocate in the movement known as Sin Maíz, No Hay País and the Colectiva del Maíz, key organizations behind the defeat of Monsanto in Mexico.

**Consuelo Crow** received her MA in anthropology from the University of Arkansas in Fayetteville and is a lifetime member of Lamda Alpha Honors Society for Anthropology. She has been researching and presenting her findings on unauthorized migration into the United States in the Border Patrol's Tucson Sector since 2008 with a focus on border checkpoints, gendered assault, and foodways. She also researches border checkpoints that bisect the Palestinian West Bank and Israel. Recently, she conducted an ethnographic study for a nonprofit in Siloam Springs, Arkansas, to assist in addressing community safety and nutrition outreach with the Latino community. She is the associate producer on the documentary film *Finding Kim*, which premiered at the 2016 Seattle International Film Festival. Crow currently lives in Seattle, Washington, with her husband, nine-year-old son, and their five cats and is building a nonprofit organization focused on environmental racism and social justice.

**Julia Curry Rodríguez** is a sociologist who teaches Chicana and Chicano Studies at San Jose State University. Her research focuses on immigrant women and children, language minorities, educational inequality, and other issues affecting Chicana/os. Before joining SJSU, Curry taught Chicana/o Studies at the University of California, Berkeley. Her more recent publications include contributions to *Mothers, Mothering and Motherhood across Cultural Differences* (2014) and the *Journal of Equity and Excellence in Education* (2012). Curry Rodríguez is the executive director of the National Association for Chicana and Chicano Studies (NACCS) and former chair of Mujeres Activas en Letras y Cambio Social (MALCS). She is currently preparing an ethnographic and historical account of women in the acequia farm families of Colorado based on the largely untapped primary archives of The Acequia Institute. She is a lifelong activist who struggles at the local and global level for humanitarian immigration reform that concedes that immigrants contribute to the quality of life, the wealth, and the social fabric of all societies.

**Lee Ann Epstein** is a doctoral student in the Culture, Literacy, and Language program at the University of Texas at San Antonio. Her research interests include food epistemologies, critical media studies, and challenging deficit paradigms toward Chicanx communities. She works at St. Philip's College in San Antonio, Texas, in the Communications and Learning Department. In addition to teaching and writing, Epstein can be found experimenting in her backyard and kitchen and wearing out the pages of seed catalogs and cookbooks.

**Catriona Rueda Esquibel** is a professor in Race and Resistance Studies in the College of Ethnic Studies at San Francisco State University (the only College of Ethnic Studies in the country). While she started out studying literature, Esquibel's academic trajectory includes poetry, drama, science fiction, queer studies, theology, food justice in communities of color, and research on traditional diets. She is coauthor of *Decolonize Your Diet: Plant-Based Mexican-American Recipes for Health and Healing* (Arsenal Pulp Press, 2015).

**Joseph C. Gallegos** was a managing partner of the oldest family farm in Colorado, the Corpus A. Gallegos Ranches (est. 1851). Before an untimely death in December 2016, he was a leading voice for acequia water rights and was involved with the environmental justice movement in Colorado since the late 1980s, when he led the campaign against the infamous Battle Mountain Gold cyanide leach mining operation in the Rito Seco. He served as mayordomo of the San Luis Peoples Ditch and as president and vice president of the Sangre de Cristo Acequia Association. At the time of his death, he was serving a third term as a Costilla County commissioner and completing work on an autobiographical history of the Gallegos family and acequia community of San Luis, Colorado.

**Rosalinda Guillen** is the founder and executive director of Community to Community Development (C2C), a 2014 Food Sovereignty Prize honoree. With a lifelong history of worker and community organizing, Guillen has contributed to organizations like the Rainbow Coalition and the United Farm Workers (UFW). She is also a member of the executive boards of Food First and The Acequia Institute.

**María L. Guillen Valdovinos** (Poesia Mariarte) is a visual creative thinker/artist and activist in environmental, food, and social justice movements. She graduated in 2010 from the University of Washington in anthropology with a minor in Womxn Studies and is currently based in occupied Coast Salish Territories. She was born in Zihuatanejo, Guerrero, Mexico, and migrated with her family in the early 1990s to the Pacific Northwest. She is also a hip-hop artist with Batallones Femeninos, a womxn's hip-hop collective based in Mexico, and does multimedia production and autonomous organizing with Shades of Silence. She is currently in the process of applying to film school and to a graduate doctoral program to extend her work of supporting womxn and Indigenous communities in defense of the earth through creative arts focused on video, audio, and music production. For more information, go to www.poesiamariarte.com.

**Rufina Juárez** is a Chicana/Mexicana/Indigena (Otomi/Hña Hñu) activist in the environmental/food justice movement. Her father was a bracero from Guanajuato and settled in the Imperial Valley, where he was a field-worker and farmer. Juárez grew up on her father's farm with her five siblings and her mother. She is the first in her family to attend college. Juarez has a bachelor's degree from the University of California, San Diego, in political science, and a master's degree from the City University of New York in public administration, and was a National Urban Fellows recipient. She is mostly known for her community involvement and infinite hours of volunteer work in the Imperial Valley, along the Texas border, in Canada, and in Los Angeles. Her work and dedication to the people's struggles and the food sovereignty movement can be seen through the documentary titled *The Garden*. She is the cofounder of the South Central Farmers, the Health & Education Fund, and the Women's Cooperative.

**Tomás Madrigal** earned his PhD in Chicana/o Studies from the University of California, Santa Barbara, and is an organizer and activist in farm worker social justice movements. He is affiliated with the organizing project of the Indigenous farm workers' movement known as Familias Unidas por la Justicia. The organization is in the midst of a three-year mobilization against Sakuma Berry Farms in Burlington, Washington. Madrigal is cofounder of the Canopy Cooperative, member of the steering committee of the Domestic Fair Trade Alliance, and a volunteer organizer at Community to Community Development. He teaches courses in Chicana/o Studies and most recently has served as a visiting lecturer at Western Washington University.

**Pancho McFarland** is a father, son, and activist and scholar in the fields of race, food and the environment, social movements, and culture. In 1999 he received his PhD in sociology from the University of Texas at Austin, where he studied the Zapatista uprising in Chiapas, México. His research on Chicana/os in hip-hop has resulted in numerous articles and the books *Chicano Rap: Gender and Violence in the Postindustrial Barrio* (University of Texas Press, 2008) and The *Chican@ Hip Hop Nation: Politics of a New Millennial Mestizaje* (Michigan State University Press, 2013). Since moving to Chicago in 2005, McFarland has taught at Chicago State University and has been involved in food justice, community gardening, and the local food movement as a scholar-farmer and executive director of the Green Lots Project. He is working on an autoethnographic study of the local food movement in Black Chicago.

**Devon G. Peña** is professor of American Ethnic Studies, Anthropology, and Program on the Environment at the University of Washington. A lifelong activist in the environmental and food justice movements, he is cofounder and president of The Acequia Institute, a nonprofit foundation dedicated to advancing these movements and promoting agroecology and permaculture farming methods through work at the institute's 181-acre acequia farm in Colorado. The institute is also home to a thirty-year seed library focused on the "Three Sisters." He also serves on the board of directors of Food First and was a member of the board of directors of the Council for Responsible Genetics (1992–2003), one of the nation's first citizen-scientist organizations to monitor and resist commercial agricultural biotechnology. A widely published and award-winning scholar, he is completing work on a thirty-year study of acequia farming communities in Colorado, *The Last Common: Endangered Lands and Disappeared People in the Politics of Place*. Among his most recent publications is the *Oxford Encyclopedia of Latinos and Latinas in Contemporary Politics, Law, and Social Movements*, senior editor and contributor (Oxford University Press, 2015).

**Adelita Sanvicente Tello** is a cofounder and director of Fundación Semillas de Vida in Mexico City and a leading organizer in the Sin Maíz, No Hay Paíz movement. She is an agronomist with a master's degree in rural development from the Universidad Autónoma Metropolitana-Xochimilco, studied economics of the agri-food system in Viterbo, Italy, and is an agroecology PhD candidate at the University of Medellin, Colombia. She has collaborated with various nongovernmental organizations and farmers, developing training programs and environmental education and planning for the sustainable management of natural resources. She has also held positions linked to the rural sector in various institutions and governments, at both the municipal and federal levels. She worked with the Regional Union of Ejidos and Communities in the Hidalgo Huasteca (URECHH) in marketing their products and participated in the Tepoztlan community's fight against a golf club, a movement that stopped the imposition of a project that was billed as the largest golf club in Latin America.

**Silvia Patricia Solís** is a doctoral candidate in the Department of Education, Culture and Society at the University of Utah. With a concentration in the anthropology of education, her research interests are coloniality and violence, decolonial feminist thought and methodologies, materiality and futurity, and indigenous knowledge systems, trauma and healing, place and movement. Her dissertation traces indigenous healing knowledges in the *saberes* of Indigenous and Mexican women living in a place called the U.S.-México border. She currently lives in Salt Lake City with her two children and partner.

**Tezozomoc** (Tezo), a former linguistics graduate student, named himself after a Mesoamerican Tepanec leader. Tezo's grandfather, a subsistence farmer in México, instilled in Tezo the techniques of subsistence farming, a deep appreciation for the land, and a respect for the long tradition of growing one's own food. Tezo's family came to Los Angeles during the green revolution of the 1970s, and Tezo and his father quickly became local leaders in South Central L.A., battling inequality through the formation of a fourteen-acre community garden located on a contested plot of land on Forty-First and Alameda. The South Central Farm, as it was known, became a cultural center and offered community members far more than just fresh produce. In 2006, after a hard-fought political battle, the community farm lost its land to a developer. Tezo was undeterred, and in 2007 he helped to form a new eighty-acre co-op, the South Central Farmers Health and Education Fund (SCFHEF), in Buttonwillow, California, about a two-hour drive north of Los Angeles. Tezozomoc is an accomplished Los Angeles Chicano poet and has been published in journals such as *CrazyQuilt, Rhino, Mind Matters Review, Left Curve, Next Phase, Minotaur Press, San Fernando Poetry Journal, Poet's Sanctuary, Black Buzzard Press, Dance of the Iguana, The Americas Review, LaHoja, Louder Than Bombs, Orale!,* and *Tight.* His poetry was included in the anthology *The Coiled Serpent: Poets Arising from the Cultural Quakes and Shifts of Los Angeles* (Tia Chucha, 2016). Tezozomoc is a *Huffington Post* blogger, where he blogs about activism and food issues. Tezozomoc's work also includes academic essays on Nahuatl indigenous languages and a chapter in the book *Teaching Indigenous Languages* (Northern Arizona University, 1997).

**Gabriel R. Valle** received his PhD in sociocultural anthropology from the University of Washington in August 2016 and serves as assistant professor of Environmental Studies at California State University San

Marcos. His research scholarship involves community-based collaborative ethnography with a focus on ethnoecology and agroecology in urban home kitchen gardens. His recently published dissertation documents how a multiethnic, multilingual community of urban gardeners in San Jose, California, utilize their home kitchen gardens to navigate uncertainty, enact autonomy, and contribute to their socioecological resilience. Since moving to San Diego, his research has expanded into the political ecology of drought and how urban gardeners practice agroecology to save water and continue to grow food in the drought-stricken state.

**Teresa de Jesus Berlinda Vigil** is a rural herbal practitioner who is familiar with natural healing methods traditional to the Upper Rio Grande bioregion. Understanding the importance of natural remedies to Hispano residents in the San Luis Valley, Vigil has for the past thirty-two years expanded her knowledge of traditional healing herbs and natural remedies common to northern New Mexico and southern Colorado. Having taken classes at Regis University for a year, including course work in herbology, Vigil's cultural perspective and lifelong medical experiences have made her a sought-after presenter. She is profiled in *Delmar's Integrative Herb Guide for Nurses* and *Fiesta's Cultural Awareness of Herbal Use*, and contributed recipes to *Cocina de la Familia*. Teresa is an instructor in the University of Washington Acequia Agroecology and Permaculture Field School and leads local and visiting college and high school students in the study of ethnobotany and ethnomedicine. She is the daughter of Jose Incarnation Gallardo and Manuelita Gallegos, whose ancestors resided in the San Luis Valley for multiple generations. She is married to Victor, and they have seven children, fourteen grandchildren, and twenty-one great-grandchildren.

# INDEX

Colectiva Legal del Pueblo, 281
Colectiva/Colectivo de Detenidos, 283–84, 286
collateralized debt obligations (CDOs), 226–27
collective action/organization, xxx, xxxii, xxxiii, 310, 368, 380
collective healing, 372
Collectivo Ecologista-Jalisco, 331
colonial wounds, xxiv–xxv, 67–70, 72, 111, 367, 393n14, 395n23
colonialism/colonization, xxvii, 5, 44, 58, 64, 70, 110, 183, 245, 289, 335, 380; ancestral diets stripped by, 142, 145–46; bio-, xvii; and capital, 316–20; and capitalism, xx, 377–78, 382; and decolonizing diets, 125–26, 138; dehumanizing effects of, 372; and dispossession, xxix, 230, 374, 382; exorcising, 360; neo-, 286; racist, 295; stories, 392–93n9; violence of, 147
coloniality: defined, xxiv; fractures/ limits of, 72; and modernity, 72; neoliberal, 366; and property, xxii; and soil conservation, xxxii–xxxiii. *See also* decoloniality; postcoloniality
*colonias,* 257, 258
colonized diets, 112, 377
Colorado, xvi, xxviii, xxix, xxxiii, 97, 355, 359–60, 388n35, 409n10, 410n34, 412n45, 442–43, 445, 447
commercialization, 200, 257–58, 329
Commission on Environmental Cooperation (CEC), 332
Committee for Immigration Reform, 281
commodification, xx, xxxii, 24, 30, 66, 336, 371
commodity foods, 138, 142, 145–46
common property resources (CPRs), 336, 407n45
communal determination, 55, 59, 372
communality, 68, 312, 335–41
communism, 300
communitas, 85
communities, 47, 184, 312, 380, 381

Community Alliance for Global Justice (CAGJ), 253, 419
community gardens, xxvii, 24–25, 41, 46–48, 55, 58, 119–22, 147, 176–82, 187, 299, 301–8, *303,* 310, 367, 432, 444, 446. *See also* urban gardens
community land reclamation project, *175*
Community Supported Agriculture (CSA), 58, 217
Community to Community Development. *See* C2C (Community to Community Development)
*compañeras,* 40
company towns, 257
Comprehensive Immigration Reform (CIR), 281–82
confined animal feeding operations (CAFOs), 377
conflict foods, xxiii, 33, 367
consumerism, 299, 386n15
Convention on Biological Diversity (CBD), xxxii, 15–16
Cook, Katsi, 140–41
Cook, Susan, 274
cooking, decolonial, xxvii, 147, 370
Cooperativa Jacal, 242–43, 247
cooperatives, xxix, xxx; autonomous, 242–43; as grassroots organization, 267; Indigenous, 368–69; women-led, 247–49; worker-run, 245
corn, 311–41. *See also* Indigenous corn belt; maíz/maize
corn civilizations, 376
Corpus A. Gallegos Ranches (San Luis, Colorado), *154, 166,* 363
counterstories, biographical, 115, 117, 122
Coyolxauhqui (Mexican goddess figure), 112, 137–38, 147, 366
Coyolxauhqui bowl (recipe), 148–49
CPRs. *See* common property resources (CPRs)
creation stories, 63–64, 392–93n9
creative spaces, 58–61
crossing over, 101, 369–70
crossing times, 369–70

Esparza, Moctesuma, 130–31
Espinoza, Alejandro, 320
Esquibel, Catriona Rueda, xxvii, 113, 125, 126–33, 135, 147, 367, 370, 372, 374, 378–79, 418, 442
Esquibel, Librada Tafoya, 131
Esteva, Gustavo, xxix, 32, 187, 206, 374, 386n15, 403n, 421
ethics: and Indigenous knowledge, 381; of place, 309; soil and land, 352
ethnicity, in classroom, xxviii, 189–206, 372, 374, 402n
ethno-archaeology, xxv, 376
ethnobotany, xxi, xxxiii, 95, 97, 106, 183, 213, 366, 379, 394n22, 447
ethnoecology, xxiii, xxiv, xxviii, 183, 371, 447
ethnoedaphology, 346, 351, 390n29
ethnography, xv–xvi, xviii, xxiii, xxv, 47, 175, 195, 368, 385n7, 441, 442, 444, 447. See also autoethnography
ethnohistoriography, 354
ethnonationalism, 377
ethnopedology, 346, 351, 353, 390n29
Eurocentrism, Western, 66, 77, 353, 385n
Europeanizing, 128–29, 373
exception, zone of, xxv, 96, 99–102, 106, 369. See also economic exception
exceptionalism: hierarchy of, 5–6; racist ideology of, 136
exploitation, xxiii, xxx–xxxi, 14–15, 30, 51, 65, 75, 84, 171, 184, 206, 228, 242, 244–45, 252, 255, 257–58, 262, 289, 298, 304, 312–13, 319, 323, 384n6
export crops, 259
expropriation, 228, 312, 317, 344
Eyzaguirre, Pablo, 47, 421, 422

Facebook, 125–26, 366, 383n3
Familias Unidas por la Justicia, xiii, 251, 253–55, 271–76, 287, 366, 444
family: blended, 191–92, 193; dysfunctional, 191; and food, 189–90, 195–96, 206
Fanon, Franz, 68, 229, 365, 393n14, 421
FAO. See Food and Agriculture Organization (FAO)
Farm Bureau, 259

farmers, as scientists, 353
farmers' markets, 58, 242, 382
farmworkers: autonomy, xxii, xxx, 235–50, 370, 376, 403n; independent union, 276–77; organizing in Washington, xxx–xxxi, 31, 251–90, 370, 376, 377, 388n43, 404–5n; racialization of, xxx–xxxi; rights of, 261, 263, 269; women, 244–47, 248
fast food, xxvii, 33, 119, 263
FDA. See Food and Drug Administration (FDA)
Federal Law on Plant Health (LFSV), 323
feminism, xxiii, xxvi–xxviii, xxx, 72, 83, 109–12, 123, 126, 130, 146–47, 184–85, 191, 441, 445–46. See also ecofeminism
FIOB. See Frente Indigena de Organizaciones Binacionales (FIOB)
first foods, xx, 377, 381, 432
First Nations Canadian health programs, 145
floating gardens, 48, 353, 354
Flores, Yvette Gisele, 54, 422
Florescano, Enrique, 314, 422
Florida, economic refugee camps in, 262
food(s): activism, xxi, xxvii, 140, 143; conflict, xxiii, 33, 367; grassroots, 379; healthy native, 147; junk, 305; junkyards, xx, 25, 297, 377; as labor, skill, heritage, 189–90, 205–6; as medicine, 184; and power of identification, 86; pyramids, 145; street, 133; studies, xi, xii, xvii, xxxiii; supply, global, 306. See also fast food; first foods; heritage food/cuisines
Food and Agriculture Organization (FAO), 318
Food and Drug Administration (FDA), 94
food autonomy, 231, 290, 348, 349, 354, 375, 377, 380
food champions, 218
food deserts, 54, 109, 118, 119; autonomy